Human-Robot Collaborations in Industrial Automation

Human-Robot Collaborations in Industrial Automation

Editor

Anne Schmitz

MDPI • Basel • Beijing • Wuhan • Barcelona • Belgrade • Manchester • Tokyo • Cluj • Tianjin

Editor
Anne Schmitz
University of Wisconsin-Stout
USA

Editorial Office
MDPI
St. Alban-Anlage 66
4052 Basel, Switzerland

This is a reprint of articles from the Special Issue published online in the open access journal *Sensors* (ISSN 1424-8220) (available at: https://www.mdpi.com/journal/sensors/special_issues/industrial_robotics).

For citation purposes, cite each article independently as indicated on the article page online and as indicated below:

LastName, A.A.; LastName, B.B.; LastName, C.C. Article Title. *Journal Name* **Year**, *Volume Number*, Page Range.

ISBN 978-3-0365-5213-2 (Hbk)
ISBN 978-3-0365-5214-9 (PDF)

Contents

About the Editor

Anne Schmitz

I received my Mechanical Engineering undergraduate degree from the University of Wisconsin-Madison. During my schooling, I explored many opportunities to apply my engineering degree. I was involved with the Formula One Racecar Team, did a semester-long co-op working on fume hoods, did a summer internship at Kimberly Clark designing a HVAC system, and did another summer internship at General Electric designing anesthesia equipment. As a senior, I became involved in research of finite element analyses of a prosthetic foot. This immediately interested me on applying engineering to medical applications. I obtained my Biomedical Engineering PhD at the University of Wisconsin-Madison. My work focused on computational biomechanics. More specifically, it developed musculoskeletal models of the body to simulate movement and see how surgery and soft tissue injury affects movement. During my graduate work, I was also a teaching assistant for Introduction to Biomechanics where I developed a love for teaching. I then did postdoctoral research at the University of Kentucky where I experimentally measured movements (e.g., running form), which provides data that can be used to validate the models I built. At Gannon University, I built computational models with a focus on the knee to optimize surgical techniques (e.g., ACL reconstruction) to restore normal function after injury. At UW-Stout, my research area is on 3D printing techniques. Specifically I am working on the reliability of the 3D printing process to create functional parts. When I am not doing research, I enjoy swimming, camping with my kids in Scouts, and playing my violin.

Editorial

Human–Robot Collaboration in Industrial Automation: Sensors and Algorithms

Anne Schmitz

Engineering and Technology Department, University of Wisconsin-Stout, Menomonie, WI 54751, USA; schmitzann@uwstout.edu

Citation: Schmitz, A. Human–Robot Collaboration in Industrial Automation: Sensors and Algorithms. *Sensors* **2022**, *22*, 5848. https://doi.org/10.3390/s22155848

Received: 1 August 2022
Accepted: 3 August 2022
Published: 5 August 2022

Publisher's Note: MDPI stays neutral with regard to jurisdictional claims in published maps and institutional affiliations.

Technology is changing the manufacturing world. For example, sensors are being used to track inventory from the manufacturing floor to a retail shelf or a customer's door, i.e., asset tracking [1]. These types of interconnected systems constitute the so-called fourth industrial revolution, i.e., Industry 4.0, and are projected to lower manufacturing costs [2]. As the manufacturing industry moves toward these integrated technologies and lower costs, engineers will need to connect these systems via the Internet of Things (IoT) [2]. These engineers will also need to design connected systems that can efficiently and safely interact with humans during the manufacturing process, e.g., a car assembly line [3]. The focal points of this Special Issue are the smart sensors that enable robots and humans to "see" each other [4–9] and the machine learning algorithms that process these complex data so the robot can make decisions [10–13].

One of the biggest challenges in human–robot collaborations is the unpredictability of human actions [14]. To address this challenge, sensors have been integrated into this collaboration to allow the robot and human operator to "see" each other. The most common way for robots to "see" humans is through three-dimensional cameras, e.g., Microsoft Kinect [15]. These data are then used to help the robots detect humans and avoid collisions. In this Special Issue, Khawaja demonstrates the use of this technology to predict human motion [5]. Based on this predicted path, the robot can follow the operator's movements and be prepared to quickly execute the next step in the task, e.g., tightening a bolt or attaching grommets. This motion prediction framework has been shown to decrease cycle time by up to 25% in the sample task studied (delivering parts and tools to a worker in an automobile assembly task). Another way for the robot to "see" the operator is through a two-dimensional camera. These cameras tend to be used in applications where robots and humans coexist. Yamakawa extended the use of two-dimensional cameras to collaborative applications [4]. A high-speed camera can be used to take images of the operator's hands, which are then processed quickly and accurately using machine learning. This imaging process has been shown to estimate the operator's grasp type in 0.07 milliseconds with 94% accuracy. One limitation of using red–green–blue (RGB) imaging is the difficulty in distinguishing between humans in the foreground and moving objects in the background. Himmelsbach addressed this limitation in the field using thermal imaging [6]. This is especially advantageous for situations where robots can "see" both the operator's workspace and walkways with roaming autonomous vehicles. These autonomous vehicles may inadvertently trigger the robot to slow down or stop. Incorporating thermal imaging allows robots to ignore these roaming robots in the background, resulting in a 50% increase in efficiency. Typically, only a single sensing modality is used to enable the robot to "see" the human operator because these data are difficult to process in real time [14,15]. Amin combined both visual and tactile sensors with the aid of machine learning to quickly process these robust data [8]. Multiple Microsoft Kinect cameras were used to detect a whole human body, while multiple cameras allowed for monitoring a larger workspace. The data from these cameras were fed into a neural network model to determine whether the operator was passing through the workspace, observing the robot, moving too close to the robot for it to work, or interacting with the robot. Tactile sensors on the robot provided additional information about the operator: no interaction, intentional contact, or incidental contact. These systems combined were able to "see" and "feel" the operator with 99% accuracy. Besides human-to-robot communication, messaging the other way from robot to human is also important because humans can become nervous around fast-moving robots. To address this, Grushko studied how robots can use haptic feedback to "talk" to a human [7]. Operators showed a 45% improvement in completion time when haptic feedback was used to inform the operator of the robot's planned trajectory. The feedback was provided through vibrations on a wearable device on the operator's glove. Another way humans and robots interact is through teaching, e.g., when the operator teaches the robot to perform a task. Typically, a robot is taught to perform a non-contact task such as spraying. Tasks that involve contact, such as picking up an object, require synchronous sensing or both traction and contact. Zhang developed a sensor that measures both of these forces using a single sensor, as opposed to a multiple-sensor arrangement [9]. This compact sensor arrangement utilizes strain gauges mounted on a cylindrical sleeve. This sensor was validated for a drawer-opening experiment where the robot was taught to approach a drawer, grab the drawer, open the drawer, and then close the drawer.

During human–robot collaborations, a robot collects data and uses them to make decisions. Due to the non-deterministic nature of these data, machine learning is used for this processing [16]. The articles in this Special Issue demonstrate the power of machine learning to optimize task scheduling, detect collisions, collaborate with more than one person, and read social cues of a person. Scheduling tasks for human–robot collaborations in a production setting can be difficult as there are uncertainties that cannot be predicted and coded a priori offline, e.g., skill differences between human operators. Pupa's online framework, which leverages the parallelism of human–robot collaboration, is one way to address this issue [10]. This novel framework has been shown to adapt to different human operator skills and reallocate task steps if the robot becomes unavailable. To accomplish this, a database was created to store the steps needed to accomplish a task. Then, a scheduler algorithm chose the most suitable task for each actor (robot or human), accounting for the operator's skill level. The task monitoring component of the framework was fed back to the database to determine which details of the task were left to accomplish. While collaborating on these tasks, there are many points on articulated robots that can collide with the operator and cause injury. The location and magnitude of these collisions can be difficult to categorize. A neural network model has been previously developed to determine when a collision has occurred [17] so the robot can adjust its force and avoid an accident. Kwon expanded this neural network to include where on the robot the collision occurred [11]. This work is important for safety, especially as robots become more complicated with more articulations. Typically, these robots collaborate with a single human. Zou used N-player game theory to extend the collaborative ability of a robot to interact with two humans [12]. This theory utilized a recursive least-squares algorithm underlying a novel controller that allowed the robot to adapt to a human's response. This controller was validated in a simulation where a robot helped two humans carry a table. Compared to a traditional linear quadratic regulator, this N-player game theory controller resulted in the humans exerting less effort. This work has the potential to extend beyond industrial robots to robots that help in homes. Akalin developed a reinforcement learning method to train robots that interact socially at home [13]. The robots were observed interacting with humans using a trial-and-error method to determine an optimal behavior. The robot learned which robot behaviors were desired through human feedback (e.g., facial expressions, vocal laughter) and stored this information in a database for later use.

In summary, human–robot collaborations are a common occurrence. The articles in this Special Issue aim to increase the efficiency and safety of these collaborations. Sensors have been incorporated into the robots and surrounding workspaces so the robot can "see" the human. Humans have been outfitted with sensors as well, so they have additional data to "see" the robot. Finally, machine learning techniques have been developed so the robots can optimize these collaborations.

Funding: This research received no external funding.

Institutional Review Board Statement: Not applicable.

Informed Consent Statement: Not applicable.

Data Availability Statement: Not applicable.

Acknowledgments: I would like to thank my colleagues Vince Wheeler and Paul Craig for their feedback on this editorial.

Conflicts of Interest: There are no conflicts of interest to disclose.

References

1. Arumugam, S.K.; Iyer, E. An Industrial IOT in Engineering and Manufacturing Industries—Benefits and Challenges. *Int. J. Mech. Prod. Eng. Res. Dev.* **2019**, *9*, 2249–6890. [CrossRef]
2. Javaid, M.; Haleem, A.; Singh, R.P.; Rab, S.; Suman, R. Significance of Sensors for Industry 4.0: Roles, Capabilities, and Applications. *Sens. Int.* **2021**, *2*, 100110. [CrossRef]
3. Krüger, J.; Lien, T.K.; Verl, A. Cooperation of Human and Machines in Assembly Lines. *CIRP Ann.* **2009**, *58*, 628–646. [CrossRef]
4. Yamakawa, Y.; Yoshida, K. Teleoperation of High-Speed Robot Hand with High-Speed Finger Position Recognition and High-Accuracy Grasp Type Estimation. *Sensors* **2022**, *22*, 3777. [CrossRef] [PubMed]
5. Khawaja, F.I.; Kanazawa, A.; Kinugawa, J.; Kosuge, K. A Human-Following Motion Planning and Control Scheme for Collaborative Robots Based on Human Motion Prediction. *Sensors* **2021**, *21*, 8229. [CrossRef] [PubMed]
6. Himmelsbach, U.B.; Wendt, T.M.; Hangst, N.; Gawron, P.; Stiglmeier, L. Human–Machine Differentiation in Speed and Separation Monitoring for Improved Efficiency in Human–Robot Collaboration. *Sensors* **2021**, *21*, 7144. [CrossRef] [PubMed]
7. Grushko, S.; Vysocký, A.; Oščádal, P.; Vocetka, M.; Novák, P.; Bobovský, Z. Improved Mutual Understanding for Human-Robot Collaboration: Combining Human-Aware Motion Planning with Haptic Feedback Devices for Communicating Planned Trajectory. *Sensors* **2021**, *21*, 3673. [CrossRef] [PubMed]
8. Mohammadi Amin, F.; Rezayati, M.; van de Venn, H.W.; Karimpour, H. A Mixed-Perception Approach for Safe Human–Robot Collaboration in Industrial Automation. *Sensors* **2020**, *20*, 6347. [CrossRef] [PubMed]
9. Zhang, Z.; Chen, Y.; Zhang, D. Development and Application of a Tandem Force Sensor. *Sensors* **2020**, *20*, 6042. [CrossRef] [PubMed]
10. Pupa, A.; Van Dijk, W.; Brekelmans, C.; Secchi, C. A Resilient and Effective Task Scheduling Approach for Industrial Human-Robot Collaboration. *Sensors* **2022**, *22*, 4901. [CrossRef] [PubMed]
11. Kwon, W.; Jin, Y.; Lee, S.J. Uncertainty-Aware Knowledge Distillation for Collision Identification of Collaborative Robots. *Sensors* **2021**, *21*, 6674. [CrossRef] [PubMed]
12. Zou, R.; Liu, Y.; Zhao, J.; Cai, H. A Framework for Human-Robot-Human Physical Interaction Based on N-Player Game Theory. *Sensors* **2020**, *20*, 5005. [CrossRef] [PubMed]
13. Akalin, N.; Loutfi, A. Reinforcement Learning Approaches in Social Robotics. *Sensors* **2021**, *21*, 1292. [CrossRef] [PubMed]
14. Cherubini, A.; Navarro-Alarcon, D. Sensor-Based Control for Collaborative Robots: Fundamentals, Challenges, and Opportunities. *Front. Neurorobot.* **2021**, *14*, 113. [CrossRef] [PubMed]
15. Arents, J.; Abolins, V.; Judvaitis, J.; Vismanis, O.; Oraby, A.; Ozols, K. Human–Robot Collaboration Trends and Safety Aspects: A Systematic Review. *J. Sens. Actuator Netw.* **2021**, *10*, 48. [CrossRef]
16. Liu, Z.; Liu, Q.; Xu, W.; Wang, L.; Zhou, Z. Robot Learning towards Smart Robotic Manufacturing: A Review. *Robot. Comput.-Integr. Manuf.* **2022**, *77*, 102360. [CrossRef]
17. Heo, Y.J.; Kim, D.; Lee, W.; Kim, H.; Park, J.; Chung, W.K. Collision Detection for Industrial Collaborative Robots: A Deep Learning Approach. *IEEE Robot. Autom. Lett.* **2019**, *4*, 740–746. [CrossRef]

Article

A Resilient and Effective Task Scheduling Approach for Industrial Human-Robot Collaboration

Andrea Pupa [1,2], **Wietse Van Dijk** [1,*], **Christiaan Brekelmans** [1] and **Cristian Secchi** [2]

[1] The Netherlands Organisation for Applied Scientific Research—TNO, 2316 ZL Leiden, The Netherlands; andrea.pupa@unimore.it (A.P.); christiaan.brekelmans@tno.nl (C.B.)

[2] Department of Sciences and Methods of Engineering, University of Modena and Reggio Emilia, 42122 Reggio Emilia, Italy; cristian.secchi@unimore.it

[*] Correspondence: wietse.vandijk@tno.nl

Abstract: Effective task scheduling in human-robot collaboration (HRC) scenarios is one of the great challenges of collaborative robotics. The shared workspace inside an industrial setting brings a lot of uncertainties that cannot be foreseen. A prior offline task scheduling strategy is ineffective in dealing with these uncertainties. In this paper, a novel online framework to achieve a resilient and reliable task schedule is presented. The framework can deal with deviations that occur during operation, different operator skills, error by the human or robot, and substitution of actors, while maintaining an efficient schedule by promoting parallel human-robot work. First, the collaborative job and the possible deviations are represented by AND/OR graphs. Subsequently, the proposed architecture chooses the most suitable path to improve the collaboration. If some failures occur, the AND/OR graph is adapted locally, allowing the collaboration to be completed. The framework is validated in an industrial assembly scenario with a Franka Emika Panda collaborative robot.

Keywords: human-robot collaboration; human-centered robotics; task planning

Citation: Pupa, A.; Van Dijk, W.; Brekelmans, C.; Secchi, C. A Resilient and Effective Task Scheduling Approach for Industrial Human-Robot Collaboration. *Sensors* **2022**, *22*, 4901. https://doi.org/10.3390/s22134901

Academic Editor: Anne Schmitz

Received: 4 June 2022
Accepted: 27 June 2022
Published: 29 June 2022

Publisher's Note: MDPI stays neutral with regard to jurisdictional claims in published maps and institutional affiliations.

1. Introduction

Industrial applications where human and robots work closely together are becoming the new paradigm of industrial settings [1]. Collaborative robots can take over repetitive, challenging, or dangerous tasks improving the well-being of the operators [2]. There are multiple challenges that emerge from this close collaboration. On one hand, the absence of barriers makes it necessary to pay close attention to how to guarantee the safety of the operator [3–5]. On the other hand, it becomes necessary to understand how to create a synergy between humans and robots that is as natural as possible, making the most out of the collaboration [6,7]. Therefore, a strategy on how to allocate and schedule tasks between humans and robots is crucial to improve the human-robot team.

The basis of this strategy is an investigation of how the different tasks that make up the work can be distributed best among the actors in the nominal situation, the task allocation problem. The individual tasks are subject to constraints that prescribe how, when, and by whom the tasks can be executed.

The characteristics of a good task allocation are captured in the fluency concept, which relates to how well the operator and the robot are adapted to each other. Fluent collaboration benefits the work execution and the job-quality of the human operator. These aspects are not captured by optimizing for task efficiency alone. There are subjective and objective metrics for fluency available of which the latter ones can directly be used as an optimization criterion. The objective metrics include the relative portion of functional and non-functional delays of the actors, and the amount of parallel work [8].

The task allocation problem can be solved during the design phase and results in the nominal schedule. This procedure has widely been investigated in other works. Refs. [9–12] focused on heterogeneous multi-agent task allocation in an industrial setting. While other authors, e.g., [13–15], propose to model the human-robot collaboration (HRC) problem as a nonlinear optimization problem. These strategies allow us to find the best nominal schedule.

Even if optimal, a nominal task schedule cannot guarantee a real improvement of the collaboration. In the task execution phase, many factors come into play that cannot be anticipated within the nominal schedule. This requires an efficient and resilient team that can anticipate and adequately respond to these abnormalities. The requirements for an efficient team of human and automated agents, i.e., robots have been formulated in [16] and contributed to a design method in [17]. The requirements and the design method were targeted at general automation challenges and applied in open ended scenarios. The industrial practice is much more constrained and the irregularities that exist have several common causes. Identifying these causes can help the design of resilient solutions for task allocation.

Firstly, it may happen that not all the actors are always available to carry out the collaboration, e.g., the robot has been dispatched to another work station. In this case it would be highly inefficient to again design the job considering only the remaining actors. It would be more convenient to consider this actor availability problem from the beginning and adapt the schedule accordingly. Secondly, the tasks may depend on each other, i.e., there could be precedence constraints, and this interdependence must be considered while ensuring the parallelism between the actors. Thirdly, the operators, as human beings, are inherently different from each other, each with their own skills, capabilities, or individual preferences. Some of them may need or prefer to be guided more during the work, e.g., newly hired workers. For others, on the other hand, an excess of information could be annoying and counterproductive from the job quality perspective. Lastly, there is always the possibility that failures will occur that prevent the correct execution. Therefore, a fallback scenario must be considered when designing a scheduling strategy for HRC scenarios.

This paper firstly presents how an industrial HRC process, with its irregularities during execution time, can be formalized into a set of interdependent tasks. Secondly, it proposes a novel adaptive task scheduling framework for collaborative cells. As a start, the framework uses a database to understand how the collaborative job is composed out of multiple interdependent tasks. Subsequently, at runtime, it monitors the task execution to understand the human operator skills and the task result, i.e., failure or success. This information is exploited to adapt the schedule online, making the framework flexible and able to face most of the situations that may arise in a real HRC scenario.

The main contributions of this paper are:

- A formulation of four primitive situations that encompass most of the scenarios that could occur in a real HRC.
- A novel adaptive task scheduling framework that is effective and applicable to the formalized situations, i.e., suitable for a real industrial application.
- The validation of the proposed framework in several variants, one for each situation, of the same experimental scenario, proving the effectiveness of the framework.

The paper is organized as follows: Section 2 presents the review of related works, while Section 3 formalizes the task scheduling problem for HRC. In Section 4, the overall proposed architecture is detailed, and in Section 5, the scheduling strategy is defined. Sections 6 and 7 address the handling of different human operators and errors, respectively. Lastly, Section 8 summarizes the experimental validation of the proposed architecture, while Section 9 sums the conclusions and presents suggestions for future research.

2. Related Works

Different approaches were presented in the literature to deal with the problem of multi-agent task scheduling. In [18], the author proposes a heuristic method in order to allocate and schedule the tasks between multiple processors. The approach is based on the communication time between the processors and the number of successor tasks, making it suitable for the problem of handling precedence constraints. In [19], a branch-and-bound procedure and a climbing discrepancy search heuristic for the parallel machine scheduling problem with precedence constraints and sequence dependent setup times is proposed. This algorithm can minimize the sum of the completion time and maximum lateness. In [20], the authors propose a Load-Balance Scheduling Algorithm, which allows for allocating and scheduling the tasks in a multiprocessor system. The idea is to use an Earliest Deadline First (EDF) heuristic to first create an n ordered tasks list. Then, based on the actual workload, to allocate the task to one processor. In general, these solutions cannot be directly applied in an HRC application, as they consider the presence of homogeneous actors.

In [21], the authors model the HRC working process as a chessboard setting, where the decision of each actor is described by the chess piece move and formulated as a Markov game model. To optimize it, they propose a decentralized Deep-Q-network based MARL (DQN-MARL) algorithm. In [22], the task scheduling is formulated as a Mixed-Integer Linear Programming Problem (MILP) inspired by the Multimode Multiprocessor Task Scheduling Problem. The cost function aims at reducing the total makespan, and the solution is obtained with a constraint programming model and the use of a Genetic Algorithm (GA). In [23], instead, the authors propose the use of a Simulated Annealing (SA) algorithm to find the optimal solution.

These works, however, do not consider the differences between individual operators. As the scheduling procedure adapts to the robot that is available, e.g., considering the robot workspace, it should also be able to modify the schedule based on the human operator that is currently going to perform the collaborative job. In [24], an integrated task allocation and task scheduling strategy for HRC is proposed. The task allocation is solved offline exploiting a two-level feedforward optimization. Furthermore, this strategy is enriched with a feedback procedure based on mutual trust to re-allocate the tasks online. Furthermore, in [25], the authors propose a multi-criteria decision-making framework for task allocation, which generates a solution that best matches the criteria you want to optimize. Moreover, in the case of unexpected events, the algorithm can be exploited for re-scheduling the remaining tasks. In [26], a two-layer dynamic rescheduling framework is presented. The first layer builds the nominal schedule solving offline an MILP problem, while the second layer exploits the real human execution time to reschedule online the tasks. In this paper, the job quality is considered inside the MILP problem both as data to be optimized and as constraints. This work has been further extended in [27] to integrate the scheduling strategy with the safety required by the robot trajectory planner. Moreover, in [28], the authors propose a genetic algorithm that exploits human and robot data, e.g., ergonomics or capabilities, to optimally schedule the tasks in an HRC scenario. The actor characteristics are given offline as input by the user. In [29], a two-level abstraction and allocation for an HRC scenario is presented. The first layer exploits the use of the A* algorithm to optimize a cost function. The second layer, instead, handles the task execution and the respective failures. If the system detects some errors, it is possible to reschedule the tasks recomputing the optimal solution.

Even if they adapt online based on what is currently happening, none of the proposed works are so general to handle all the considered primitive situations at the same time:

1. Scarce resources;
2. Parallelism between the actors;
3. Different human operator skills;
4. Errors during the execution.

The proposed approach focuses on developing a resilient scheduling framework, that is general and applicable to real industrial scenarios. Differently from other works, e.g., ref. [26,30], that are more oriented on the optimization and reduction of the total makespan. The framework does not generate the nominal schedule, it is therefore, to a large extent, supplementary to existing methods.

3. Problem Statement

Industrial Human-Robot Collaboration is characterized by multiple agents that work toward a common goal. In this paper, a situation where a human operator H and a robot R collaborate in a shared workspace, namely the collaborative cell, is considered. The human and the robot have a pre-defined task distribution, which is defined by a set of nominal task schedules. The agents must perform their respective tasks in order to complete the collaborative job. A job typically represents an industrial process, such as the assembly of a product. In this work, it is assumed that the nominal task schedules have already been computed, e.g., exploiting [6,31].

An effective representation of a general collaborative job can be achieved with an AND/OR graph $\mathcal{G} = (T, E)$, as shown in Figure 1 [32]. Each node represents the task, T_i, while each directed edge E_{ij} means that the parent task T_i must be executed after the child task T_j. Multiple unlinked edges sharing the same parent represent an OR constraint. This constraint imposes that the parent task can only be executed if at least one of the children has been completed, e.g., T_4 and T_5 in Figure 1. Thanks to the OR constraint, it is possible to define multiple paths that lead from the same starting point to the same end point, namely *equivalent paths*, e.g., $T_2 + T_5 \equiv T_1 + T_3 + T_4$ in Figure 1. In this work, the first tasks that belong to equivalent paths are defined as *equivalent tasks*, e.g., T_1 is equivalent to T_2. When multiple edges share the same parent and are connected with an arc, they model an AND constraint. This constraint imposes that the parent task can be executed only if all the children have been completed, e.g., T_3 and T_4. The job is finished when the final task T_6 is completed; this does not require that all tasks in the job are completed.

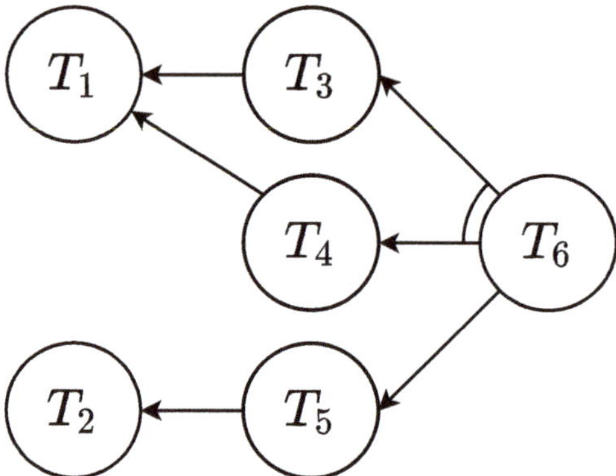

Figure 1. AND/OR graph representation. The unlinked edges represent the OR relations, while the ones connected with an arc represent the AND relations. For example, in the figure, T_6 can be executed after both T_3 and T_4 or after only T_5.

Thanks to their structure, the AND/OR graphs intrinsically model both the parallelism between the actors, which is required to reduce the waiting times, and the precedence constraints between the tasks. Furthermore, through the OR constraints it is possible to represent already in the design phase all the different situations that may arise due to the available actor capabilities. This might be the temporary absence of an actor such as a robot that is in maintenance, or an inexperienced operator that needs additional task instructions. In this paper, only human operators are considered to have different skill levels. To handle this, each human task is defined with the minimum expertise level that is required from the operator to be able to perform that task. The robot is assumed to have a fixed set of capabilities that are known during design time, which is common in industrial settings.

Scheduling the task does not ensure that the actor will always perform it correctly. For this reason, each task T_i is associated with recovery actions that allow to restore the nominal behavior of the human-robot team. This set of actions can be very complex and, in turn, represented with a AND/OR graph. For ease of reading, this set of actions will be represented in the paper as a single task. Moreover, the collaborative cell is enriched with a task monitoring strategy that allows to check if the task execution has been successful and, in case of failure, it provides information about the error. To achieve this, many algorithms are already available in the literature [33–35].

In this work, we aim at designing a scheduling framework that:

- Takes as input the skills of the human operator and its knowledge in the collaborative job, automatically schedules online tasks between the different actors choosing the most suitable path for the situation on the AND/OR tree.
- Automatically handles most situations and the failures that may occur within an industrial scenario, making it suitable for HRC in industrial settings.

4. Architecture

The proposed framework is shown in Figure 2, where different components may be distinguished:

- The *Database* block is responsible for storing all the information regarding the tasks composing the collaborative job, through the *Task Details* block, and analyzing all the data to evaluate the desired metrics, this is achieved in the *Data Evaluation* block.
- The *Scheduler* block, which is the core of the framework, takes care of choosing the most suitable task for each actor online, considering the human operator capabilities and the parallelism in the collaborative job.
- The *Task Monitoring* block oversees the task execution and communicates with the Database if the task has been completed or not. Moreover, in case of failure, it gives information about the error so that proper recovery actions can be scheduled.
- The *Human Capabilities* block takes the result of the data analysis to estimate the human operator capabilities and knowledge.

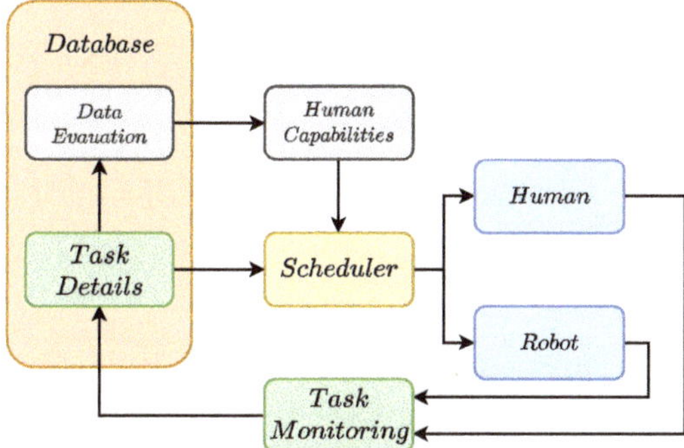

Figure 2. The overall architecture.

It is worth noting that in Figure 2 the overall framework has been presented. However, the grey blocks are out of the scope of this paper and have only been included for completeness.

The overall procedure starts offline with the design of the AND/OR graph and inserting all the tasks inside the database. Each task T_i is associated with its description and requirements, i.e., the actor that must perform it, the precedence constraints as AND/OR constraints, and the minimal required expertise level. For example, T_6 in Figure 1 can be defined as the following:

$$T_6 : \{description : "Final",$$
$$requirements :$$
$$\{actor : H,$$
$$precedence : "(T_3 \wedge T_4) \vee T_5'',$$
$$level : 1\}\}$$

Subsequently, all these task definitions are passed to the scheduler, which searches along the AND/OR tree to choose, for each actor, the tasks that best suit the current scenario, e.g., operator experience, actor availability. This is achieved, exploiting the data coming from the others blocks. Once a task has been chosen, it is forwarded to the respective actor who must perform it.

At this point, the task monitoring block continuously checks which tasks have been concluded and what the final task result is, i.e., success or failure. This information is also stored inside the database and used for two different purposes. Firstly, in the case of error, it is exploited to choose a proper set of recovery tasks that may be performed in order to continue the collaboration. Secondly, it is used by the data evaluation strategy in order to increase or reduce the human operator expertise level. Finally, the scheduler is triggered again to assign another task, until the collaborative job is concluded.

5. Scheduler

The scheduler is the core of the proposed framework and has the goal of distributing the tasks among the available actors, i.e., humans and robots. It requires as input both the AND/OR graph of the collaborative job, already defined in a previous design phase, and the availability of actors along with their skills and capabilities. These two inputs are exploited to choose at runtime the best path for the current scenario, tailoring the specific needs of the human operator during the collaboration. Thus, the scheduler aims at improving the synergy between the actors with a consequent improvement of the HRC.

The entire scheduling pipeline is implemented according to the pseudo-code reported in Algorithm 1.

The scheduler needs as input the AND/OR graph $\mathcal{G}$ coming from the database, the set of available actors $\mathcal{A}$, and the actual human operator expertise level H_L (Line 1). It immediately sets to false the variable End_J, which is used to identify when the collaborative job is concluded and instantiate to true the list of Boolean variables $\mathcal{F}_A$, which indicates if each actor is free in order to start a new task (Lines 2–5).

Then, the algorithm enters a while loop where it continuously executes the overall pipeline until the collaborative job has been finished. Firstly, the scheduler checks for each task T if the requirements have been satisfied (Line 9). This is translated in checking whether the respective agent is available to accept a new task and if the precedence tasks have been executed. If this is the case, the algorithm checks if the level of the human operator is sufficient to execute the task (Line 10). This condition allows the scheduler to handle the "Different Operator Skills" situation detailed in Section 6. It is worth noting that if the actor of task T is not the human operator, the function **checkLevel**() always returns true. This is because the robot always has the required skills to perform the tasks assigned to it. If the level check is also successful, the task T is scheduled to the respective actor, who is immediately marked as unavailable, and it is added to the list of the tasks to be monitored $\mathcal{T}_A$ (Lines 11 and 12). At this point, all the equivalent tasks of T are discharged and removed from the graph. Since the scheduler works online, this procedure is necessary because otherwise it could happen that the algorithm runs through two parallel paths, which is unnecessary to reach the goal.

Algorithm 1 Scheduler()

1: **Require:** $\mathcal{G}, \mathcal{A}, H_L$
2: $End_J \leftarrow false$
3: **for** $a \in \mathcal{A}$ **do**
4:    $Free_a \leftarrow true$
5:    $\mathcal{F_A} \leftarrow$ **pushback**$(Free_a)$
6: **end for**
7: **while** $End_J = false$ **do**
8:    **for** $T \in \mathcal{G}$ **do**
9:      **if checkRequirements**$(T, \mathcal{F_A})$ **then**
10:       **if checkLevel**(T, G, H_L) **then**
11:         $\mathcal{F_A} \leftarrow$ **setBusy**$(T, \mathcal{F_A})$
12:         $\mathcal{T_A} \leftarrow$ **pushback**(T)
13:         $\mathcal{G} \leftarrow$ **discardEqTasks**(T)
14:       **end if**
15:      **end if**
16:    **end for**
17:    **for** $T \in \mathcal{T_A}$ **do**
18:      $R, E \leftarrow$ **taskMonitor**(T)
19:      **if** $R = \varnothing$ **then**
20:       *continue*
21:      **else if** $R = Executed$ **then**
22:       **if isFinal**(T) **then** $End_J \leftarrow true$
23:      **end if**
24:      **else if** $R = Failed$ **then**
25:       $\mathcal{G} \leftarrow$ **applyRecovery**$(T, E, \mathcal{G})$
26:      **end if**
27:      $\mathcal{T_A} \leftarrow$ **remove**$(T, \mathcal{T_A})$
28:      $\mathcal{F_A} \leftarrow$ **setFree**$(T, \mathcal{F_A})$
29:    **end for**
30: **end while**

Then, the scheduler uses the database to check, for all the active tasks, the information regarding the task monitoring (Line 18). The task monitoring strategy can be implemented using several strategies available in the literature as, e.g., ref. [36,37], and allows to detect critical deviations in the collaboration. In turn, critical deviations from the nominal behavior are translated into a failure, which is stored inside the database. If no results are available, the task is not concluded, and the scheduler continues to inspect the other tasks (Line 20). In the other cases, the behavior of the scheduler depends on the type of the result. If the task has been concluded, the algorithm only checks if this is a final task, and the job is finished (Line 22). If the actor has failed, the scheduler locally adapts the graph in order to generate a recovery procedure (Line 22). This local adaptation depends on the type of error E coming from the monitoring and it is further detailed in Section 7. Lastly, the concluded task is removed from $\mathcal{T_A}$ and the actor is marked as free (Lines 27 and 28).

6. Different Operator Skills

By definition, an HRC application is characterized by the presence of both humans and robots. The differences between these two actors are quite intuitive. Human operators are capable of very complex tasks, improving execution every time, but they can hardly reach or work in hazardous environments. Robots, on the other hand, are less affected by hazards in the surrounding environment, but they are not able to intrinsically learn from their last task execution.

Unlike robots, the human is never a constant factor, human operators have different skills, which can be improved or acquired. When starting on a new type of job, it is very likely that a human operator needs detailed instructions on the tasks that need to be performed. Therefore, it becomes useful to divide the work into several tasks that are easy to comprehend, allowing the operator to acquire the necessary skills and knowledge. At some point, the operator will acquire much more experience in the collaborative work, and it may be more convenient to combine some tasks into one, avoiding unnecessary fragmentation. An example can be a complex wiring activity: a new human operator may require wire-by-wire instructions, while for expert operators a single instruction with an overview of all the wires is sufficient.

The framework handles this situation in two different phases: First, during the design of the AND/OR graph and the insertion of the tasks inside the database. Second, adding the check level step inside the scheduler; see Algorithm 1 in Section 5. This constraint is implemented according to the pseudocode in Algorithm ?

Algorithm 2 checkLevel()

1: **Require:** $T, \mathcal{G}, H_L$
2:   $a \leftarrow \textbf{getActor}(T)$
3:   **if** $a \neq H$ **then**
4:     **return** $true$
5:   **else**
6:     $\mathcal{T}_E \leftarrow getEqTasks(T, \mathcal{G})$
7:     **for** $t \in \mathcal{T}_E$ **do**
8:       **if** $t.level > T.level$ **and** $t.level \leq H_L$ **then**
9:         **return** $false$
10:       **end if**
11:     **end for**
12:     **if** $T.level \leq H_L$ **then**
13:       **return** $true$
14:     **end if**
15:     **return** $false$
16: **end if**

The level checking needs as input the task to be analyzed T, the AND/OR graph $\mathcal{G}$, and the actual human operator expertise level H_L (Line 1). It immediately gets the actor a that must perform the task and, if it is not a human operator, it returns $true$ allowing to schedule the task (Lines 2–4). If the actor is the human, it is necessary to investigate more to decide if T is the most suitable task. The algorithm firstly builds a list of all the equivalent tasks $\mathcal{T}_E$ analyzing the AND/OR graph (Line 6). Then, for each equivalent task, it checks if there is a task that requires a greater knowledge than the one that is required by the current analyzed task and if this knowledge is still admissible for the current human operator. If this is the case, the algorithm returns $false$ (Line 9). It is worth noting that this part of the code allows to always choose the task that best suits the human knowledge level, improving the collaboration and the job quality for the human operator. If no better tasks are available, the algorithm checks if this task can be performed by the human operator before allowing its execution (Line 13).

7. Error Representation

During the collaboration, it is very unlikely that everything happens exactly as planned. Human operators and robots will make errors during the task execution, preventing the nominal behavior. The framework must be able to handle these errors and adapt the behavior of the actors accordingly.

Errors can be classified into two categories based on the way the error is handled:

- *Restorable Error*, which is an error that does not preclude the task and, after some checking and restoring actions, it is possible to retry the execution. This may happen when the robot accidentally hits something and, after the human operator confirms that there are no damages or safety problems, the task is assigned again to the robot, i.e., the repair task brings the product to the state *before* the erroneous task.

- *Non-Restorable Error*, which is an error that precludes the task execution, and it is necessary to execute another task to continue the collaborative job. This may happen when the robot places an object with a wrong orientation and the scheduler assigns a new task to replace the object in the correct pose, i.e., the repair task brings the product to the state *after* the erroneous task.

In the proposed framework, both categories of errors are handled by locally adapting the AND/OR graph. This allows the scheduler to continue working without losing its generality. It is worth noting that some errors are of a type that do not allow the collaboration to continue, e.g., a safety violation. Since these cases require more severe intervention of an operator and a consequent reset of the collaborative job, they are not covered by the proposed framework.

7.1. Restorable Error

After a restorable error occurs, the nominal behavior of the actors could be ideally restored since the execution of the task is not precluded. The way the framework handles this situation is illustrated in Figure 3b.

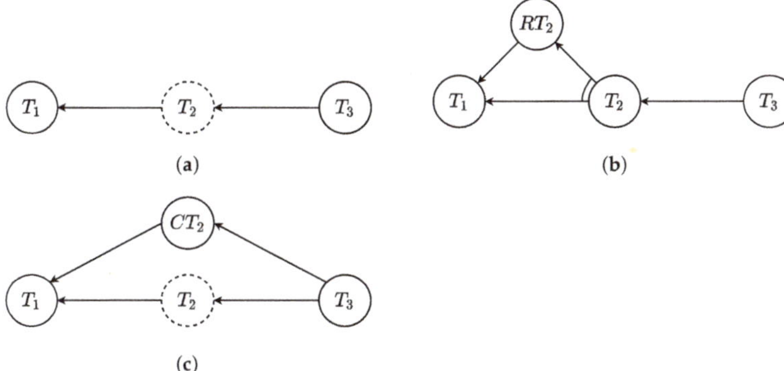

Figure 3. Representation of the errors and the adaptation of the AND/OR graph. The dashed line in (a) Means that T_2 failed. (b) Shows the restoring path, while (c) shows the corrective one.

A new graph path that goes from the previous task to the failed one is generated. This path contains all the tasks that must be executed to restore the nominal behavior, e.g., ask the human if it is safe, and is attached to the original AND/OR graph with an AND constraint. Then, the task that has previously failed is reset and marked as a task that must be still scheduled. According to Algorithm 1 in Section 5, with this strategy the scheduler is forced to go through this new restoring path before scheduling again the task that had previously failed.

7.2. Non-Restorable Error

When the error compromises the correct execution of the collaborative job, the restoring procedure may not be enough. In this case, it is necessary to ask one of the actors to execute a set of tasks to recover from the failure and continue the collaboration with the next tasks. The procedure implemented in the framework is illustrated in Figure 3c.

A new graph path is inserted that goes from the previous task to the task that follow the failed one. This path is composed by all the tasks necessary to correct from the failure, e.g., adjust the orientation, and it is attached with an OR constraint. According to Algorithm 1 in Section 5, this new path allows the scheduler to continue the collaborative job, omitting the previously failed task.

8. Experiments

The proposed framework has been experimentally validated in a human-robot collaborative scenario, where the human operator works together with a Franka Emika Panda, a 7-DoF collaborative robot. Several experiments have been carried out, focusing on four situations of which a video is available as supplementary material:

- Different human operator skills;
- Substitution of human actions with robot action;
- Error handling;
- Parallelism of the actors with resource sharing.

During the experiments, the human operator has been guided exploiting the Arkite's Human Interface Mate (HIM), which also enables the interaction. The complete setup for the experiment is shown in Figure 4.

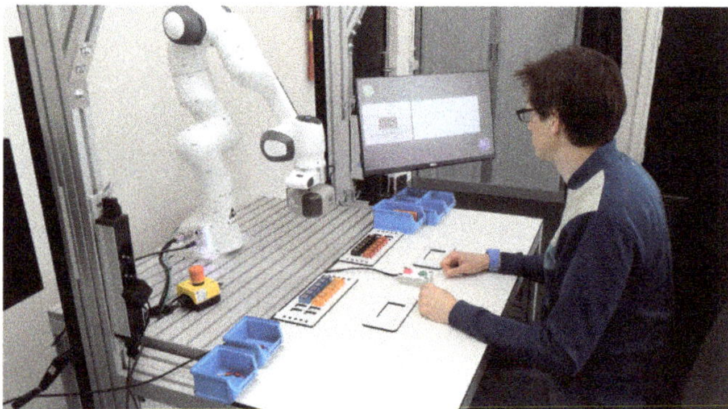

Figure 4. Setup of the experiment. The image shows all the equipment used during the experiments.

All the software components were developed in Python 3.8 and exploiting Apache Kafka, while the Franka Emika Panda is position controlled using the MoveIt Motion Planning Framework and the standard ROS libraries.

The first three situations can be represented with the AND/OR tree shown in Figure 5a, where different paths can be distinguished. The yellow one requires only the human operator to perform the job, while on the blue path, the robot performs some of the tasks. In the last part, the tree divides based on the expertise of the human operator. A detailed description of all the tasks composing the collaborative job is shown in Table 1. It is worth noting that the tasks related to the pick and place of both the casing and the connectors are part of the first phase, while the ones related to the wiring are part of the second phase. The parallel work situation, instead, is schematized in Figure 5b. This represents a double assembly job where both a robot and an expert human operator are available. For ease of reading, the tasks that have been doubled are denoted with the subscripts A and B.

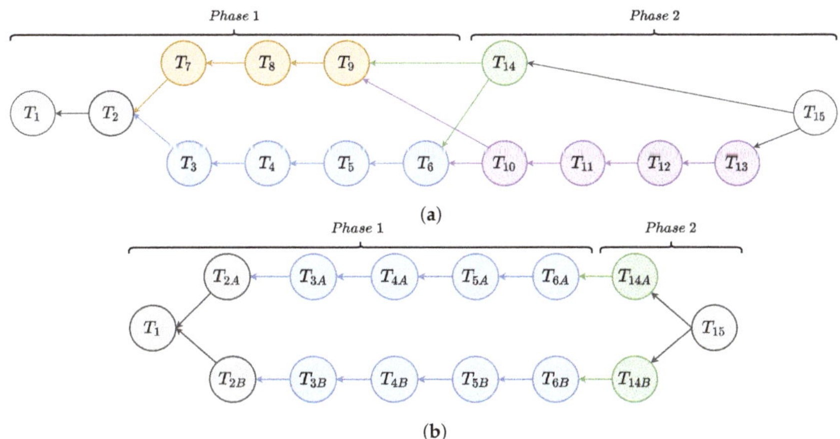

Figure 5. AND/OR graphs representing all the experiments performed. (**a**) Representation of the first three scenarios; (**b**) representation of the parallel work.

Table 1. Tasks description of AND/OR graph in Figure 5a.

Task Index	Description	Agent	Phase
1	Start the job.	H	
2	Pick and Place casing.	H	
3	Move in Home.	R	Phase 1
4–6	Robot Pick and Place 3 connectors.	R	
7–9	Human Pick and Place 3 connectors.	H	
10–13	Connect 4 wires one by one—not expert user.	H	
14	Connect all the wires—expert user.	H	Phase 2
15	Confirm completion.	H	

8.1. Different Skills

These experiments mainly focus on the second phase of the collaborative scenario. Initially, the operator is inexperienced in the collaborative work to be pursued. The collaboration starts when the operator confirms being ready and the scheduler immediately asks to place the casing in the correct spot, i.e., T_1 and T_2. At this point, the robot starts to pick and place all the connectors, while the human operator is waiting. This is because no parallel tasks are available, see the blue path in Figure 5a. Once the robot has completed the tasks, the collaborative job continues with the second phase. Since the operator is learning, the operator receives detailed step-by-step instructions about the wiring, represented by the purple path in Figure 5b. Exploiting the input coming from the human capabilities block, the scheduler is aware of the human expertise level.

After multiple executions of the collaborative job, the human operator became more expert, and the *Human Capabilities* block detects an upgrade of the user level. In this context, displaying the wiring instructions step by step may be tedious and annoying, with a consequent reduction of the job quality. The scheduler can adapt and chooses the green path in Figure 5a, which merges $T_{10} - T_{13}$ in one single task T_{14} so the operator only receives high level instructions.

This experiment validates the functionality of the **checkLevel**() constraint presented in Section 6, demonstrating that the overall framework can adapt to different human operators.

8.2. Actor Substitution

This experiment simulates the situation where, for whatever reason, the robot is unavailable. This may happen when the tool of the robot is under maintenance, and a fallback strategy is required. In the analyzed scenario, this applies to the robot tasks during the first phase. To simulate such unavailability, the robot actor is removed from the actors list and, without changing the tree, the job is started. As before, the human operator confirms and places the casing in the correct position. At this point, the scheduler can only go through the only human path, the yellow one in Figure 5a, asking the human operator to pick and place the connectors.

8.3. Error Handling

In this experiment, the human operator intentionally triggers robot errors, which are subsequently handled by the framework. The framework uses the task monitoring block of Section 7 to detect the correct execution of the task. The first time a task fails, a restorable error is triggered. If the same task fails another time, a non-restoring action is required.

During the execution of T_3, the human operator hinders the robot, which immediately stops for safety reasons. At this point, the tree is locally adapted by inserting a restoring task. Thanks to this task, the scheduler firstly asks the human operator if they are safe and, if possible, moves the robot to the home position ready to retry the execution. This is shown in Figure 6b,c. Subsequently, the human operator hinders the robot again, causing another failure. Since the task failed twice, the task monitoring generates a not restorable error. This is because it would be better to ask the human operator to execute this task. For this reason, the AND/OR graph is modified again adding a new path to reach T_4. As before, the human operator must confirm that everything is fine and the robot goes back in a home position, but this time the scheduler asks the human to pick and place the connector in the correct place. All the steps are illustrated in Figure 6d,e. From this point, the robot can resume its work.

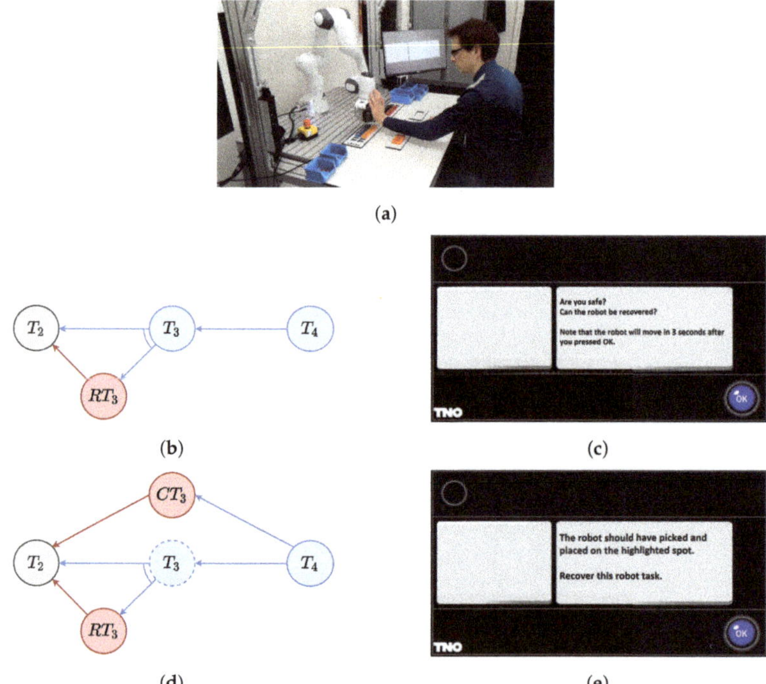

(a)

(b)

(c)

(d)

(e)

Figure 6. Failure of of T_3. (**a**) The human hindering the robot. (**b,d**) The AND/OR graph locally adapted. In (**c**), the Arkite system asks if the human is safe before moving again, while in (**e**), the instruction to guide the human operator in correcting the task is shown.

8.4. Parallel Work

The previous situations validate the effectiveness of the framework and its features in a sequential scenario. The framework is also capable of scheduling parallel tasks. In this case, the robot works in parallel, sharing resources with the human operator. This is shown in the last scenario, where the human and the robot work on two products at the same time while sharing a workspace. Without losing generality, the scenario has been simplified avoiding the possibility of different paths.

When the collaborative job starts, the scheduler immediately asks the human operator to pick and place the first casing. Subsequently, the robot starts to insert the three connectors while the human operator places the second casing. Once the robot has finished working on the first casing, the scheduler asks the human to make the wiring. It is worth noting that since the human operator is an expert user, the wiring is composed by a single task. In the meantime, the robot finishes putting the connectors inside the second casing, allowing the human to conclude the job with the second wiring.

9. Conclusions and Future Works

In this paper, a framework for resilient and effective task scheduling in a collaborative industrial scenario has been built. With this framework, industrial collaborative human-robot processes can be modeled and executed. This includes the nominal execution behavior as well as alternative execution behaviors that might be caused by common situations at the workplace. These situations include adjusting to the skill level of the operator, actor substitution, and error handling.

Starting from the definition of the tasks composing the collaborative job, namely the AND/OR graph, the scheduler is responsible for deciding what is the most suitable task for the actors to execute at each point in time. The task monitoring strategy is exploited to understand if the task has been correctly executed and, if necessary, to locally adapt the AND/OR graph to prevent that a task failure will result in a halted process. Moreover, the scheduler exploits the information about the human expertise to improve the HRC. In the paper, the methodologies used to online estimate the human expertise have not been investigated. In this paper we have used a description that included a single human operator and robot, but the framework can be extended to larger teams without major modifications.

The experimental validation has been conducted in a scenario with different execution variations, showing and proving that the proposed framework handles most situations that may occur in a real industrial setting.

Future works aim at improving the scheduling strategy in order to also choose the path that optimizes a desired cost function, e.g., exploiting the A* algorithm over the AND/OR graph [12]. Moreover, the framework should be tested in a user study that involves a real-world industrial scenario. This would lead to a further validation and improvement of the overall framework. Lastly, the framework could be extended to also handle the synchronization between the actors. In this scenario, some strategies such as [38,39] could be used to enable the mutual communication between the two actors.

Supplementary Materials: The following supporting information can be downloaded at: https://www.mdpi.com/article/10.3390/s22134901/s1, Video S1: A Resilient and Effective Task Scheduling Approach for Industrial Human-Robot Collaboration.

Author Contributions: Conceptualization, A.P., W.V.D. and C.B.; methodology, A.P. and W.V.D.; software, A.P., W.V.D. and C.B.; validation, A.P.; writing—original draft preparation, A.P.; writing—review and editing, W.V.D., C.B. and C.S.; visualization, A.P.; supervision, W.V.D. and C.S. All authors have read and agreed to the published version of the manuscript.

Funding: This project has received funding from the European Union's Horizon 2020 research and innovation program under grant agreement No. 818087 (ROSSINI) and SMITZH (the smart industry hub of the greater Rotterdam—The Hague area) coordinated by InnovationQuarter and is supported by Rotterdam-The Hague Metropolitan Region, the Province of South Holland, and the Ministry of Economic Affairs and Climate Policy.

Institutional Review Board Statement: Not applicable.

Informed Consent Statement: Not applicable.

Data Availability Statement: Not applicable.

Conflicts of Interest: The authors declare no conflict of interest.

References

1. Villani, V.; Pini, F.; Leali, F.; Secchi, C. Survey on human—Robot collaboration in industrial settings: Safety, intuitive interfaces and applications. *Mechatronics* **2018**, *55*, 248–266. [CrossRef]
2. Pham, Q.; Madhavan, R.; Righetti, L.; Smart, W.; Chatila, R. The impact of robotics and automation on working conditions and employment. *IEEE Robot. Autom. Mag.* **2018**, *25*, 126–128. [CrossRef]
3. Pupa, A.; Arrfou, M.; Andreoni, G.; Secchi, C. A Safety-Aware Kinodynamic Architecture for Human-Robot Collaboration. *IEEE Robot. Autom. Lett.* **2021**, *6*, 4465–4471. [CrossRef]
4. Benzi, F.; Secchi, C. An Optimization Approach for a Robust and Flexible Control in Collaborative Applications. In Proceedings of the 2021 IEEE International Conference on Robotics and Automation (ICRA), Xi'an, China, 30 May–5 June 2021; pp. 3575–3581. [CrossRef]
5. Merckaert, K.; Convens, B.; Wu, C.J.; Roncone, A.; Nicotra, M.M.; Vanderborght, B. Real-time motion control of robotic manipulators for safe human—Robot coexistence. *Robot. Comput.-Integr. Manuf.* **2022**, *73*, 102223. [CrossRef]
6. Pupa, A.; Landi, C.T.; Bertolani, M.; Secchi, C. A Dynamic Architecture for Task Assignment and Scheduling for Collaborative Robotic Cells. In *Human-Friendly Robotics 2020*; Saveriano, M., Renaudo, E., Rodríguez-Sánchez, A., Piater, J., Eds.; Springer International Publishing: Cham, Switzerland, 2021; pp. 74–88.
7. Sheridan, T.B. Human-Robot Interaction: Status and Challenges. *Hum. Factors J. Hum. Factors Ergon. Soc.* **2016**, *58*, 525–532. [CrossRef]
8. Hoffman, G. Evaluating Fluency in Human-Robot Collaboration. *IEEE Trans. Hum.-Mach. Syst.* **2019**, *49*, 209–218. [CrossRef]
9. He, Y.; Shao, Z.; Xiao, B.; Zhuge, Q.; Sha, E. Reliability driven task scheduling for heterogeneous systems. In Proceedings of the Fifteenth IASTED International Conference on Parallel and Distributed Computing and Systems, Marina del Rey, CA, USA, 3–5 November 2003; Volume 1, pp. 465–470.
10. Qin, X.; Xie, T. An availability-aware task scheduling strategy for heterogeneous systems. *IEEE Trans. Comput.* **2008**, *57*, 188–199. [CrossRef]
11. Makrini, I.E.; Merckaert, K.; Winter, J.D.; Lefeber, D.; Vanderborght, B. Task allocation for improved ergonomics in Human-Robot Collaborative Assembly. *Interact. Stud.* **2019**, *20*, 102–133. [CrossRef]
12. Pratissoli, F.; Battilani, N.; Fantuzzi, C.; Sabattini, L. Hierarchical and Flexible Traffic Management of Multi-AGV Systems Applied to Industrial Environments. In Proceedings of the 2021 IEEE International Conference on Robotics and Automation (ICRA), Xi'an, China, 30 May–5 June 2021; pp. 10009–10015. [CrossRef]
13. Michalos, G.; Spiliotopoulos, J.; Makris, S.; Chryssolouris, G. A method for planning human robot shared tasks. *CIRP J. Manuf. Sci. Technol.* **2018**, *22*, 76–90. [CrossRef]
14. Li, K.; Liu, Q.; Xu, W.; Liu, J.; Zhou, Z.; Feng, H. Sequence Planning Considering Human Fatigue for Human-Robot Collaboration in Disassembly. *Procedia CIRP* **2019**, *83*, 95–104. [CrossRef]
15. Ayough, A.; Zandieh, M.; Farhadi, F. Balancing, Sequencing, and Job Rotation Scheduling of a U-Shaped Lean Cell with Dynamic Operator Performance. *Comput. Ind. Eng.* **2020**, *143*, 106363. [CrossRef]
16. Klein, G.; Woods, D.D.; Bradshaw, J.M.; Hoffman, R.R.; Feltovich, P.J. Ten challenges for making automation a "team player" in joint human-agent activity. *IEEE Intell. Syst.* **2004**, *19*, 91–95. [CrossRef]
17. Johnson, M.; Bradshaw, J.M.; Feltovich, P.J.; Jonker, C.M.; Van Riemsdijk, M.B.; Sierhuis, M. Coactive Design: Designing Support for Interdependence in Joint Activity. *J. Hum.-Robot Interact.* **2014**, *3*, 43. [CrossRef]
18. Ramamritham, K. Allocation and scheduling of precedence-related periodic tasks. *IEEE Trans. Parallel Distrib. Syst.* **1995**, *6*, 412–420. [CrossRef]
19. Gacias, B.; Artigues, C.; Lopez, P. Parallel machine scheduling with precedence constraints and setup times. *Comput. Oper. Res.* **2010**, *37*, 2141–2151. [CrossRef]
20. Zhang, K.; Qi, B.; Jiang, Q.; Tang, L. Real-time periodic task scheduling considering load-balance in multiprocessor environment. In Proceedings of the 2012 3rd IEEE International Conference on Network Infrastructure and Digital Content, Beijing, China, 21–23 September 2012; pp. 247–250. [CrossRef]
21. Yu, T.; Huang, J.; Chang, Q. Optimizing task scheduling in human-robot collaboration with deep multi-agent reinforcement learning. *J. Manuf. Syst.* **2021**, *60*, 487–499. [CrossRef]
22. Ferreira, C.; Figueira, G.; Amorim, P. Scheduling Human-Robot Teams in collaborative working cells. *Int. J. Prod. Econ.* **2021**, *235*, 108094. [CrossRef]

23. Nourmohammadi, A.; Fathi, M.; Ng, A.H. Balancing and scheduling assembly lines with human-robot collaboration tasks. *Comput. Oper. Res.* **2022**, *140*, 105674. [CrossRef]

24. Rahman, S.M.; Wang, Y. Mutual trust-based subtask allocation for human–robot collaboration in flexible lightweight assembly in manufacturing. *Mechatronics* **2018**, *54*, 94–109. [CrossRef]

25. Nikolakis, N.; Kousi, N.; Michalos, G.; Makris, S. Dynamic scheduling of shared human-robot manufacturing operations. *Procedia CIRP* **2018**, *72*, 9–14. [CrossRef]

26. Pupa, A.; Van Dijk, W.; Secchi, C. A Human-Centered Dynamic Scheduling Architecture for Collaborative Application. *IEEE Robot. Autom. Lett.* **2021**, *6*, 4736–4743. [CrossRef]

27. Pupa, A.; Secchi, C. A Safety-Aware Architecture for Task Scheduling and Execution for Human-Robot Collaboration. In Proceedings of the 2021 IEEE/RSJ International Conference on Intelligent Robots and Systems (IROS), Prague, Czech Republic, 27 September–1 October 2021; pp. 1895–1902. [CrossRef]

28. Raatz, A.; Blankemeyer, S.; Recker, T.; Pischke, D.; Nyhuis, P. Task scheduling method for HRC workplaces based on capabilities and execution time assumptions for robots. *CIRP Ann.* **2020**, *69*, 13–16. [CrossRef]

29. Johannsmeier, L.; Haddadin, S. A hierarchical human-robot interaction-planning framework for task allocation in collaborative industrial assembly processes. *IEEE Robot. Autom. Lett.* **2016**, *2*, 41–48. [CrossRef]

30. Zhang, M.; Li, C.; Shang, Y.; Liu, Z. Cycle Time and Human Fatigue Minimization for Human-Robot Collaborative Assembly Cell. *IEEE Robot. Autom. Lett.* **2022**, *7*, 6147–6154. [CrossRef]

31. De Mello, L.H.; Sanderson, A.C. A correct and complete algorithm for the generation of mechanical assembly sequences. In Proceedings of the 1989 IEEE International Conference on Robotics and Automation. IEEE Computer Society, Scottsdale, AZ, USA, 14–19 May 1989; pp. 56–57.

32. Homem de Mello, L.; Sanderson, A. AND/OR graph representation of assembly plans. *IEEE Trans. Robot. Autom.* **1990**, *6*, 188–199. [CrossRef]

33. Canham, R.; Jackson, A.; Tyrrell, A. Robot error detection using an artificial immune system. In Proceedings of the NASA/DoD Conference on Evolvable Hardware, Chicago, IL, USA, 9–11 July 2003; pp. 199–207. [CrossRef]

34. Pettersson, O. Execution monitoring in robotics: A survey. *Robot. Auton. Syst.* **2005**, *53*, 73–88. [CrossRef]

35. Trung, P.; Giuliani, M.; Miksch, M.; Stollnberger, G.; Stadler, S.; Mirnig, N.; Tscheligi, M. Head and shoulders: automatic error detection in human-robot interaction. In Proceedings of the 19th ACM International Conference on Multimodal Interaction, Glasgow, UK, 13–17 November 2017; pp. 181–188.

36. Vo, N.N.; Bobick, A.F. From stochastic grammar to bayes network: Probabilistic parsing of complex activity. In Proceedings of the IEEE Conference on Computer Vision and Pattern Recognition, Columbus, OH, USA, 23–28 June 2014; pp. 2641–2648.

37. Maeda, G.; Ewerton, M.; Neumann, G.; Lioutikov, R.; Peters, J. Phase estimation for fast action recognition and trajectory generation in human—Robot collaboration. *Int. J. Robot. Res.* **2017**, *36*, 1579–1594. [CrossRef]

38. Muthugala, M.; Srimal, P.; Jayasekara, A. Improving robot's perception of uncertain spatial descriptors in navigational instructions by evaluating influential gesture notions. *J. Multimodal User Interfaces* **2021**, *15*, 11–24. [CrossRef]

39. Capelli, B.; Villani, V.; Secchi, C.; Sabattini, L. Understanding multi-robot systems: on the concept of legibility. In Proceedings of the 2019 IEEE/RSJ International Conference on Intelligent Robots and Systems (IROS), Macau, China, 3–8 November 2019; pp. 7355–7361.

Article

Teleoperation of High-Speed Robot Hand with High-Speed Finger Position Recognition and High-Accuracy Grasp Type Estimation

Yuji Yamakawa [1,*,†] **and Koki Yoshida** [2,†]

1   Interfaculty Initiative in Information Studies, The University of Tokyo, Tokyo 153-8505, Japan
2   School of Engineering, The University of Tokyo, Tokyo 153-8505, Japan; k.yoshida.890@gmail.com
*   Correspondence: y-ymkw@iis.u-tokyo.ac.jp
†   These authors contributed equally to this work.

Abstract: This paper focuses on the teleoperation of a robot hand on the basis of finger position recognition and grasp type estimation. For the finger position recognition, we propose a new method that fuses machine learning and high-speed image-processing techniques. Furthermore, we propose a grasp type estimation method according to the results of the finger position recognition by using decision tree. We developed a teleoperation system with high speed and high responsiveness according to the results of the finger position recognition and grasp type estimation. By using the proposed method and system, we achieved teleoperation of a high-speed robot hand. In particular, we achieved teleoperated robot hand control beyond the speed of human hand motion.

Keywords: teleoperation; high-speed image processing; machine learning; finger position recognition; grasp type estimation

Citation: Yamakawa, Y.; Yoshida, K. Teleoperation of High-Speed Robot Hand with High-Speed Finger Position Recognition and High-Accuracy Grasp Type Estimation. *Sensors* **2022**, *22*, 3777. https://doi.org/10.3390/s22103777

Academic Editor: Anne Schmitz

Received: 26 April 2022
Accepted: 12 May 2022
Published: 16 May 2022

Publisher's Note: MDPI stays neutral with regard to jurisdictional claims in published maps and institutional affiliations.

1. Introduction

Technology for realizing remote systems such as teleoperation, telerobotics, telexistence, etc., has been an important issue [1–3], and much research has been actively carried out. In the recent situation, in particular with the effects of COVID-19, remote work (telework) by office workers has become commonplace. In the future, teleoperation using robot technology will be applied to industrial fields, and object handling and manipulation using remote systems are considered to be essential and critical tasks. In order to achieve this, we consider that telerobotics technology based on sensing human hand motion and controlling a robot hand will be essential. Thus, this research focuses on the teleoperation of a robot hand on the basis of visual information about human hand motion. The reason why we use visual information is that it is troublesome for users to have to put on contact devices [4–7] before operating the system, and non-contact-type systems are considered to be more suitable for users.

Here, we describe related work in the fields of teleoperation and telerobotics based on visual information. Interfaces based on non-contact sensing generally recognize human hand gestures and control a slave robot based on these gestures [8,9]. In the related work in the field of humanoid robotics, a low-cost teleoperated control system for a humanoid robot has been developed [10]. In wearable robotics, semantic segmentation has been performed by using Convolutional Neural Networks (CNNs) [11]. Such interfaces are intuitive for users and do not involve the restrictions involved with contact-type input devices. Some devices for recognizing human hand gestures have been developed, and some systems have been also constructed [12,13]. Lien et al. proposed a high-speed (10,000 Hz) gesture recognition method based on the position change of the hand and fingers by radar [14]. This method can recognize the rough hand motion, but not its detail. Zhang et al. performed human hand and finger tracking using a machine learning technique based on RGB images, but the operating speed was limited to 30 fps [15]. Tamaki et al. created a database consisting of finger joint angles obtained by using a data glove, hand contour information, and nail positions obtained from images, and they also proposed a method of estimating hand and finger positions by searching the database at 100 fps [16].

Premeratne discussed some techniques for hand gesture recognition for Human–Computer Interaction (HCI) [17]. Furthermore, Ankit described recent activities on hand gesture recognition for robot hand control [18]. Hoshino et al. [19] and Griffin et al. [20] proposed methods of mapping between human hand and robot hand motions. On the other hand, Meeker et al. [21] created a mapping algorithm experimentally. Sean et al. developed a system that can operate a robot arm according to human intention [22]. Niwa et al. proposed "Tsumori" control, which can achieve a unique robot operation for an operator based on learning the correspondence of a human operation and robot motion [23]. Fallahinia and Mascaro proposed a method of estimating hand grasping power based on the nail color [24].

Summarizing the above, we can conclude that the disadvantages of the previous approaches are as follows:

1. Low speed: The sampling rate is course, and the gain of the robot controller becomes small, resulting in low responsiveness.
2. Low responsiveness: The latency from the human motion to the robot motion is long, making it difficult to remotely operate the robot. Furthermore, the system cannot respond to rapid and random human motion.

Regarding the low speed and low responsiveness, Anvari et al. [25] and Lum et al. [26] discussed the system latency in surgical robotics, and they claimed that the latency affects the task completion and performance. Thus, it is strongly desirable for teleoperation systems to have as low a system latency as possible.

To overcome these disadvantages, we also developed a high-speed telemanipulation robot hand system consisting of a stereo high-speed vision system, a high-speed robot hand, and a real-time controller [27,28]. In the stereo high-speed vision system, which is composed of two high-speed cameras and an image-processing PC, the 3D positions of the fingertips of a human subject were calculated by a triangulation method. Then, mapping between the human hand and the robot hand was performed. Finally, robot hand motion was generated to duplicate the human hand motion. With this high-speed system, we achieved a system latency so low that a human being cannot recognize the latency from the human hand motion to the robot hand motion [29,30].

In the present research, we aim to achieve even lower latency so that an intelligent system with vision cannot recognize the latency. Realizing such an extremely low-latency teleoperated system will contribute to solutions for overcoming latency issues in cases where the latency of telemanipulated systems may occur in more distant places. In addition, this technology will enable high-level image processing using the remaining processing time. In this paper, we propose a new method that fuses machine learning and high-speed image-processing techniques to obtain visual information about human hand motion. In general, the speed of machine learning methods is considered to be very low, and therefore, we consider that it is not suitable to adapt machine learning methods for real-time and real-world interactions between a human and a robot. By using our proposed method, we can overcome the issue with the low speed of the machine learning processing. Concretely speaking, the low-speed characteristics of machine learning can be improved by using high-speed image processing and interpolating the results of the machine learning with the results of the high-speed image processing. Although the finger position is estimated by machine learning using a CNN and high-speed image-processing technologies in this research, the integration of machine learning and high-speed image-processing technologies can be considered to be applicable to other target tracking tasks. Thus, our proposed method with high speed and high intelligence possesses the generality of the target-tracking method.

In addition, since our proposed method does not require three-dimensional measurement and camera calibration is also not needed, it is easy to set up the system. Moreover, motion mapping from the human hand motion to the robot hand motion is not performed in our proposed method. Therefore, kinematic models of the human hand and robot hand are not needed either. As a result, it is considered to be easy to implement our developed teleoperation system in actual situations.

The contributions of this paper are the following:

1. Integration of a machine learning technique and high-speed image processing;
2. High-speed finger tracking using the integrated image processing;
3. High-accuracy grasp type estimation;
4. Real-time teleoperation of a high-speed robot hand system;
5. Evaluation of the developed teleoperation system.

Furthermore, Table 1 shows the positioning of this research. The characteristics of our proposed method are "non-contact", "intention extraction", and "high-speed".

Table 1. Positioning of this research.

Evaluation Index	Conventional Method	Proposed Method
Comfortable operation	Contact	Non-contact
Application to various robots	Motion mapping	Intention extraction
Fine-motion recognition	Low-speed	High-speed

The rest of this paper is organized as follows: Section 2 describes an experimental system for teleoperation. Section 3 explains a new method for achieving grasp type estimation based on high-speed finger position recognition. Section 4 shows evaluations of the proposed method and the experimental results of teleoperation. Section 5 concludes with a summary of this research and future work.

2. Experimental System

This section explains our experimental system for the teleoperation of a high-speed robot hand based on finger position recognition and grasp type estimation. The experimental system, as shown in Figure 1, consists of a high-speed vision system (Section 2.1), a high-speed robot hand (Section 2.2), and a real-time controller (Section 2.3). All of the components were placed in the same experimental environment.

2.1. High-Speed Vision System

This subsection explains the high-speed vision system, consisting of a high-speed camera and an image-processing PC. As the high-speed camera, we used a commercial product (MQ013MG-ON) manufactured by Ximea. The full image size was 1280 pixels (width) × 1024 pixels (height), and the frame rate at the full image size was 210 frames per second (fps). In this research, since we decreased the image size, we increased the frame rate from 210 fps to 1000 fps. The reason why we set the frame rate at 1000 fps is that the servo control systems for the robot and machine system were both operated at 1000 Hz. In general, the raw image acquired by the high-speed camera was dark because of the significantly short exposure time. Therefore, we used an LED light to obtain brighter raw images from the high-speed camera.

The raw image data acquired by this high-speed camera were transferred to the image-processing PC. The image-processing PC ran high-speed image processing to track the finger position and to estimate the grasp type. The details of the image processing are explained in Section 3. The results of the image processing were sent to a real-time controller, described in Section 2.3. By performing real-time, high-speed (1000 Hz) image processing, we could control the high-speed robot hand described in Section 2.2 at 1 kHz. The sampling frequency of 1 kHz was the same as the sampling frequency of the servo-motor control.

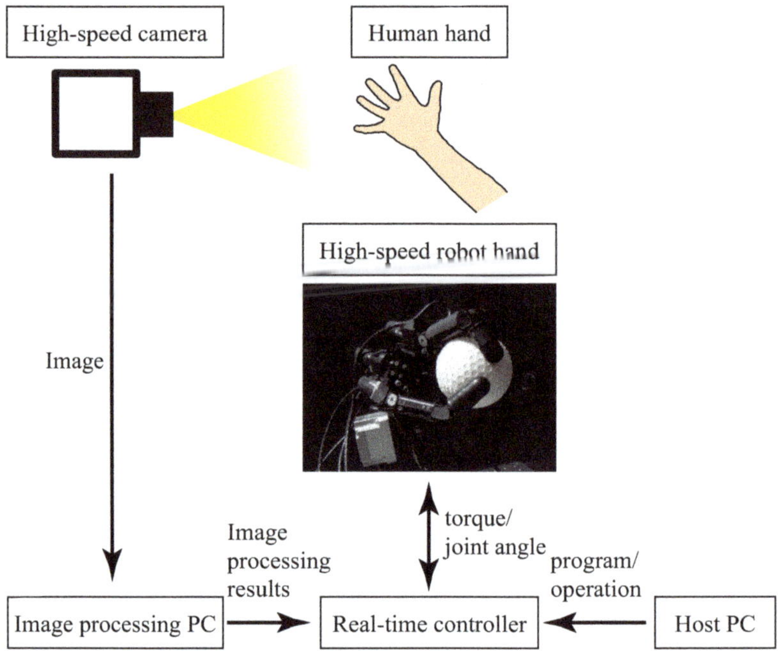

Figure 1. Structure of the experimental system.

The specifications of the image-processing PC are as follows: Dell XPS 13 9360, CPU: Intel® Core (™) i7-8550U @1.80 GHz, RAM: 16.0 GB, OS: Windows 10 Pro, 64 bit.

2.2. High-Speed Robot Hand

This subsection describes the high-speed robot hand, which was composed of three fingers [31]. A photograph of the high-speed robot hand is shown in the center of Figure 1. The number of degrees of freedom (DoF) of the robot hand was 10; the middle finger had 2 DoF, the left and right fingers 3 DoF, and the wrist 2 DoF. The joints of the robot hand could be closed by 180 degrees in 0.1 s, which is fast motion performance beyond that possible by a human. Each joint angle of the robot hand was controlled using a Proportional and Derivative (PD) control law, given by

$$\tau = k_p(\theta_d - \theta) + k_d(\dot{\theta}_d - \dot{\theta}), \tag{1}$$

where τ is the torque input as the control input for the high-speed robot hand control, θ_d and θ are the reference and actual joint angles of the finger of the robot hand, and k_p and k_d are the proportional and derivative gains of the PD controller.

2.3. Real-Time Controller

As the real-time controller, we used a commercial product manufactured by dSPACE. The real-time controller had a counter board (reading encoder attached to the motors of the robot hand), digital-to-analog (DA) output, and two Ethernet connections (one was connected to the host PC and the other to the image-processing PC). We operated the real-time controller through the host PC, and we also implemented the program of the proposed method in the host PC.

The real-time controller received the results of image processing via Ethernet communication. Then, the real-time controller generated a control signal for the robot hand to appropriately control the robot hand according to the results of the image processing and output the control signal to the robot hand.

3. Grasp Type Estimation Based on High-Speed Finger Position Recognition

This section explains a new method for estimating grasp type, such as power grasp or precision grasp, based on high-speed finger position recognition using machine learning and high-speed image processing, and our proposed method can be mainly divided into two components: high-speed finger position recognition described in Section 3.1 and grasp type estimation described in Section 3.2.

Figure 2 shows the overall flow of the proposed teleoperation method, detailed below:

1. Acquisition of the image by the high-speed camera:
    First, images can be captured by the high-speed camera at 1000 fps.
2. Estimation of finger position by CNN and finger tracking by high-speed image processing:
    The CNN and finger tracking are executed on the images. The calculation process of the CNN is run at 100 Hz, and finger tracking is run at 1000 Hz; the results of the CNN are interpolated by using the results of finger tracking. As a result, the finger positions are recognized at 1000 Hz.
3. Estimation of grasp type by decision tree classifier:
    Based on the finger positions, grasp type estimation is performed by using a decision tree classifier.
4. Grasping motion of the high-speed robot hand:
    According to the estimated grasp type, the high-speed robot hand is controlled to grasp the object.

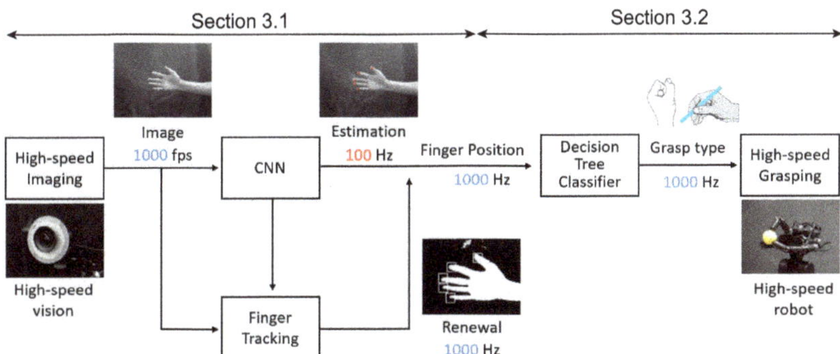

Figure 2. Overall flow of the proposed teleoperation method.

3.1. High-Speed Finger Position Recognition with CNN

This subsection explains the method for high-speed finger position recognition. By recognizing the finger positions at high speed (for instance, 1000 fps), we can reduce the latency from the human hand motion to the robot hand motion and estimate the grasp type with high accuracy. Conventional image-processing methods are too slow (around 30 fps) for actual application. This research can solve the speed issue with the image processing conventionally used.

The proposed method was implemented by using machine learning and high-speed image-processing technologies. As the machine learning method, we used a Convolutional Neural Network (CNN). As the high-speed image processing method, we used tracking of a Region Of Interest (ROI) and image processing of the ROI. Here, the ROI was extremely small for the full image size and was set at the position of the result of the CNN, namely roughly at the position of the fingers.

3.1.1. Estimation of Finger Position by CNN

By using the CNN, we estimated the positions of six points (five fingertips and the center position of the palm) in the 2D image captured by the high-speed camera. The advantages of using the CNN to recognize hand positions include robustness against finger-to-finger occlusion and robustness against background effects.

The model architecture of the CNN is as follows:

- Input: an array of $128 \times 128 \times 1$;
- Output: 12 values;
- Alternating layers: six Convolution layers and six Max Pooling layers;
- Dropout layer placed before the output layer;
- The filter size of the Convolution layers was 3×3, the number of filters 32, and the stride 1;
- The pool size of Max Pooling was 2×2.

This architecture was created by referring to a model [32] used for image classification and modifying it according to handling multiple-output regression problems.

In addition, the value of Dropout was set at 0.1, and the activation function and parameter optimization were Relu and RMSprop [33], respectively. The loss was calculated using the Mean-Squared Error (MSE). Adding the Dropout layer was expected to suppress overlearning, and reducing the number of layers was expected to be effective in suppressing overlearning and reducing inference and learning times [34].

Figure 3 shows an example of annotation on a hand image, where the annotated positions indicating the center points of the fingertips and palms of the five fingers are shown as blue dots.

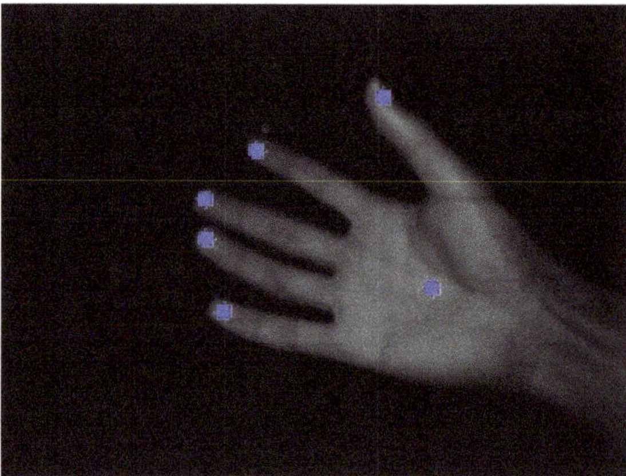

Figure 3. Annotation result of fingertip positions.

3.1.2. Finger Tracking by High-Speed Image Processing

Since the estimation by the CNN is much slower than imaging by the high-speed camera and there are many frames where the hand position cannot be recognized during the estimation by the CNN, the CNN processing becomes the rate-limiting step of the system. While compensating for the frames where the CNN estimation is not performed, we achieved real-time acquisition of the hand position. Figure 4 [35] shows a schematic of the method that combines CNN estimation performed at low frequency with high-frequency hand tracking to obtain the hand position. In Figure 4, the orange dots indicate the execution of the CNN, and the gray dots indicate the execution of hand tracking. By performing hand tracking in frames where the CNN is not performed, the hand position can be obtained at high frequency.

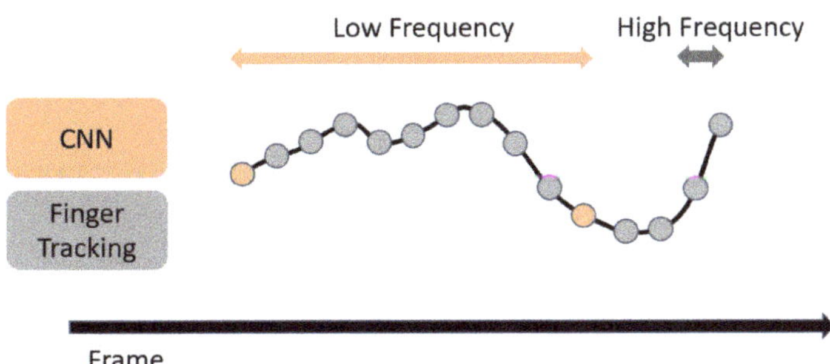

Figure 4. Concept of fusing the CNN and finger tracking.

Next, we explain the requirements that the hand-tracking method should satisfy. First, in order to perform real-time hand position recognition, information obtained in frames after the frame to be tracked cannot be used. For example, if we perform linear interpolation of two CNN results to interpolate the hand position in the frame between the CNN steps, we need to wait for the second CNN to be executed, which impairs the real-time performance of the system. Therefore, it is necessary to track the hand based on the information obtained from the frame to be processed and the earlier frames.

In addition, it is desirable to obtain the data by measurement rather than by prediction. This is because it is not always possible to accurately recognize high-speed and minute hand movements if the current hand position is inferred from the trend of past hand positions. By calculating the current hand position from the information from the current frame, instead of predicting based on the information from the past frames, we can achieve more accurate recognition of sudden hand movements.

In this study, we propose a real-time, measurement-based recognition method for hand positions in frames where the CNN is not performed. This method involves hand tracking using frame-to-frame differences of fingertip center-of-gravity positions. The proposed hand-tracking method calculates the hand position in the corresponding frame by using three data sets: the hand position from the past CNN estimation results, the image of the frame in which the CNN was executed, and the image of the corresponding frame. As the amount of hand movement between frames, we calculated the difference in hand positions from the two images and added it to the hand position estimated by the CNN to treat it as the hand position in the corresponding frame.

The following is the specific method of calculating the hand position when the most recent frame in which the CNN is executed is n, the frame in which the tracking process is performed is $n + k$, and the frame in which the CNN is executed again is $n + T$:

1.   *n*-th frame: CNN

   In the n-th frame, let an estimated fingertip position obtained by the CNN be P_n. Using Equation (2) below, the image is binarized, the ROI with the center position P_n is extracted, and the center of the fingertip in the ROI is assumed to be C_n (Figure 5a). In the image binarization, the original image and the binarized image are $src(i, j)$ and $f(i.j)$, respectively. Furthermore, the threshold of the image binarization is set at *thre*.

$$f(i,j) = \begin{cases} 0 & \text{if } src(i,j) < thre \\ 1 & \text{otherwise} \end{cases} \tag{2}$$

The image moment is represented by m_{pq} (Equation (3)), and the center position (C_n) of the fingertip in the ROI is ($m_{10}/m_{00}, m_{01}/m_{00}$):

$$m_{pq} = \sum_i \sum_j i^p j^q f(i,j) \tag{3}$$

The value of P_n is substituted for the fingertip position Q_n in the n-th frame.

$$Q_n = P_n \tag{4}$$

2. $(n + k)$-th frame: Finger tracking
After binarizing the image in the $(n + k)$-th frame $(0 < k < T)$, the ROI with the center position P_n is extracted, and the center of the fingertip in the ROI is assumed to be C_{n+k} (Figure 5b). At that time, let the finger position in the $(n + k)$-th frame be Q_{n+k}, calculated by the following equation:

$$Q_{n+k} = P_n + C_{n+k} - C_n \tag{5}$$

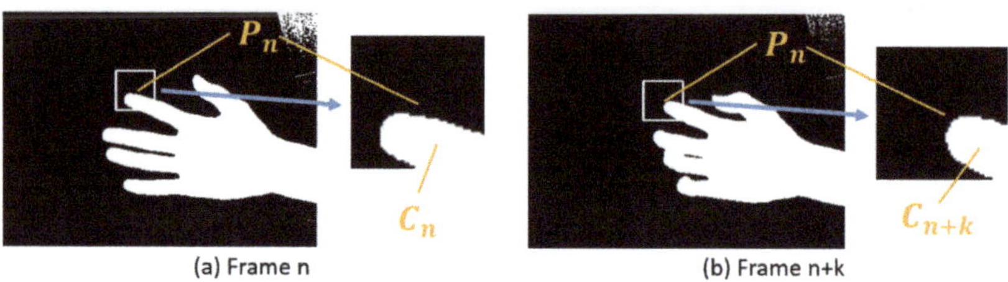

Figure 5. An example of finger tracking: (**a**,**b**) show images at the n-th and $(n + k)$-th frames, respectively.

If the hand-tracking module receives a new CNN estimation result from the CNN module every T frames, where T is a predefined constant, and updates the result for processing, the hand-tracking process will have to wait for every frame that exceeds T for inference by the CNN. When the number of frames required for inference by the CNN exceeds T, the hand-tracking process needs to wait. This increases the latency of hand tracking, since the inference time of the CNN may vary in actual execution and the inference result may not be sent by the CNN module even after T frames have passed. On the other hand, if the CNN results are updated in frames received from the CNN module instead of in frames at regular intervals, the latency is reduced because the last received CNN result is used even if the CNN result is delayed. The latency is reduced because the last received CNN result is used even if the transmission of the CNN result is delayed.

Based on the above, we devised two different methods for hand tracking with and without a waiting time for the CNN estimation results: a low-latency mode with a variable T value and a high-accuracy mode with a constant T value. The low-latency mode is effective for applications where low latency is more important than accuracy, such as anticipating human actions. On the other hand, the high-accuracy mode is suitable for applications where accurate acquisition of hand positions is more important than low latency, such as precise mechanical operations. In this research, we adopted the low-latency mode to track the human hand motion, because we aimed at the development of a teleoperation system with high speed and low latency. An overview of the algorithm for the low-latency mode is shown in Algorithm 1. The algorithm for the high-accuracy mode is shown in Algorithm A1 in Appendix A.

The characteristics of the two modes are summarized below:

Algorithm 1. Low-latency mode:

As the result of the CNN, which is used for finger tracking, the latest result is utilized. The advantage is that the latency is reduced because no time is required to wait for the CNN results. The disadvantage is that if the CNN processing is delayed, the tracking process will be based on the CNN results for distant frames, which will reduce the accuracy.

Algorithm A1. High-accuracy mode:

By fixing the interval T of the number of frames at which the CNN is executed, the process of updating the estimate by CNN is performed at fixed intervals. The advantage is that hand tracking is based on frequently acquired CNNs, which improves accuracy. The disadvantage is that when the CNN processing is delayed, the latency increases because there is a waiting time for updating the CNN results before the tracking process starts.

Algorithm 1 Finger tracking with low latency.

1: *resultCNN* {CNN result to receive}
2: *resultFT* {Finger tracking result to send}
3: **while** True **do**
4:    **if** *resultCNN* is received **then**
5:       calculate C_n
6:       *resultFT* $\leftarrow$ *resultCNN*
7:    **else**
8:       calculate C_{n+k}
9:       *resultFT* $\leftarrow$ *resultCNN* $+ C_{n+k} - C_n$
10:   **end if**
11:   send *resultFT*
12: **end while**

3.2. Grasp Type Estimation

This subsection explains the method for estimating the grasp type on the basis of the results of the high-speed finger tracking. Furthermore, we explain the robot hand motion according to the estimated grasp type.

3.2.1. Estimation of Grasp Type by Decision Tree

We also used machine learning technology to estimate the grasp type on the basis of the finger position and the center position of the palm, which are estimated by the CNN and hand tracking at high speed. As representative grasp types to be estimated, we considered two grasp types: (1) a power grasp using the palm of the hand and (2) a precision grasp using only the fingertips. As a result, we categorized the grasps to be estimated into three types: "power grasp", "precision grasp", and "non-grasp", as shown in Figure 6. In the "power grasp", all four fingers, and not the thumb, move in the same way and tend to face the thumb, whereas in the "precision grasp", the positions of the thumb and index finger tend to separate from those of the little finger and ring finger. The "non-grasp" state corresponds to the extended state of the fingers.

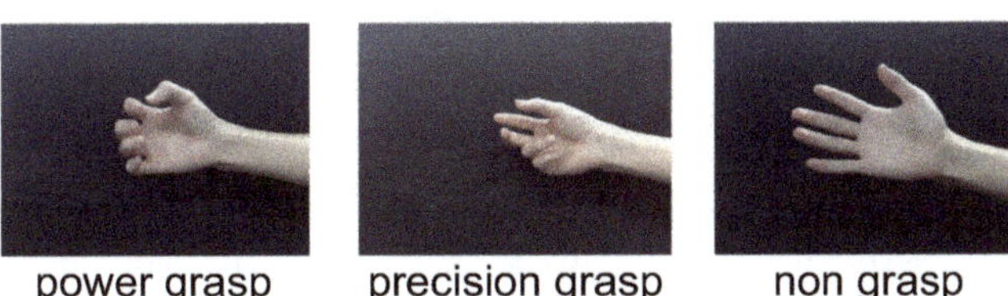

Figure 6. Differences among power grasp (**left**), precision grasp (**middle**), and non-grasp (**right**).

To accurately estimate the grasp type, we used decision trees (decision tree classifier) as the machine learning method. Decision trees have the features of fast classification and readability of the estimation criteria. In particular, the ease of interpretation of the estimation criteria and the possibility of creating algorithms with adjustments are reasons for using decision trees as a classification method.

<u>Preprocessing of hand position data:</u>

In order to improve the accuracy of the decision tree, we preprocessed the input data, namely the hand positions (Figure 7). In the preprocessing, we first calculate the distance between the middle finger and the palm of the hand in the frame with the fingers extended as a hand size reference to calibrate the hand size. When the coordinates of the middle finger and the palm of the hand in the image of the frame to be calibrated are (x_{m0}, y_{m0}) and (x_{p0}, y_{p0}), respectively, the distance between the two points can be given by

$$r_0 = \sqrt{(x_{m0} - x_{p0})^2 + (y_{m0} - y_{p0})^2}. \tag{6}$$

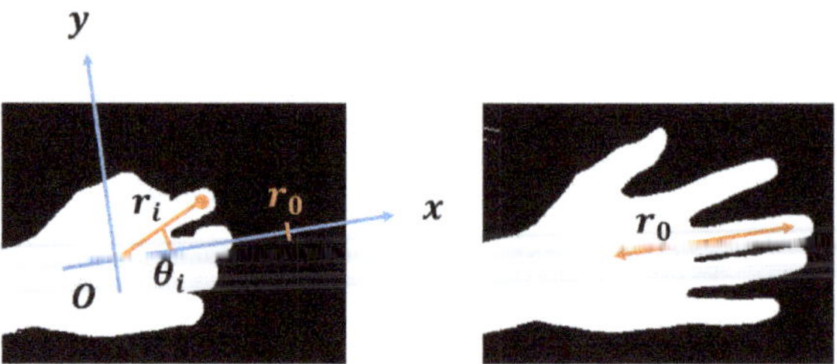

Figure 7. Direction of middle finger and angle between each finger and middle finger.

Next, we calculate the position of the hand, (r, θ), in the polar coordinate system with the palm of the hand serving as the origin and the direction of extension of the middle finger serving as the x-axis, based on the position of the hand represented by the coordinates in the image. When the coordinates of the middle finger and the palm in the image are (x_m, y_m) and (x_p, y_p), respectively, the declination angle of the polar coordinate of the middle finger, θ_m, is expressed by the following formula:

$$\theta_m = \arctan\left(\frac{y_m - y_p}{x_m - x_p}\right). \tag{7}$$

Furthermore, the polar coordinate (r_i, θ_i) corresponding to the coordinate (x_i, y_i) of finger i in the image is expressed by the following equation with the direction of the middle fingertip serving as the positive direction of the x-axis. To calibrate the size of the hand, we divide r_i by the hand size reference r_0.

$$r_i = \frac{1}{r_0}\sqrt{(x_i - x_p)^2 + (y_i - y_p)^2} \tag{8}$$

$$\theta_i = \arctan\left(\frac{y_i - y_p}{x_i - x_p}\right) - \theta_m \tag{9}$$

The vector **r** containing the distance (r_i) of each finger and the vector Θ containing the declination angle (θ_i) obtained in this way for each frame are used as inputs to the decision tree. By using the relative positions of the fingers with respect to the palm in polar coordinates as inputs, the data to be focused on are clarified, and by dividing the data by the size of the hand with the fingers extended, the characteristics of the hand morphology can be extracted while suppressing the effects of differences in hand size between individuals and differences in the distance between fingers and camera lens for each execution.

3.2.2. Grasping Motion of High-Speed Robot Hand

The object is grasped by the high-speed robot hand according to the result of the high-speed grasp type estimation described above.

The middle finger of the robot hand has two joints, that is the root and tip links, and the left and right fingers also have three joints, the root, tip, and rotation around the palm. The root and tip links operate in the vertical direction to bend and stretch the fingers, allowing them to wrap around objects. The rotation joint around the palm moves horizontally and can change its angle to face the middle finger, which enables stable grasping. The wrist part of the robot hand has two joints, that is flexion/extension and rotation joints, which enables the finger to move closer to the object to be grasped.

Since the time required to rotate the finger joint 180 deg is 0.1 s, it takes approximately 50 ms to close the finger from the open position to 90 deg for grasping. From our research using a high-speed control system, the latency from the image input to the torque input of the robot hand is about 3 ms [36]. If the value of the estimated grasping configuration oscillates, the target angle is frequently changed, and the robot hand becomes unstable, so the grasping operation is started when the same grasp type is received continuously for a certain number of frames.

4. Experiments and Evaluations

This section explains the experiments, experimental results, and evaluations for finger position recognition (Section 4.1—Exp. 1-A), grasp type estimation (Section 4.2—Exp. 1-B), and teleoperated grasping by a robot hand (Section 4.3—Exp. 2), respectively. Figure 8 shows an overview of the experiments and evaluations for each part.

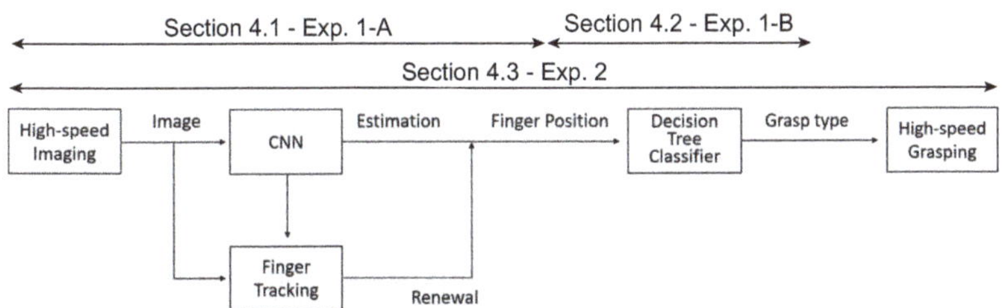

Figure 8. Overview of experiments and evaluations.

4.1. Finger Position Recognition

This subsection explains the experiment and evaluation for finger position recognition based on the proposed method with high-speed image processing and the CNN.

4.1.1. Preparation for Experiment

We trained the CNN model described in Section 3, which estimated hand positions from images. To increase the amount of data for training, we performed data augmentation by random scaling (0.7~1.0) and rotation (−60~60 deg.) operations. As a result, we could obtain 9000 images from 1000 images by data augmentation. The training process was performed for 200 epochs, with 70% of the prepared data used as the training data and 30% used as the validation data. The slope of the loss function calculated from the Mean-Squared Error (MSE) became lower around epoch 30. When we calculated the Mean Absolute Error (MAE) of the estimation results for the validation data, the MAE was less than 10 pixels. The width of the fingers in the image was about 15 pixels, which means that the hand position estimation was accurate enough for the hand-tracking process. As a result of 5 trials of 100 consecutive inferences, the mean and standard deviation of the inference times were 7.03 ms and 1.46 ms, respectively. Furthermore, the longest was 24.6 ms, and the shortest was 3.75 ms. Thus, an average of seven hand-tracking runs was taken for each update of the CNN results for 1000 fps of image acquisition by the high-speed camera.

4.1.2. Experiment—1-A

The exposure time of the high-speed camera was set to 0.5 ms, the image size to 400 pixels wide and 300 pixels high, the square ROI for the hand-tracking process to 40 pixels by 40 pixels, and the threshold for the binarization process to $thre = 20$. In such a situation, we captured the hand opening and closing in 1000 frames during 1 s and applied the proposed hand tracking for hand position recognition. The images captured in the experiment were stored, the CNN was run offline on all images, and the results were used as reference data for the comparison method.

4.1.3. Results

The output speed of the hand position was 1000 Hz, which was the same speed as the imaging. The latency between the end of imaging and the output of the hand position was also 1 ms.

Hand images are shown in Figure 9i: starting with the finger extended (Figure 9(i-a)), bending the finger (Figure 9(i-b)), folding it back around Frame 400 (Figure 9(i-c)), and stretched again (Figure 9(i-d)–(i-f)). Figure 9ii is a graph of the hand positions estimated over 1000 frames by the proposed method. The image coordinates of the five fingers and the palm center point are represented as light blue for the index finger, orange for the middle finger, gray for the ring finger, yellow for the pinky finger, dark blue for the thumb, and green for the palm.

The errors in the proposed method and the errors in the comparison method are shown in Table 2. The average error of the five fingers in the proposed method was 1.06 pixels. Furthermore, the error in the comparison method was 1.27 pixels, which is 17% bigger than that in the proposed method. For all fingers, the error in the proposed method was smaller than that of the comparison method. Note that 1 pixel in the hand image corresponds to approximately 0.7 mm in the real world.

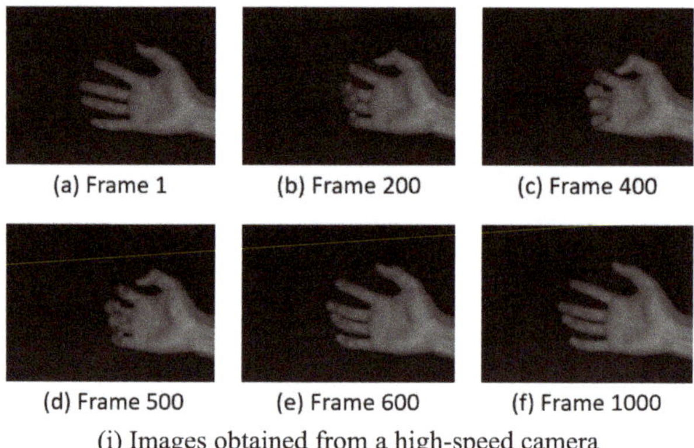

(a) Frame 1   (b) Frame 200   (c) Frame 400

(d) Frame 500   (e) Frame 600   (f) Frame 1000

(i) Images obtained from a high-speed camera

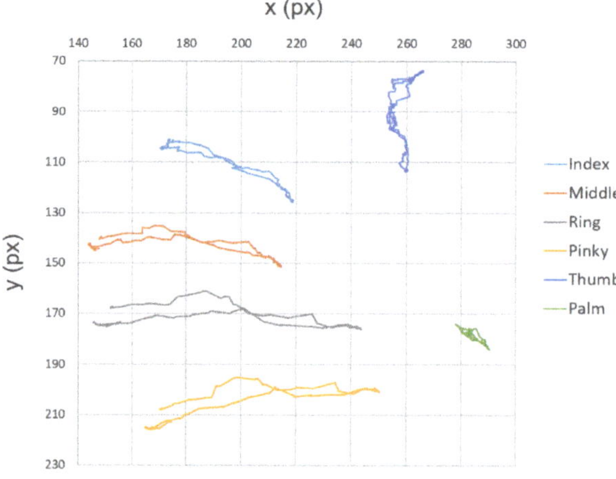

(ii) (x, y) coordinates of each finger and palm

Figure 9. Result of finger tracking: (i) images obtained from the high-speed camera and (ii) (x, y) coordinates of each finger and the palm.

Table 2. Mean-Squared Error (MSE) of finger positions estimated by CNN w/ and w/o finger tracking.

Finger	MSE	
	with Finger Tracking/Pixel	**without Finger Tracking/Pixel**
Index	0.95	1.03
Middle	0.97	1.24
Ring	1.19	1.60
Pinky	1.18	1.47
Thumb	0.99	1.02
Average	1.06	1.27

4.1.4. Discussion

First, we consider the execution speed of hand position recognition. From the experimental results, the output speed of the hand position was 1000 Hz, which is the same as that of the image capturing, indicating that the hand position recognition is fast enough. In addition, the latency from the end of imaging to the output of the hand position was 1 ms, indicating that the total execution time of the inter-process data sharing and hand-tracking process itself was 1 ms. Thus, the effectiveness of the proposed method described in Section 3 was shown.

Next, we discuss the reason why the error in the proposed method is smaller than that in the comparison method. In the proposed method, even for the frames where the CNN is not executed, the hand position recognition at 1000 Hz by the tracking process can output values close to the reference data. On the other hand, the comparison method outputs the last CNN estimation result without updating it for the frames where the CNN is not executed, and thus, the updating in the hand position recognition is limited to 100 Hz. The effectiveness of the proposed method for fast hand tracking is the reason why the proposed method has superior accuracy.

4.2. Grasp Type Estimation

This subsection explains the experiment and evaluation for grasp type estimation based on the finger position recognition described above.

4.2.1. Preparation for the Experiment

We trained a decision tree that outputs a grasp type label using the hand position as input. First, we captured 49 images of a power grasp, 63 images of a precision grasp, and 52 images of a non-grasp and annotated them with the grasp type label. Next, we annotated the hand positions in the images by estimating the CNN trained in the above subsection. After preprocessing, we transformed the hand positions represented in the Cartesian coordinate system into a polar coordinate system centered on the palm of the hand and normalized them by hand size to obtain 10 variables. The number of variables in the decision tree was two: length r and angle θ for each of the five types of fingers (index, middle, ring, pinky, and thumb). The length r is the ratio of the distance from the center of the palm to the tip of the middle finger with the fingers extended to the distance from the palm to each fingertip. The angle θ is the angle between the finger and the middle finger, with the thumb direction being positive and the little finger direction being negative.

Based on the above variables as inputs, a decision tree was trained using the leave-one-out method [37]. The model was trained with the depth of the decision tree from 1 to 10. As a result, when the depth was three, the accuracy was about 0.94, which can be considered to be sufficient. Thus, we decided that the depth of the tree structure should be set at three. Furthermore, the parameters of the decision tree such as the classification conditions, Gini coefficient, and depth were obtained.

4.2.2. Experiment—1-B

We used a series of images of the grasping motion to evaluate the learned decision tree. We took a series of 500 frames of hand images of each type of grasping motion, starting from the open fingers, performing a power or precision grasp, and then, opening the fingers again. The exposure time of the high-speed camera was 0.5 ms, the image size 400 pixels (width) × 300 pixels (height), and the frame rate 1000 fps. Based on the hand positions in the images recognized by the CNN and hand tracking (accurate mode), we performed the grasp type estimation using the decision tree and calculated the correct answer rate.

4.2.3. Result

Figures 10 and 11 show the results of grasp type estimation for a series of images and the hand image in a representative frame. The horizontal axes in Figures 10 and 11 represent the frame number, and the vertical axis also represents the label of the grasp type, where 0 indicates non-grasp, 1 indicates a power grasp, and 2 indicates a precision grasp. Figure 10 is the result for the power grasp motion, which is a non-grasp at Frame 0 (a), judged as a power grasp at Frame 82 (b), a continued power grasp (c), and a non-grasp again at Frame 440 (d). Figure 11 is the result for the precision grasp motion, which is a non-grasp at Frame 0 (a), judged as a precision grasp at Frame 72 (b), a continued precision grasp (c), and a non-grasp again at Frame 440 (d). No misjudgments occurred in either the power grasp or the precision grasp experiments.

4.2.4. Discussion

First, we discuss the stability of the grasp type estimation by the proposed method. From the results shown in Figures 10 and 11, there was no misjudgment of the grasp type estimation. In addition, high-accuracy grasp type estimation was achieved successfully.

Next, we describe the processing speed of the grasp type estimation. The mean and standard deviation of the inference speed for three trials of 1000 consecutive inferences were 0.07 ms and 0.02 ms, respectively. This is much shorter than 1 ms and is fast enough for a system operating at 1000 Hz.

Finally, the hand position recognition and grasp type estimation for the hand images evaluated in Experiment 1-A and 1-B are shown in Figure 12. From this result, we can conclude that the effectiveness of the proposed method for the hand position recognition and grasp type estimation is confirmed.

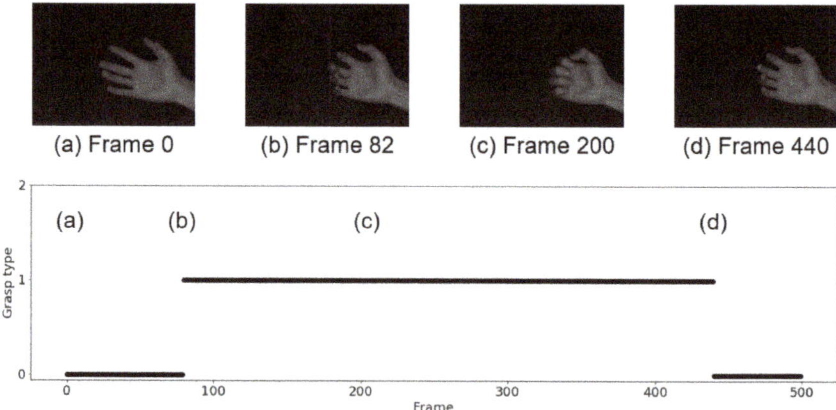

Figure 10. Estimation result in the case of power grasp.

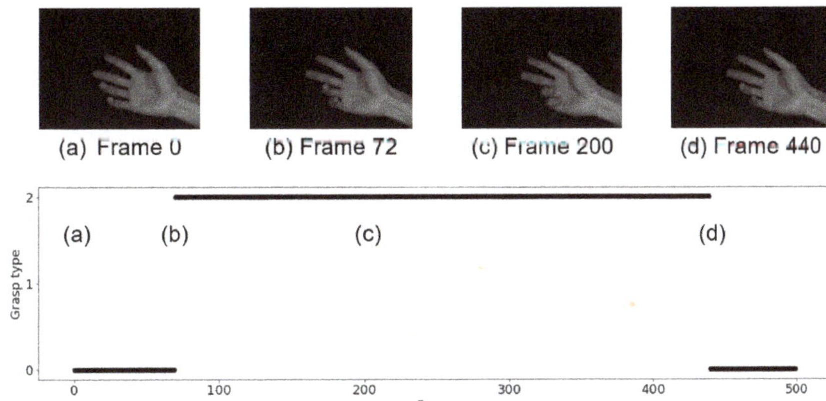

Figure 11. Estimation result in the case of precision grasp.

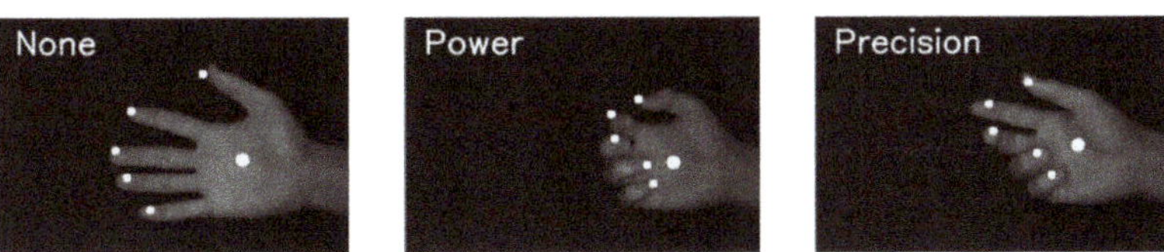

Figure 12. Difference among non-grasp, power grasp, and precision grasp and fingertip and palm positions (white circles).

4.3. Teleoperated Grasp by Robot Hand

This subsection explains the experiment and evaluation for teleoperated grasping by a robot hand on the basis of human hand motion sensing.

4.3.1. Experiment—2

In this experiment, the robot hand grasped a Styrofoam stick-shaped object with a diameter of 0.05 m and a length of 0.3 m, which was suspended by a thread in the robot's range of motion. The human operator performed a "power grasp" or a "precision grasp" with his/her hand in the field of view of the high-speed camera. The real-time controller calculated the reference joint angles of the robot hand for the grasping operation according to the received grasp type and provided the reference joint angles as step inputs to the robot hand through the real-time control system.

4.3.2. Results

A video of the grasping process of the robot hand can be seen at our web site [38], and the hand fingers of the robot hand and the operator are shown in Figure 13. From this result, we confirmed the effectiveness of the proposed teleoperation of the robot hand based on the hand tracking and grasp type estimation.

4.3.3. Discussion

Since the operating frequency of both image processing and robot control was 1000 Hz in the experiment, the operating frequency of the entire system was 1000 Hz. Therefore, the robot hand manipulation with high-speed image processing proposed in this study achieved the target operating frequency of 1000 Hz.

Next, we consider the latency of the entire system. We define the latency as the time from the imaging to the completion of the robot hand motion. That is to say, the latency can be evaluated from the total time for the image acquisition, image processing, including hand tracking and grasp type estimation, transmitting the results from the image-processing PC to the real-time controller, and implementing robot hand motion.

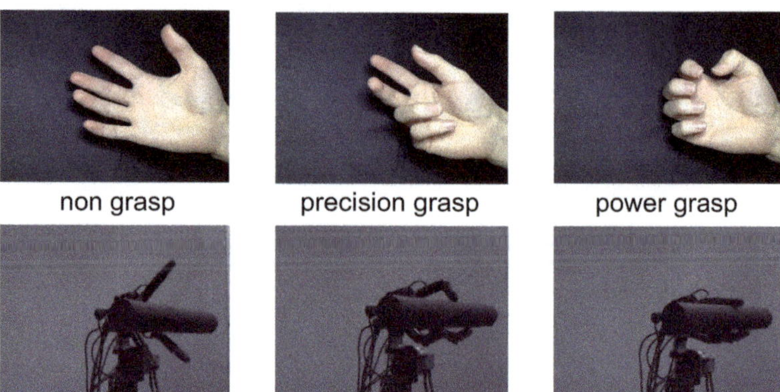

Figure 13. Experimental result of teleoperation; left, middle, and right show non-grasp, precision grasp, and power grasp, respectively.

First, the latency from the end of imaging to the end of transmission of the image processing result was 1 ms. Second, the latency from the transmission of the result by the image-processing PC to the output of the control input by the real-time controller was from 2~3 ms, and the worst case was 3 ms. Next, we need to consider the latency from the output of the control input to the completion of the robot hand motion. The time required to converge to ±10% of the reference joint angle is shown in Table 3. From the top in Table 3, "Joint" means the root and tip links of the middle finger and the root, tip, and rotation of the left and right thumbs around the palm, respectively. Furthermore, Figure 14 shows the step response of the tip link of the left and right thumbs from an initial value of 0.0 rad to a reference joint angle of 0.8 rad. The dashed line depicts the range of ±10% of the reference joint angle 0.8 rad, and the slowest convergence time was 36 ms. Therefore, it took 36 ms to complete the grasp after the real-time controller received the grasp type. As described above, the latency of the entire system was 1 ms for the imaging and the grasp type estimation, 3 ms for the communication between the image-processing PC and the real-time controller, and 36 ms for the robot operation, totaling 40 ms for the teleoperation of the robot hand. Since this value (40 ms) is close to the sampling time of the human eye (around 33 ms), our developed system is fast enough for robot teleoperation.

Table 3. Time for convergence of robot hand joint angle to ±10% of the reference angle.

Finger	Joint	Time/ms
Middle finger	root	25
	top	25
Left and right thumbs	root	26
	top	36
	rotation around palm	18

Figure 15 shows the timeline of the human and robot grasping motions. From the experimental results, it took about 80 ms from the time the hand starts the grasping motion to the time the hand form that is estimated to be a specific grasping form is captured and 150 ms until the time the motion is completed. On the other hand, it took 40 ms for the robotic hand to complete the grasp after the hand configuration that is estimated to be a specific grasp configuration is captured. In other words, the grasping by the robotic hand is completed when the grasping motion by the fingers is completed at 80 ms, and the remote grasping operation using the robotic hand is realized by anticipating the motion from the human hand morphology during the grasping motion, called pre-shaping. Consequently, we achieved teleoperated robot hand control beyond the speed of human hand motion by using the proposed method and system. This result may contribute to compensate the latency due to the network in the teleoperation.

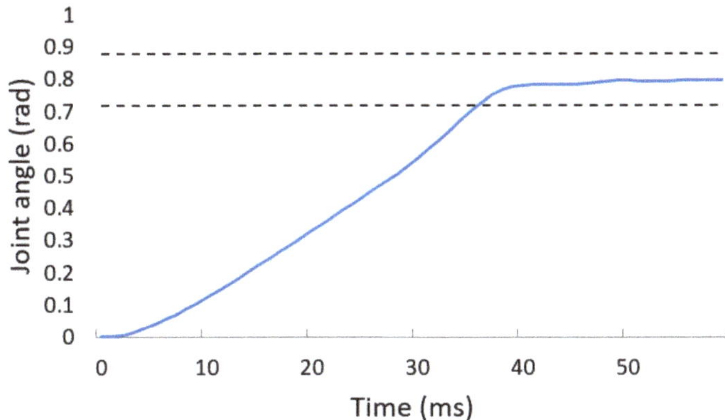

Figure 14. Joint response of high-speed robot hand.

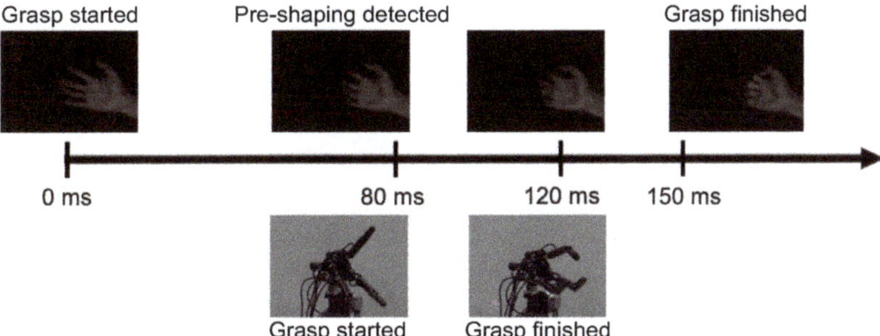

Figure 15. Time series of human hand motion and robot hand motion.

5. Conclusions

The purpose of the work described in this paper was to develop an intuitive and fast telerobot grasping and manipulation system, which requires fast recognition of the intended grasping method from the operator's gestures. In this paper, we proposed a method for fast recognition of grasping intentions by obtaining hand positions from gesture images and estimating the grasp type from the hand positions by machine learning. In particular, we combined machine learning and tracking to achieve both high speed and accuracy in hand position acquisition. In the evaluation experiments of the hand position recognition method, we achieved a mean-squared error of 1.06 pixel, an operating frequency of 1000 Hz, and a latency of 1 ms. In the evaluation experiment of the grasp type estimation, we also achieved an accuracy of 94%, and the inference time was 0.07 ms. These results show that the operating frequency of the system from gesture capturing to grasping form estimation was 1000 Hz and the latency was 1 ms, which confirms the effectiveness of the proposed method. As a result of the remote grasping operation experiment by the high-speed robot hand using the high-speed grasping form estimation system, the grasping operation was completed in 40 ms after the hand image was captured. This is the time when the grasping operation by the fingers was completed at 80%, and the high-speed tele-grasping operation was realized successfully.

The first application of the system proposed in this study is HMI, which uses high-speed gesture recognition. In this study, the grasping form was obtained from the hand position. However, the proposed method can be applied to HMI that supports various hand gestures because it is easy to obtain other types of gestures. Next, human–machine coordination using fast and accurate hand position recognition can be considered as an application. Since the hand positions are acquired with high speed and high accuracy, the system can be applied to human–machine coordination using not only gestures, but also hand positions, and remote master–slave operation of robots by mapping.

There are still some issues to be solved in this research. One of them is the 3D measurement of the hand position. Currently, the hand position is only measured in two dimensions, which restricts the orientation of the operator's hand, but we believe that three-dimensional measurement will become possible by using multiple cameras, color information, and depth information. The other is hand position recognition against a miscellaneous background. In this study, the background of the hand image was black, but by training the machine learning model using images with various backgrounds, the hand position can be recognized without being affected by the background, and the applicability of the system will be enhanced. These issues described above will be solved in the future.

Author Contributions: Conceptualization, Y.Y.; methodology, K.Y. and Y.Y.; software, K.Y.; validation, K.Y. and Y.Y.; formal analysis, K.Y.; investigation, K.Y. and Y.Y.; writing—original draft preparation, Y.Y., K.Y.; writing—review and editing, Y.Y., K.Y.; project administration, Y.Y.; funding acquisition, Y.Y. All authors have read and agreed to the published version of the manuscript.

Funding: This research was supported in part by the JST, PRESTO Grant Number JPMJPR17J9, Japan.

Institutional Review Board Statement: Not applicable.

Informed Consent Statement: Not applicable.

Data Availability Statement: Not applicable.

Conflicts of Interest: The authors declare no conflicts of interest.

Appendix A

In this Appendix, we show an algorithm of another mode, which is a high-accuracy mode for finger tracking, shown in Algorithm A1.

Algorithm A1 Finger tracking with high accuracy

1: $frame\,number \leftarrow 0$
2: **while** True **do**
3:   **if** $frame\,number\,(mod\,T) \equiv 0$ **then**
4:     **repeat**
5:       wait
6:     **until** $resultCNN$ is received
7:     calculate C_n
8:     $resultFT \leftarrow resultCNN$
9:   **else**
10:     calculate C_{n+k}
11:     $resultFT \leftarrow resultCNN + C_{n+k} - C_n$
12:   **end if**
13:   send $resultFT$
14:   $frame\,number \leftarrow frame\,number + 1$
15: **end while**

References

1. Sheridan, T.B. *Telerobotics, Automation and Human Supervisory Control*; The MIT Press: Cambridge, MA, USA, 1992.
2. Tachi, S. *TeTelecommunication, Teleimmersion and Telexistence*; Ohmsha: Tokyo, Japan, 2003.
3. Tachi, S. *TeTelecommunication, Teleimmersion and Telexistence II*; Ohmsha: Tokyo, Japan, 2005.

4. Cui, J.; Tosunoglu, S.; Roberts, R.; Moore, C.; Repperger, D.W. A review of teleoperation system control. In Proceedings of the Florida Conference on Recent Advances in Robotics, Ft. Lauderdale, FL, USA, 8–9 May 2003; pp. 1–12.

5. Tadakuma, R.; Asahara, Y.; Kajimoto, H.; Kawakami, N.; Tachi, S. Development of anthropomorphic multi-DOF master-slave arm for mutual telexistence. *IEEE Trans. Vis. Comput. Graph.* **2005**, *11*, 626–636. [CrossRef] [PubMed]

6. Angelika, P.; Einenkel, S.; Buss, M. Multi-fingered telemanipulation—Mapping of a human hand to a three finger gripper. In Proceedings of the 17th IEEE International Symposium on Robot and Human Interactive Communication, Munich, Germany, 1–3 August 2008; pp. 465–470.

7. Kofman, J.; Wu, X.; Luu, T.J. Teleoperation of a robot manipulator using a vision-based human-robot interface. *IEEE Trans. Ind. Electron.* **2005**, *52*, 1206–1219. [CrossRef]

8. Chen, N.; Chew, C.-M.; Tee, K.P.; Han, B.S. Human-aided robotic grasping. In Proceedings of the 21st IEEE International Symposium on Robot and Human Interactive Communication, Paris, France, 9–13 September 2012; pp. 75–80.

9. Hu, C.; Meng, M.Q.; Liu, P.X.; Wang, X. Visual gesture recognition for human-machine interface of robot teleoperation. In Proceedings of the 2003 IEEE/RSJ International Conference Intelligent Robots and Systems, Las Vegas, NV, USA, 37–31 October 2003; pp. 1560–1565.

10. Cela, A.; Yebes, J.J.; Arroyo, R.; Bergasa, L.M.; Barea, R.; López, E. Complete Low-Cost Implementation of a Teleoperated Control System for a Humanoid Robot. *Sensors* **2013**, *13*, 1385–1401. [CrossRef] [PubMed]

11. Yang, K.; Bergasa, L.M.; Romera, E.; Huang, X.; Wang, K. Predicting Polarization Beyond Semantics for Wearable Robotics. In Proceedings of the 2018 IEEE-RAS 18th International Conference on Humanoid Robots, Beijing, China, 6–9 November 2018; pp. 96–103.

12. Zhang, Z. Microsoft Kinect Sensor and Its Effect. *IEEE Multimed.* **2012**, *19*, 4–10. [CrossRef]

13. Weichert, F.; Bachmann, D.; Rudak, B.; Fisseler, D. Analysis of the Accuracy and Robustness of the Leap Motion Controller. *IEEE Sens. J.* **2013**, *13*, 6380–6393. [CrossRef] [PubMed]

14. Lien, J.; Gillian, N.; Karagozler, M.E.; Amihood, P.; Schwesig, C.; Olson, E.; Raja, H.; Poupyrev, I. Soli: Ubiquitous Gesture Sensing with Millimeter Wave Radar. *ACM Trans. Graph* **2016**, *35*, 1–19. [CrossRef]

15. Zhang, F.; Bazarevsky, V.; Vakunov, A.; Tkachenka, A.; Sung, G.; Chang, C.; Grundmann, M. MediaPipe hands: On-device real-time hand tracking. *arXiv* **2020**, arXiv:2006.10214.

16. Tamaki, E.; Miyake, T.; Rekimoto, J. A Robust and Accurate 3D Hand Posture Estimation Method for Interactive Systems. *Trans. Inf. Process. Soc. Jpn.* **2010**, *51*, 229–239.

17. Premaratne, P. *Human Computer Interaction Using Hand Gestures*; Springer: Berlin/Heidelberg, Germany, 2014.

18. Ankit, C. *Robust Hand Gesture Recognition for Robotic Hand Control*; Springer: Berlin/Heidelberg, Germany, 2018.

19. Hoshino, K.; Tamaki, E.; Tanimoto, T. Copycat hand—Robot hand generating imitative behavior. In Proceedings of the IECON 2007—33rd Annual Conference of the IEEE Industrial Electronics Society, Taipei, Taiwan, 5–8 November 2007; pp. 2876–288.

20. Griffin, W.; Findley, R.; Turner, M.; Cutkosky, M. Calibration and mapping of a human hand for dexterous telemanipulation. In Proceedings of the ASME IMECE Symp. Haptic Interfaces for VirtualEnvironments and Teleoperator Systeme, Orlando, FL, USA, 5–10 November 2000; pp. 1–8.

21. Meeker, C.; Haas-Heger, M.; Ciocarlie, M. A Continuous Teleoperation Subspace with Empirical and Algorithmic Mapping Algorithms for Non-Anthropomorphic Hands. *arXiv* **2019**, arXiv:1911.09565v1.

22. Saen, M.; Ito, K.; Osada, K. Action-intention-based grasp control with fine finger-force adjustment using combined optical-mechanical tactile sensor. *IEEE Sens. J.* **2014**, *14*, 4026–4033. [CrossRef]

23. Niwa, M.; Iizuka, H.; Ando, H.; Maeda, T. Tumori Control: Manipulate Robot with detection and transmission of an Archetype of Behavioral Intention. *Trans. Virtual Real. Soc. Jpn.* **2012**, *17*, 3–10.

24. Fallahinia, N.; Mascaro, S.A. Comparison of Constrained and Unconstrained Human Grasp Forces Using Fingernail Imaging and Visual Servoing. In Proceedings of the IEEE Conference on Robotics and Automation 2020 (ICRA 2020) Camera-Ready, Paris, France, 31 May–31 August 2020; pp. 2668–2674.

25. Anvari, M.; Broderick, T.; Stein, H.; Chapman, T.; Ghodoussi, M.; Birch, D.W.; Mckinley, C.; Trudeau, P.; Dutta, S.; Goldsmith, C.H. The impact of latency on surgical precision and task completion during robotic-assisted remote telepresence surgery. *Comput. Aided Surg.* **2005**, *10*, 93–99. [CrossRef] [PubMed]

26. Lum, M.J.H.; Rosen, J.; King, H.; Friedman, D.C.W.; Lendvay, T.S.; Wright, A.S.; Sinanan, M.N.; Hannaford, B. Teleoperation in surgical robotics—Network latency effects on surgical performance. In Proceedings of the 31st Annual International Conference of the IEEE Engineering in Medicine and Biology Society, Minneapolis, MI, USA, 3–6 September 2009; pp. 6860–6863.

27. Katsuki, Y.; Yamakawa, Y.; Watanabe, Y.; Ishikawa, M. Development of fast-response master-slave system using high-speed non-contact 3D sensing and high-speed robot hand. In Proceedings of the 2015 IEEE/RSJ International Conference on Intelligent Robots and Systems (IROS), Hamburg, Germany, 28 September–2 October 2015; pp. 1236–1241.

28. Yamakawa, Y.; Katsuki, Y.; Watanabe, Y.; Ishikawa, M. Development of a High-Speed, Low-Latency Telemanipulated Robot Hand System. *Robotics* **2021**, *10*, 41. [CrossRef]

29. Ito, K.; Sueishi, T.; Yamakawa, Y.; Ishikawa, M. Tracking and recognition of a human hand in dynamic motion for Janken (rock-paper-scissors) robot. In Proceedings of the 2016 IEEE International Conference on Automation Science and Engineering (CASE), Fort Worth, TX, USA, 21–25 August 2016; pp. 891–896.

30. Available online: http://www.hfr.iis.u-tokyo.ac.jp/research/Janken/index-e.html (accessed on 20 April 2022).

31. Namiki, A.; Imai, Y.; Ishikawa, M.; Kaneko, M. Development of a high-speed multifingered hand system and its application to catching. In Proceedings of the 2003 IEEE/RSJ International Conference on Intelligent Robots and Systems, Las Vegas, NV, USA, 27 October–1 November 2003; pp. 2666–2671.
32. The Keras Blog—Building Powerful Image Classification Models Using Very Little Data. Available online: https://blog.keras.io/building-powerful-image-classification-models-using-very-little-data.html (accessed on 20 April 2022).
33. Neural Networks for Machine Learning Lecture 6a Overview of Mini-Batch Gradient Descent. Available online: https://www.cs.toronto.edu/~tijmen/csc321/slides/lecture_slides_lec6.pdf (accessed on 20 April 2022).
34. Canziani, A.; Paszke, A.; Culurciello, E. An Analysis of Deep Neural Network Models for Practical Applications. *arXiv* **2017**, arXiv:1605.07678.
35. Yoshida, K.; Yamakawa, Y. 2D Position Estimation of Finger Joints with High Spatio-Temporal Resolution Using a High-speed Vision. In Proceedings of the 21nd Conference of the Systems Integration Division of the Society of Instrument and Control Engineers (SI2020), Fukuoka, Japan, 8–10 September 2020; pp. 1439–1440. (In Japanese)
36. Huang, S.; Shinya, K.; Bergstrom, N.; Yamakawa, Y.; Yamazaki, T.; Ishikawa, M. Dynamic compensation robot with a new high-speed vision system for flexible manufacturing. *Int. J. Adv. Manuf. Technol.* **2018**, *95*, 4523–4533. [CrossRef]
37. Wong, T. Performance evaluation of classification algorithms by k-fold and leave-one-out cross validation. *Pattern Recognit.* **2015**, *48*, 2839–2846. [CrossRef]
38. Available online: http://www.hfr.iis.u-tokyo.ac.jp/research/Teleoperation/index-e.html (accessed on 20 April 2022).

Article

A Human-Following Motion Planning and Control Scheme for Collaborative Robots Based on Human Motion Prediction

Fahad Iqbal Khawaja [1,2,*], **Akira Kanazawa** [1], **Jun Kinugawa** [1] and **Kazuhiro Kosuge** [1,3]

[1] Center for Transformative AI and Robotics, Graduate School of Engineering, Tohoku University, Sendai 980-8579, Japan; kanazawa@irs.mech.tohoku.ac.jp (A.K.); kinugawa@irs.mech.tohoku.ac.jp (J.K.); kosuge@tohoku.ac.jp or kosuge@hku.hk (K.K.)

[2] Robotics and Intelligent Systems Engineering (RISE) Laboratory, Department of Robotics and Artificial Intelligence, School of Mechanical and Manufacturing Engineering (SMME), National University of Sciences and Technology (NUST), Sector H-12, Islamabad 44000, Pakistan

[3] Department of Electrical and Electronic Engineering, The University of Hong Kong, Pokfulam, Hong Kong

[*] Correspondence: fahad.iqbal@smme.nust.edu.pk or fahad@irs.mech.tohoku.ac.jp

Citation: Khawaja, F.I.; Kanazawa, A.; Kinugawa, J.; Kosuge, K. A Human-Following Motion Planning and Control Scheme for Collaborative Robots Based on Human Motion Prediction. *Sensors* **2021**, *21*, 8229. https://doi.org/10.3390/s21248229

Academic Editor: Anne Schmitz

Received: 8 November 2021
Accepted: 4 December 2021
Published: 9 December 2021

Publisher's Note: MDPI stays neutral with regard to jurisdictional claims in published maps and institutional affiliations.

Abstract: Human–Robot Interaction (HRI) for collaborative robots has become an active research topic recently. Collaborative robots assist human workers in their tasks and improve their efficiency. However, the worker should also feel safe and comfortable while interacting with the robot. In this paper, we propose a human-following motion planning and control scheme for a collaborative robot which supplies the necessary parts and tools to a worker in an assembly process in a factory. In our proposed scheme, a 3-D sensing system is employed to measure the skeletal data of the worker. At each sampling time of the sensing system, an optimal delivery position is estimated using the real-time worker data. At the same time, the future positions of the worker are predicted as probabilistic distributions. A Model Predictive Control (MPC)-based trajectory planner is used to calculate a robot trajectory that supplies the required parts and tools to the worker and follows the predicted future positions of the worker. We have installed our proposed scheme in a collaborative robot system with a 2-DOF planar manipulator. Experimental results show that the proposed scheme enables the robot to provide anytime assistance to a worker who is moving around in the workspace while ensuring the safety and comfort of the worker.

Keywords: human–robot interaction; human–robot collaboration; collaborative robots; motion planning; robot control; human motion prediction; human-following robots

1. Introduction

The concept of collaborative robots was introduced in the early 1990s. The first collaborative system was proposed by Troccaz et al. in 1993 [1]. This system uses a passive robot arm to ensure safe operation during medical procedures. In 1996, Colgate et al. developed a passive collaborative robot system and applied it to the vehicle's door assembly process carried out by a human worker [2]. In 1999, Yamada et al. proposed a skill-assist system to help a human worker carry a heavy load [3].

The collaborative robot systems are being actively introduced in the manufacturing industry. The International Organization of Standardization (ISO) amended its robot safety standards ISO 10128-1 [4] and 10128-2 [5] in 2011 to include the safety guidelines for human-robot collaboration. This led to an exponential rise in collaborative robot research and development. Today, many companies are manufacturing their own versions of collaborative robots, and these robots are being used in industries all over the world. Collaborative robots are expected to play a major role in the Industry 5.0 environments where people will work together with robots and smart machines [6].

In 2010, a 2-DOF co-worker robot "PaDY" (in-time Parts and tools Delivery to You robot) was developed in our lab to assist a factory worker in an automobile assembly process [7]. This process comprises a set of assembly tasks that are carried out by a worker while moving around the car body. PaDY assists the worker by delivering the necessary parts and tools to him/her for each task. The original control system of PaDY was developed based on a statistical analysis of the worker's movements [7].

Many studies have been carried out on the human–robot collaborative system. Hawkins et al. proposed an inference mechanism of human action based on a probabilistic model to achieve wait-sensitive robot motion planning in 2013 [8]. D'Arpino et al. proposed fast target prediction of human reaching motion for human–robot collaborative tasks in 2015 [9]. Unhelkar et al. designed a human-aware robotic system, in which human motion prediction is used to achieve a safe and efficient part-delivery task between the robot and the stationary human worker in 2018 [10]. A recent survey on the sensors and techniques used for human detection and action recognition in industrial environments can be seen in [11]. The studies cited above [8–10] have improved efficiency of the collaborative tasks by incorporating human motion prediction into robot motion planning.

Human motion prediction was also introduced to PaDY. In 2012, the delivery operation delay of the robot–human handover tasks was reduced by utilizing prediction of the worker's arrival time at the predetermined working position [12]. In 2019, a motion planning system was developed which optimized a robot trajectory by taking the prediction uncertainty of the worker's movement into account [13]. In those studies [12,13], the robot repeats the delivery motion from its home position to each predetermined assembly position. If the robot can follow the worker during the assembly process, the worker can pick up necessary parts and tools from the robot at any time. Thus, more efficient collaborative work could be expected by introducing human-following motion.

In this paper, human-following motion of the collaborative robot is proposed for delivery of parts and tools to the worker. The human-following collaborative robot system needs to stay close to the human worker, while avoiding collision with the worker under the velocity and acceleration constraints. The contribution of this paper is summarized as follows:

1. The proposed human-following motion planning and control scheme enables the worker to pick up the necessary parts and tools when needed.
2. The proposed scheme achieves the human-following motion with a sufficiently small tracking error without adversely affecting the safety and comfort of the worker.
3. Experiments conducted in an environment similar to a real automobile assembly process illustrate the effectiveness of the proposed scheme.

This is a quantitative study where we conducted the experiments ourselves and analyzed the data collected from these experiments to deduce the results. The proposed scheme predicts the motion of the worker and calculates an optimal delivery position for the handover of parts and tools from the worker to the robot for each task of the assembly process. This scheme has been designed for a single worker operating within his/her workspace. It is not designed for the cases when multiple workers are operating in the same workspace, or when the worker moves beyond the workspace.

The rest of the paper is organized as follows. Section 2 describes the related works. Section 3 gives an overview of the proposed scheme, including the delivery position determination, the worker's motion prediction, and trajectory planning and control scheme. The experimental results are discussed in Section 4. Section 5 concludes this paper.

2. Related Works

In this section, we present a review of the existing research on human–robot handover tasks, human-following robots, and motion/task planning based on human motion prediction.

2.1. Human–Robot Handover

Some studies have considered the problem of psychological comfort of the human receiver during the handover task. Baraglia et al. addressed the issue of whether and when a robot should take the initiative [14]. Cakmak et al. advocated the inclusion of user preferences while calculating handover position [15]. They also identified that a major cause of delay in the handover action is the failure to convey the intention and timing of handing over the object from the robot to the human [16]. Although these studies deal with important issues for improving the human–robot collaboration, it is still difficult to apply them in actual applications because psychological factors cannot be directly observed.

Some other studies used observable physical characteristics of the human worker for planning a robot motion that is safe and comfortable for the worker. Mainprice et al. proposed a motion planning scheme for human–robot collaboration considering HRI constraints such as constraints of distance, visibility and arm comfort of the worker [17]. Aleotti et al. devised a scheme in which the object is delivered in such a way that its most easily graspable part is directed towards the worker [18]. Sisbot et al. proposed a human-aware motion planner that is safe, comfortable and socially acceptable for the human worker [19].

The techniques and algorithms mentioned above operate with the assumption that the worker remains stationary in the environment. To solve the problem of providing assistance to a worker who moves around in the environment, we propose a human-following approach with HRI constraints in this paper.

2.2. Human-Following Robots

Several techniques have been proposed to carry out human-following motion in various robot applications. One of the first human-following approaches was proposed by Nagumo et al., in which an LED device carried by the human was detected and tracked by the robot using a camera [20].

Hirai et al. performed visual tracking of the human back and shoulder in order to follow a person [21]. Yoshimi et al. used several parameters including the distance, speed, color and texture of human clothes to achieve stable tracking in complex situations [22]. Morioka et al. used the reference velocities for human-following control calculated from estimated human position under the uncertainty of the recognition [23]. Suda et al. proposed a human–robot cooperative handling control using force and moment information [24].

The techniques cited in this section focus on performing human-following motion of the robot to achieve safe and continuous tracking. However, these schemes use the feedback of the observed/estimated current position of the worker. This makes it difficult for the robot to keep up with the worker who is continuously moving around in the workspace. In this paper, we solve this problem by applying human motion prediction and MPC.

2.3. Motion/Task Planning Based on Human Motion Prediction

In recent years, many studies have proposed motion planning using human motion prediction. The predicted human motion is used to generate a safe robot trajectory. Mainprice et al. proposed to plan a motion that avoids the predicted occupancy of the 3D human body [25]. Fridovich-Keil et al. proposed to plan a motion that avoids the risky region calculated by the confidence-aware human motion prediction [26]. Park et al. proposed a collision-free motion planner using the probabilistic collision checker [27].

Several studies have proposed robot task planning to achieve collaborative work based on human motion prediction. Maeda et al. achieved a fluid human–robot handover by estimating the phase of human motion [28]. Liu et al. presented a probabilistic model for human motion prediction for task-level human–robot collaborative assembly [29]. Cheng et al. proposed an integrated framework for human–robot collaboration in which the robot perceives and adapts to human actions [30].

Human motion prediction has been effectively used in various problems of human–robot interaction. In this paper, we apply the human motion prediction to human-following motion of the collaborative robot for delivery of parts/tools to a worker.

3. Proposed Motion Planning and Control Scheme

3.1. System Architecture

Figure 1 shows the system architecture of our proposed scheme. This scheme consists of three major parts:

1. Delivery position determination;
2. Worker's motion prediction;
3. Trajectory planning and control.

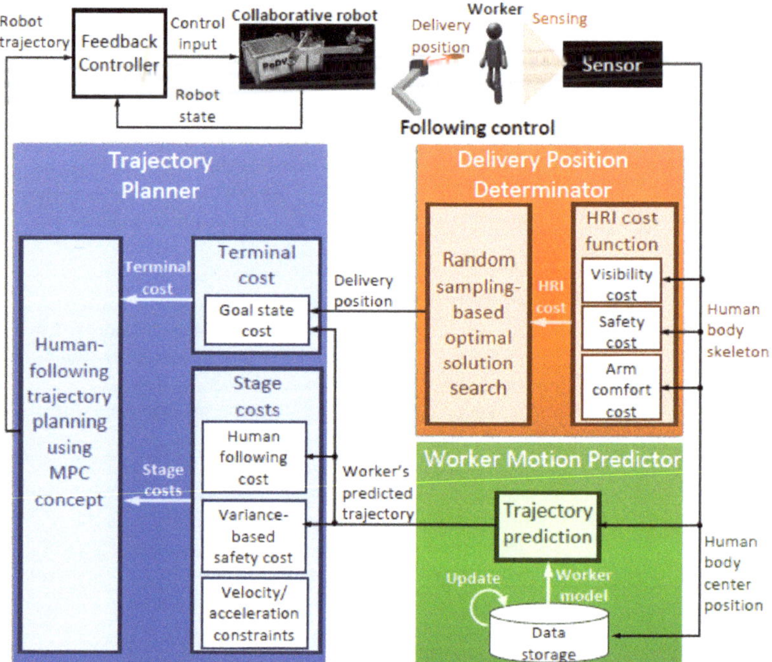

Figure 1. System architecture.

In delivery position determination, an optimal delivery position is estimated using an HRI-based cost function. This cost function is calculated using the skeletal data of the worker measured by the 3-D vision sensor. In workers' motion prediction, the position data obtained from the vision sensor are used to predict the motion of the worker. Moreover, after the completion of each work cycle, the worker's model is updated using the stored position data. In the trajectory planning step, an optimal trajectory from the robot's current position to the goal position is calculated using the receding horizon principle of Model Predictive Control (MPC). The robot motion controller ensures that the robot follows the calculated trajectory. The detailed description of these three parts of our scheme is given in the subsequent subsections.

3.2. Delivery Position Determination

In our proposed scheme, the delivery position is determined by optimizing an HRI-based cost function. This cost function includes terms related to the safety, visibility and arm comfort of the worker. These terms are calculated from the worker's skeletal data observed by the 3-D vision sensor in real-time. This concept was first introduced by Sisbot et al. for motion planning of mobile robots [31]. The analytical form of the HRI-based cost function was proposed in our previous study [32]. Here, we provide a brief description of the cost function and solver for determining the optimal delivery position.

Let $p_{\text{del}} \in \mathbb{R}^n$ be the n dimensional delivery position, then the total cost $Cost(p_{\text{del}}, s_{\text{w}})$ is expressed as:

$$Cost(p_{\text{del}}, s_{\text{w}}) = C_V(p_{\text{del}}, s_{\text{w}}) + C_S(p_{\text{del}}, s_{\text{w}}) + C_A(p_{\text{del}}, s_{\text{w}})$$

where s_{w} is latest sample of the worker's skeletal data obtained from the sensor $C_V(p_{\text{del}}, s_{\text{w}})$ is the visibility cost that maintains the delivery position within the visual range of the worker. This cost is expressed as a function of the difference between the worker's body orientation and the direction of the delivery position with respect to the worker's body center. $C_S(p_{\text{del}}, s_{\text{w}})$ is the safety cost that prevents the robot from colliding with the worker. This cost is expressed as a function of the distance between the worker's body center and the delivery position. $C_A(p_{\text{del}}, s_{\text{w}})$ is the arm comfort cost that maintains the delivery position within the suitable distance and orientation for the worker. This cost is a function of the joint angles of the worker's arm. In addition, this cost penalizes the delivery position where the worker needs to use his/her non-dominant hand.

The optimal delivery position is calculated by minimizing the cost function $Cost(p_{\text{del}}, s_{\text{w}})$. Since $Cost(p_{\text{del}}, s_{\text{w}})$ is a non-convex function, we use Transition-based Rapidly-exploring Random Tree(T-RRT) method [33] to find the globally optimal solution. We apply T-RRT only in the vicinity of the worker to calculate the optimal solution in real-time. The process of determining the optimal delivery position is summarized in Algorithm 1.

Algorithm 1 Determination of Optimal Delivery Position using T-RRT

Input: Worker's position p_{w},
    Current sample of the worker's skeleton s_{w},
    Sampling range r_{s},
    HRI cost function $Cost(p_{\text{del}}, s_{\text{w}})$
Output: Optimal delivery position p_{del}
  1: Set the sampling area S_{near} using p_{w} and r_{s}
  2: $p_{\text{cur}} \leftarrow Sample(S_{\text{near}})$
  3: $Cost_{\text{cur}} \leftarrow Cost(p_{\text{cur}}, s_{\text{w}})$
  4: $Counter \leftarrow 0$
  5: **while** $Counter \leq Counter_{\text{max}}$ **do**
  6:    $p_{\text{rand}} \leftarrow Sample(S_{\text{near}})$
  7:    $p_{\text{new}} \leftarrow p_{\text{cur}} + \delta(p_{\text{rand}} - p_{\text{cur}})$
  8:    $Cost_{\text{new}} \leftarrow Cost(p_{\text{new}}, s_{\text{w}})$
  9:    **if** $TransitionTest(Cost_{\text{new}}, Cost_{\text{cur}}, d_{\text{new}-\text{cur}})$ **then**
10:      $p_{\text{cur}} \leftarrow p_{\text{new}}$
11:      $Cost_{\text{cur}} \leftarrow Cost_{\text{new}}$
12:      $Counter \leftarrow 0$
13:    **else**
14:      $Counter \leftarrow Counter + 1$
15:    **end if**
16: **end while**
17: $p_{\text{del}} \leftarrow p_{\text{cur}}$
18: **return** p_{del}

Figure 2 shows an example of the cost map in the workspace around the worker calculated from the HRI constraints. The worker's shoulder positions (red squares) and the calculated delivery position (green circle) are shown in the figure. We can see that the proposed solver can calculate the delivery position that has the minimum cost in the cost map.

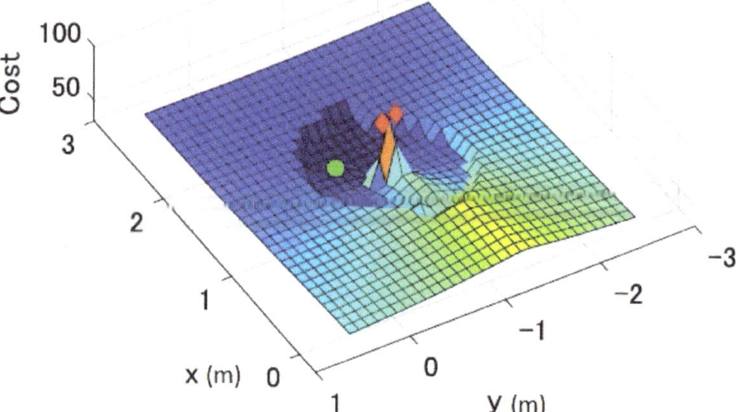

Figure 2. Example of the cost map calculated from the HRI constraints and its optimal delivery position.

3.3. Worker's Motion Prediction

The worker's motion is predicted by using Gaussian Mixture Regression (GMR) proposed in our previous work [34]. GMR models the worker's past movements and predicts his/her future movements in the workspace. Here, we provide a brief description of the motion prediction using GMR.

Suppose that $p_c = p_w^{(t)} \in \mathbb{R}^n$ is the worker's current position at time step t, $p_h = \left(p_w^{(t-1)} \ p_w^{(t-2)} \ \cdots p_w^{(t-d)} \right)^T \in \mathbb{R}^{n \times (d-1)}$ is the position history, and d is the length of the position history. GMR models the conditional probability density function $p_r(p_c|p_h)$ whose expectation $E[p_c|p_h]$ means the worker's future position and variance $V[p_c|p_h]$ is the uncertainty of the prediction. The details of GMR calculation are shown in Appendix A.

The procedure for the long-term motion prediction using GMR is summarized in Algorithm 2. The calculation to predict the worker's position at the next time step is repeated until the length of the predicted trajectory becomes equal to the maximum prediction length T_p. The worker's predicted motion is expressed as the sequence of Gaussian distributions $\left(\mathcal{N}_w^{(t_c)}, \mathcal{N}_w^{(t_c+1)}, \cdots \mathcal{N}_w^{(t_c+T_p)} \right)$ starting from the current time t_c. $\mathcal{N}_w^{(t)}$ is the worker's predicted position distribution at step t expressed as:

$$\mathcal{N}_w^{(t)} = \mathcal{N}(\boldsymbol{\mu}_w^{(t)}, \boldsymbol{\Sigma}_w^{(t)}) \tag{1}$$

where $\boldsymbol{\mu}_w^{(t)}$ is the mean vector and $\boldsymbol{\Sigma}_w^{(t)}$ is the covariance matrix of worker's predicted position at step t.

If the worker repeats his/her normal movement, which is indicated in the process chart of the assembly process, our prediction system can predict the worker's movement accurately enough for the system. According to our previous research, the RMSE (Root Mean Square Error) of the worker's movement was about 0.3 m [34]. The RMSE was calculated by the comparison between the initial predicted worker's movement and the observed worker's movement when the worker started to move to the next working position.

When the worker moves differently from his/her normal movement, it is not easy to ensure the accuracy of the prediction. However, the proposed system operates safely even in this case, since the variance of the predicted position of the worker is included in the cost function used in the motion planning as shown in our previous study [13].

Algorithm 2 Worker's motion prediction using GMR

Input: Current time t_c,
    Current position $p_c^{(t_c)}$,
    Position history $p_h^{(t_c)}$,
    Max prediction length T_p
Output: Predicted trajectory $\left(\mathcal{N}^{(t_c)}, \mathcal{N}^{(t_c+1)}, \cdots, \mathcal{N}^{(t_c+T_p)}\right)$
  1: **while** $k = 1$ *to* T_p **do**
  2:    $p_h^{(t_c+k)} = \left(p_c^{(t_c+k-1)} \; p_c^{(t_c+k-2)} \; \cdots \; p_c^{(t_c+k-d-1)}\right)^{\mathrm{T}}$
  3:    $\mu_w^{(t_c+k)} = E[p_c^{(t_c+k)} | p_h^{(t_c+k)}]$
  4:    $\Sigma_w^{(t_c+k)} = V[p_c^{(t_c+k)} | p_h^{(t_c+k)}]$
  5:    $p_c^{(t_c+k)} = E[p_c^{(t_c+k)} | p_h^{(t_c+k)}]$
  6: **end while**

3.4. Trajectory Planning and Control

Figure 3 shows the concept of human-following motion planning using the worker's predicted motion. The sequence of the worker's predicted position distributions $(\mathcal{N}_w^{(t_c)}, \mathcal{N}_w^{(t_c+1)}, \cdots \mathcal{N}_w^{(t_c+T_p)})$ is given to the trajectory planner. The sequence of the robot states, that is the robot trajectory $\left(q^{(t_c)}, q^{(t_c+1)}, \cdots q^{(t_c+T_0)}\right)$, is calculated so that the robot's state at each time step follows the corresponding predicted position of the worker.

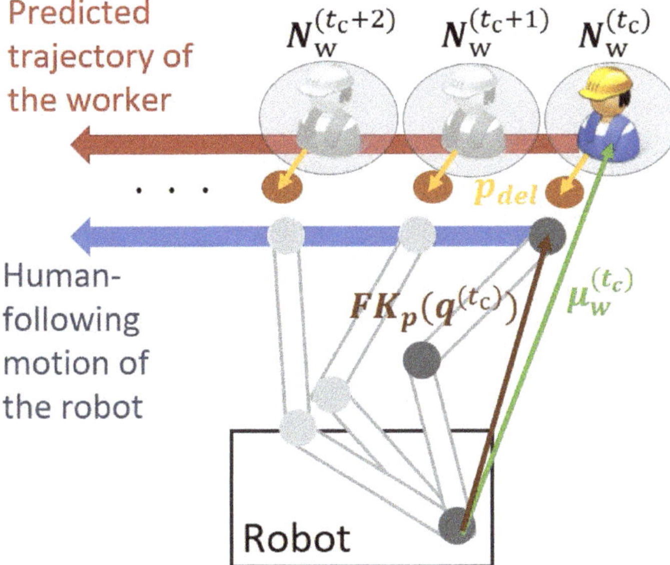

Figure 3. Concept of human-following motion planning based on the predicted trajectory of the worker.

To achieve the prediction-based human-following robot motion, we use an MPC-based planner to consider the evaluation function for finite time future robot states. This is a well-known strategy and is often used in real-time robot applications such as task-parametrized motion planning [35] and multi-agent motion planning [36].

The cost function used in MPC consists of terminal cost and stage cost. The terminal cost deals with the cost at the terminal state of the robot, which is the delivery position in our case. Stage cost considers the state of the robot during the whole trajectory from the current configuration to the goal configuration. A distinct feature of our scheme is that the optimal delivery position, found by optimizing the HRI-based cost function, is used to calculate the terminal cost. In addition, the predicted trajectory of the worker is used to calculate the stage cost. This scheme plans the collision-free robot trajectory that follows the moving worker efficiently under the safety cost constraint and the robot's velocity and acceleration constraints.

The cost function J used for the optimization of the proposed trajectory planner is expressed as:

$$J = \varphi(q(t_c + T_o)) + \int_{t_c}^{t_c+T_o} (L_1(\dot{q}(k)) + L_2(q(k)) + L_3(q(k)))dk \tag{2}$$

where $q = (\theta, \dot{\theta})^T \in \mathbb{R}^{2N_j}$ is the state vector of the manipulator, $\theta = (\theta_1, \theta_2, \cdots, \theta_{N_j})^T \in \mathbb{R}^{N_j}$ is the vector composed of the joint angles of the manipulator, N_j is the degrees of freedom of the manipulator, $T_o(T_o \leq T_p)$ is the length of the trajectory (in our experiments, we used $T_o = T_p$ as a rule of thumb) and $\varphi(q(t_c + T_o))$ is the terminal cost which prevents the calculated trajectory of the robot from diverging. It is expressed as:

$$\varphi(q(t_c + T_o)) = \frac{1}{2}\left(FK_{N_j}(q(t_c + T_o)) - x_{\text{del}}\right)^T R\left(FK_{N_j}(q(t_c + T_o)) - x_{\text{del}}\right) \tag{3}$$

where FK_j is the forward kinematics of the robot that transform the robot state q from joint coordinates to position p_j and velocity v_j in the workspace coordinates. R is the diagonal positive definite weighting matrix. x_{del} is the terminal state of the robot which is calculated based on the optimal delivery position and the predicted mean position of the worker. In this study, x_{del} becomes $x_{\text{del}} = \left(\mu_w^{(t_c+T_o)} + p_{\text{del}}, 0\right)^T$, where $\mu_w^{(t_c+T_o)}$ is the worker's predicted position at the end of the trajectory $(t_c + T_o)$, and p_{del} is the calculated delivery position for the worker's observed position. We calculate p_{del} after each sampling interval and assume that the variation in p_{del} is negligibly small during the sampling interval of the sensing system, which is 30 ms.

$L_1(\dot{q}(k))$, $L_2(q(k))$ and $L_3(q(k))$ are the stage costs which are expressed as:

$$L_1(\dot{q}(k)) = \frac{1}{2}\sum_{j=1}^{N} r_j(B_{\text{vel},j}(\dot{\theta}_j(k)) + B_{\text{acc},j}(\ddot{\theta}_j(k))) \tag{4}$$

$$L_2(q(k)) = w\sum_{j=1}^{N_j} \frac{1}{D_M\left(FK_{p,j}(q(k)), \mu_w^{(k)}, \Sigma_w^{(k)}\right)}. \tag{5}$$

$$L_3(q(k)) = \frac{1}{2}\sum_{j=1}^{N} (FK_{p,j}(q(k)) - (\mu_w^{(k)} + p_{\text{del}})^T)\, Q\, (FK_{p,j}(q(k)) - (\mu_w^{(k)} + p_{\text{del}})) \tag{6}$$

$L_1(\dot{q}(k))$ is the stage cost to maintain the robot velocity and acceleration within their maximum limits. $B_{\text{vel},j}(\dot{\theta}_j(k))$ and $B_{\text{acc},j}(\ddot{\theta}_j(k))$ are defined as:

$$B_{\text{vel},j}(\dot{\theta}_j(k)) = \begin{cases} 0 & (||\dot{\theta}_j|| \leq \dot{\theta}_{\max,j}) \\ (||\dot{\theta}_j|| - \dot{\theta}_{\max,j})^2 & (||\dot{\theta}_j|| > \dot{\theta}_{\max,j}) \end{cases} \tag{7}$$

$$B_{\text{acc},j}(\ddot{\theta}_j(k)) = \begin{cases} 0 & (||\ddot{\theta}_j|| \leq \ddot{\theta}_{\max,j}) \\ (||\ddot{\theta}_j|| - \ddot{\theta}_{\max,j})^2 & (||\ddot{\theta}_j|| > \ddot{\theta}_{\max,j}) \end{cases} \tag{8}$$

where $\dot{\theta}_{\mathrm{max},j}$ and $\ddot{\theta}_{\mathrm{max},j}$ are the maximum velocity and maximum acceleration of the jth joint, respectively.

$L_2(q(k))$ is the stage cost that prevents the robot from hitting the worker. w is a weighting coefficient of this cost function. $D_{\mathrm{M}}(x, \mu, \Sigma) = \sqrt{(x - \mu)^{\mathrm{T}} \Sigma^{-1} (x - \mu)}$ is the Mahalanobis distance that considers the variance of the probabilistic density distribution. Using the Mahalanobis distance between the predicted worker's position distribution $\mathcal{N}(\mu_{\mathrm{w}}^{(k)}, \Sigma_{\mathrm{w}}^{(k)})$ and the end-effector position $FK_{p,j}(q(k))$ at step k, an artificial potential field is constituted according to the predicted variance. The artificial potential becomes wider in the direction of larger variance in the predicted position.

$L_3(q)$ is the stage cost to ensure that the robot follows the worker's motion. Q is the diagonal positive definite weighting matrix. This cost function is responsible for the human-following motion of the robot based on the worker's predicted trajectory as shown in Figure 3. For each time step of the predicted position distribution of the worker $\mathcal{N}(\mu_{\mathrm{w}}^{(k)}, \Sigma_{\mathrm{w}}^{(k)})$, the desirable state of the robot is calculated so that the robot's endpoint follows the predicted mean position of the worker $\mu_{\mathrm{w}}^{(k)}$ offset by the calculated delivery position p_{del}.

Now we can define the optimization problem that will be solved by our proposed system.

$$\begin{aligned}
\text{minimize} \quad & J \\
\text{subject to} \quad & \dot{q} = f(q, u) \\
& q(t) = q_{\mathrm{cur}}
\end{aligned}$$

where f denotes the nonlinear term of the robot's dynamics, u is the input vector, and $q(t)$ is the initial state of the trajectory which corresponds to the current state q_{cur} of the robot. To solve this optimization problem with the equality constraints described above, we use the calculus of variations. The discretized Euler–Lagrange equations that the optimal solution should satisfy are expressed as:

$$q^{(k+1)} = q^{(k)} + f(q^{(k)}, u^{(k)}) \Delta t_{\mathrm{s}}, \tag{9}$$

$$q^{(t)} = q_{\mathrm{cur}}, \tag{10}$$

$$\lambda^{(k)} = \lambda^{(k+1)} - \left(\frac{\partial H}{\partial q}\right)^{\mathrm{T}} (q^{(k+1)}, u^{(k)}, \lambda^{(k+1)}), \tag{11}$$

$$\lambda^{(t+T_o)} = \left(\frac{\partial \varphi}{\partial q}\right)^{\mathrm{T}} (q^{(t+T_o)}), \tag{12}$$

$$\frac{\partial H}{\partial u}(q^{(k)}, u^{(k)}, \lambda^{(k)}) = 0, \tag{13}$$

where H is the Hamiltonian and is defined as:

$$H(q, u, \lambda) = L_1(\dot{q}(k)) + L_2(q(k)) + L_3(q(k)) + \lambda^{\mathrm{T}} f(q, u). \tag{14}$$

The procedure for calculating the online trajectory is shown in Algorithm 3. After the sequential optimization based on the gradient decent, we obtain the optimal trajectory $\left(q^{(t)}, q^{(t+1)}, \cdots, q^{(t+T_o)}\right)$. For detailed calculations, please refer to our previous study [13].

Algorithm 3 Robot Trajectory Generator

Input: Target delivery position p_{del},
    Predicted worker's trajectory $\left(\mathcal{N}^{(t)}, \mathcal{N}^{(t+1)}, \cdots \mathcal{N}^{(t+T_p)}\right)$,
    Current state of the robot q_{cur},
    Max length of the robot trajectory T_o
Output: Optimal trajectory is $\left(q^{(t)}, q^{(t+1)}, \cdots q^{(t+T_o)}\right)$

1: Initialize the set of input vectors u
2: $q^{(t)} \Leftarrow q_{\text{cur}}$
3: **while** $\Sigma_{k=t}^{t+T_o} |\frac{\partial H}{\partial u}(q^{(k+1)}, u^{(k)}, \lambda^{(k+1)})| < \epsilon$ **do**
4:     **while** $k = 1$ *to* T_o **do**
5:         $q^{(t+k)} \Leftarrow q^{(t+k-1)} + f(q^{(t+k-1)}, u^{(t+k-1)})\Delta t_s$
6:     **end while**
7:     **while** $k = T_o$ *to* 1 **do**
8:         $\lambda^{(t+k-1)} \Leftarrow \lambda^{(t+k)} - \left(\frac{\partial H}{\partial q}\right)^{\text{T}} (q^{(t+k)}, u^{(t+k)}, \lambda^{(t+k+1)})$
9:     **end while**
10:    **while** $i = 1$ *to* T_o **do**
11:        $s_i \Leftarrow \left(\frac{\partial H}{\partial u}\right)^{\text{T}} (q^{(t+i-1)}, u^{(t+i-1)}, \lambda^{(t+i)})$
12:    **end while**
13:    $u \Leftarrow u + cs$
14: **end while**
15: **while** $k = 1$ *to* T_o **do**
16:    $q^{(t+k)} \Leftarrow q^{(t+k-1)} + f(q^{(t+k-1)}, u^{(t+k-1)})\Delta t_s$
17: **end while**

---

4. Experiment

4.1. Experimental Setup

To evaluate the performance of the proposed scheme in a real-world environment, we used the planar manipulator PaDY proposed in our previous study [7]. PaDY was designed to assist the workers of an automobile factory. A parts tray and a tool holder were attached to the end-effector of PaDY to store the parts and tools required for car assembly tasks. The robot delivers the parts and tools to the worker during the assembly process. For the details of the hardware design of PaDY, please refer to [7].

The proposed scheme was installed in a computer with an Intel Core i7-3740QM (Quad-core processor, 2.7 GHz) with 16GB memory. All calculations were done within 30 ms, the sampling interval of the sensing system that tracks the position of the human worker.

We designed an experiment to demonstrate the effectiveness of the worker's motion prediction in the human-following behavior of our proposed scheme. Figure 4 shows the experimental workspace and Figure 5 shows the top view of the setup for this experiment. In this experiment, the worker needs to perform the following six tasks:

1. Tightening a bolt (Task 1);
2. Attaching three grommets (Task 2);
3. Attaching one grommet (Task 3, Task 4, Task 5, Task 6).

Each task is performed at a separate working position in the workspace. The experiment is carried out as shown in Figure 6. The experiment is carried out as follows.

1. The experiment begins when the robot starts to approach the worker standing at the working position for Task 1. The worker takes a bolt and the bolt tightening tool from the robot (Figure 6a).
2. The worker performs Task 1 (Figure 6b).
3. The worker moves to the working position for Task 2 and the robot follows him. The worker returns the bolt tightening tool to the tool holder (Figure 6c) and picks up three grommets from the parts tray.

4. The worker performs Task 2 (Figure 6d).
5. The worker moves to the working position for Task 3 and picks up a grommet from the tray (Figure 6e).
6. The worker performs Task 3 (Figure 6f).
7. The worker moves to the working position for Task 4 and picks up a grommet from the tray (Figure 6g).
8. The worker performs Task 4 (Figure 6h).
9. The worker moves to the working position for Task 5 and picks up another grommet from the parts tray (Figure 6i).
10. The worker performs Task 5 (Figure 6j).
11. The worker moves to the working position Task 6 and picks up the last grommet from the parts tray (Figure 6k).
12. The worker performs Task 6 (Figure 6l) and this concludes the experiment.

We performed this experiment with four different participants (A, B, C and D) to evaluate the robustness of the system for different workers. Each participant is asked to perform the complete work cycle ten times. The first trial is performed without using the predicted motion of the worker. Whereas, in all other trials, the predicted motion of the worker is used and the worker model is sequentially updated after completing each trial.

For more details about the experiment, please see the Supplementary Materials.

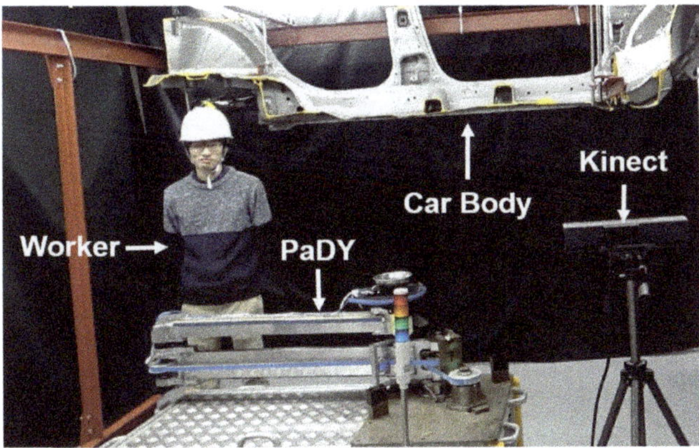

Figure 4. Experimental workspace.

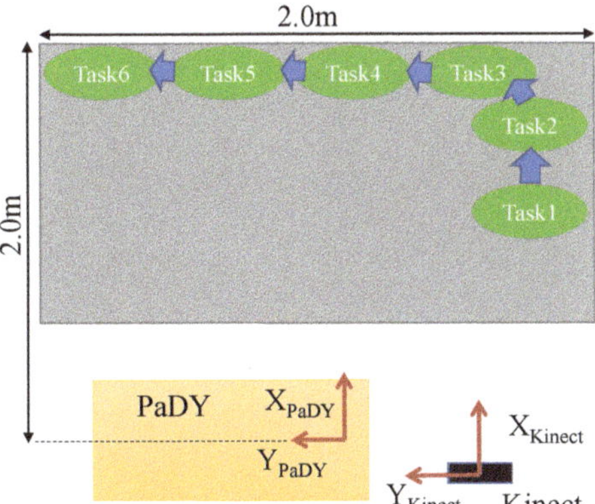

Figure 5. Top view of the experimental setup.

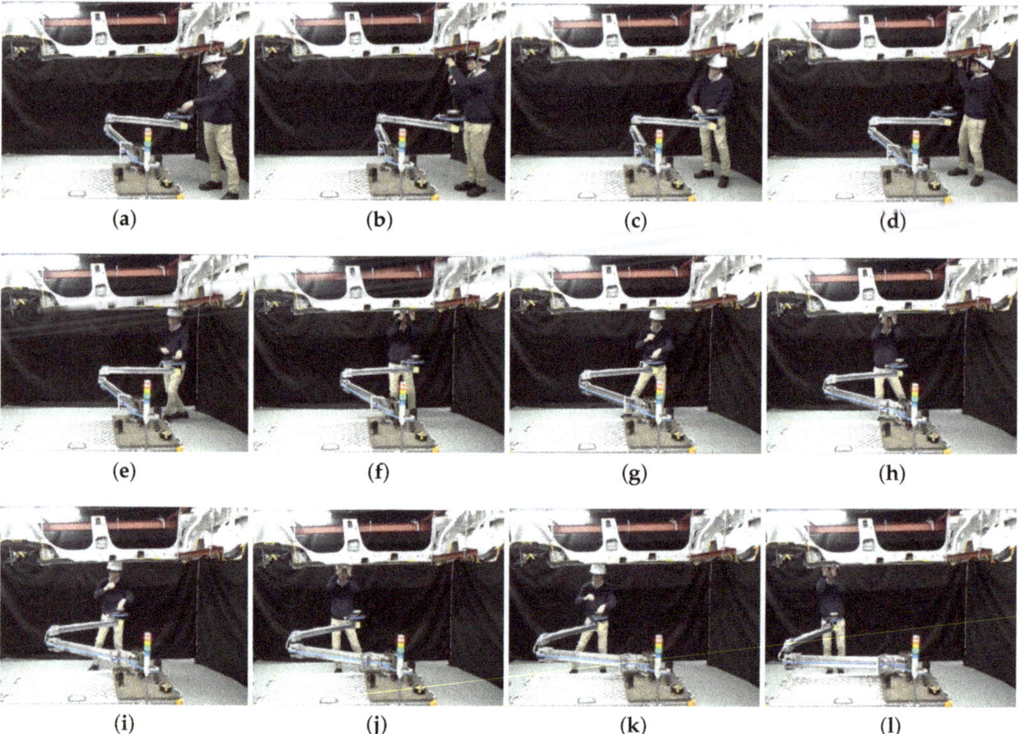

Figure 6. Experiment showing a complete work cycle where six tasks are performed. (**a**) A bolt and the tool are picked up; (**b**) Task 1 is performed; (**c**) The tool is returned and 3 grommets are picked up; (**d**) Task 2 is performed; (**e**) A grommet is picked up; (**f**) Task 3 is performed; (**g**) A grommet is picked up; (**h**) Task 4 is performed; (**i**) A grommet is picked up; (**j**) Task 5 is performed; (**k**) A grommet is picked up; (**l**) Task 6 is performed.

4.2. Tracking Performance

Figure 7a shows the estimated delivery position and the robot's end-effector position for trial 1 of a participant when the robot's motion is calculated based on the observed position of the worker without using the motion prediction. The black vertical lines show the time when the worker performs each assembly task. Figure 7b shows the estimated delivery position and the robot's end-effector position for trial 10 of the same participant when the robot's motion is calculated based on the predicted position of the worker using the proposed scheme.

At the beginning of the experiment, the robot is at its home position and the participant is at the working position for Task 1. The robot starts its human-following motion after arriving at the delivery position for Task 1 (at around 6 s in Figure 7a,b). We can see that the robot keeps following the participant during the whole experiment in both schemes (with and without the use of motion prediction).

The green line in Figure 7a,b shows the tracking error which is the difference between the delivery position and the end-effector position. We can see that the maximum tracking error is reduced from about 0.5 m to 0.3 m by using the motion prediction.

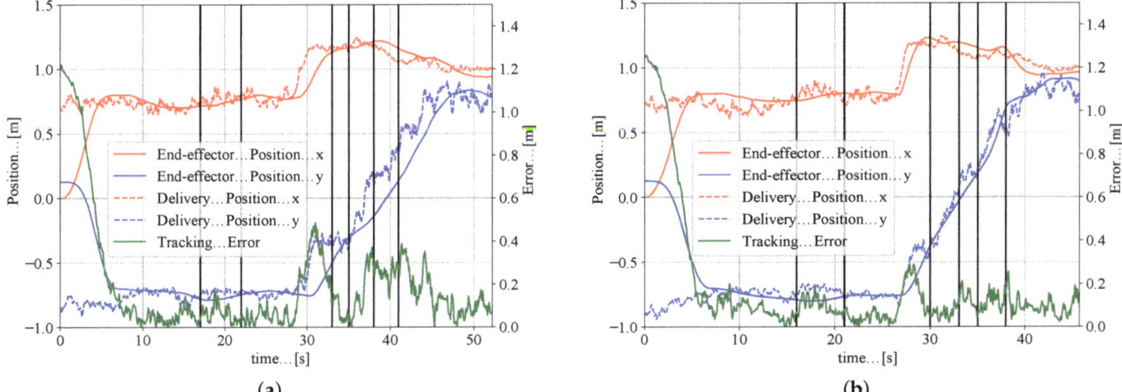

Figure 7. Tracking performance. (**a**) When motion prediction is not used; (**b**) When motion prediction is used.

It is not possible to completely eliminate the tracking error since the manipulator used for the experiments has a mechanical torque limiter at each joint and the maximum angular acceleration without activating the torque limiter is 90 deg/s^2. In both Figure 7a,b, a large tracking error around 30 s can be observed. This is because the participant makes a large movement around 30 s and the robot cannot follow the participant because of its acceleration limit.

4.3. Cycle Time

Figure 8 shows the comparison of the cycle time of the four participants in each trial. We define cycle time as the time required for a participant to complete all six tasks of the assembly process. In Figure 8, the cycle time of each trial is normalized by the time of trial 1. Remember that motion prediction was not used in trial 1.

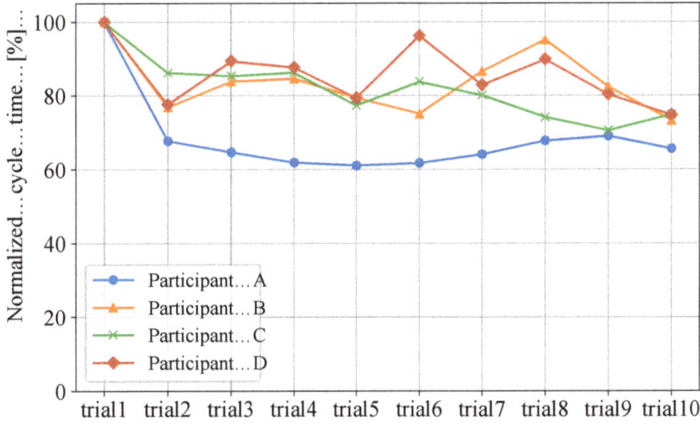

Figure 8. Comparison of Cycle Time.

We can see that the cycle time for each participant decreases as the number of trials increases. The cycle time of trial 10 is reduced to 65.6–74.8% of the cycle time of trial 1. This shows that motion prediction can improve the performance of participants and help them complete the assembly process faster.

Note that the proposed system ignores the dynamics of the interaction between the robot and the worker, assuming that the worker is well trained and the behavioral dynamics of the worker with respect to the robot's movements can be ignored. If the effects of the robot's motion on the worker can be modeled, the system can better deal with the effect of the interaction between the worker and the robot and a further improvement in the worker's time efficiency could be expected.

4.4. HRI-Based Cost

Table 1 shows the average and maximum HRI-based costs for each participant during the human-following motion of the robot. Since the HRI-based cost increases as the safety and comfort of the worker decreases, it is desirable to have a low HRI-based cost in human–robot collaboration.

Table 1. Summary of HRI-based costs during the human-following motion for each worker.

Worker	Average Cost (without Prediction)	Average Cost (with Prediction)	Max Cost (without Prediction)	Max Cost (with Prediction)
Worker A	8.99	11.79	36.34	35.82
Worker B	12.73	9.90	38.17	34.44
Worker C	18.35	13.56	39.65	31.33
Worker D	16.30	17.50	31.47	31.26

In Table 1, we see that there are no significant differences in the average and maximum HRI-based costs between trial 1 (when motion prediction is not used) and trial 10 (when motion prediction is used) for all four participants. Therefore, we conclude that the proposed prediction-based human-following control reduces the work cycle time without adversely affecting the safety and comfort of the workers.

5. Conclusions

We proposed a human-following motion planning and control scheme for a collaborative robot which supplies the necessary parts and tools to a worker in an automobile assembly process. The human-following motion of the collaborative robot makes it possible to provide anytime assistance to the worker who is moving around in the workspace.

The proposed scheme calculates an optimal delivery position for the current position of the worker by performing non-convex optimization of an HRI-based cost function. Whenever the worker's position changes, the new optimal delivery position is calculated. Based on the observed movement of the worker, the motion of the worker is predicted and the robot's trajectory is updated in real-time using model predictive control to ensure a smooth transition between the previous and new trajectories.

The proposed scheme was applied to a planar collaborative robot called PaDY. Experiments were conducted in a real environment where a worker performed a car assembly process with the assistance of the robot. The results of the experiments confirmed that our proposed scheme provides better assistance to the worker, improves the work efficiency, and ensures the safety and comfort of the worker.

This scheme has been designed for a single worker operating within his/her workspace. It is not designed for the cases when multiple workers are operating in the same workspace, or when the worker moves beyond the workspace. Moreover, we did not consider the dynamics of interaction between the robot and the human, assuming that the human worker in the factory is well trained and his/her behavior dynamics to the robot motion is negligible. If the effects of the robot's motion on the human can be modeled, the system can better deal with the effect of the interaction and further improvement in time efficiency could be expected.

We believe that the human-following approach has tremendous potential in the field of collaborative robotics. The ability to provide anytime assistance is a key feature of our proposed method, and we believe it will be very useful in many other collaborative robot applications.

Supplementary Materials: The supplementary material is available online at https://youtu.be/-jkPoK5URdw (accessed on 4 December 2021), Video: Human-Following Motion Planning and Control Scheme for Collaborative Robots.

Author Contributions: Conceptualization, J.K. and K.K.; methodology, A.K. and K.K.; software, F.I.K. and A.K.; validation, F.I.K. and A.K.; formal analysis, F.I.K. and A.K.; investigation, A.K. and K.K.; resources, J.K. and K.K.; data curation, F.I.K. and A.K.; writing—original draft preparation, F.I.K. and A.K.; writing—review and editing, F.I.K., A.K. and K.K.; visualization, F.I.K.; supervision, K.K.; project administration, K.K.; funding acquisition, J.K. and K.K. All authors have read and agreed to the published version of the manuscript.

Funding: This research received no external funding.

Institutional Review Board Statement: Not applicable.

Informed Consent Statement: Informed consent was obtained from all subjects involved in the study.

Data Availability Statement: The dataset generated from the experiments in this study can be found at https://github.com/kf-iqbal-29/Dataset-HumanFollowingCollaborativeRobot.git.

Conflicts of Interest: The authors declare no conflict of interest.

Abbreviations

The following abbreviations are used in this manuscript:

HRI	Human–Robot Interaction
MPC	Model Predictive Control
ISO	International Organization of Standardization
DOF	Degrees Of Freedom
PaDY	In-time Parts and tools Delivery to You robot
T-RRT	Transition-based Rapidly exploring Random Trees
GMR	Gaussian Mixture Regression
RMSE	Root Mean Square Error

Appendix A. Detail Calculation of Gaussian Mixture Regression

Suppose that $p_c = p_w^{(t)} \in \mathbb{R}^n$ is the worker's current position at time step t, $p_h = \left(p_w^{(t-1)} \; p_w^{(t-2)} \; \cdots \; p_w^{(t-d)} \right)^{\mathrm{T}} \in \mathbb{R}^{n \times (d-1)}$ is the position history, and d is the order of the autoregressive model. Then the joint distribution p_r of p_c and p_h can be expressed as

$$p_r(p_h, p_c) = \sum_{m=1}^{M} \pi_m \mathcal{N}(p_h, p_c | \mu_m, \Sigma_m), \tag{A1}$$

where

$$\mu_m = \begin{bmatrix} \mu_m^{p_h} \\ \mu_m^{p_c} \end{bmatrix}, \tag{A2}$$

$$\Sigma_m = \begin{bmatrix} \Sigma_m^{p_h p_h} & \Sigma_m^{p_h p_c} \\ \Sigma_m^{p_c p_h} & \Sigma_m^{p_c p_c} \end{bmatrix}. \tag{A3}$$

The expectation $E[p_c | p_h]$ and the variance $V[p_c | p_h]$ of the conditional probability density function $p_r(p_c | p_h)$ are expressed as

$$E[p_c | p_h] = \sum_{m=1}^{M} h^m(p_h) \mu', \tag{A4}$$

$$V[p_c | p_h] = \sum_{m=1}^{M} h^m(p_h) \left(\Sigma' + \mu' \mu'^{\mathrm{T}} \right) - E[p_c | p_h] E[p_c | p_h]^{\mathrm{T}}, \tag{A5}$$

where

$$h^m(\boldsymbol{p}_\mathrm{h}) = \frac{\pi_m \mathcal{N}(\boldsymbol{p}_\mathrm{h}|\boldsymbol{\mu}_m^{\boldsymbol{p}_\mathrm{h}}, \boldsymbol{\Sigma}_m^{\boldsymbol{p}_\mathrm{h}\boldsymbol{p}_\mathrm{h}})}{\sum_{k=1}^{K} \pi_k \mathcal{N}(\boldsymbol{p}_\mathrm{h}|\boldsymbol{\mu}_k^{\boldsymbol{p}_\mathrm{h}}, \boldsymbol{\Sigma}_k^{\boldsymbol{p}_\mathrm{h}\boldsymbol{p}_\mathrm{h}})}, \tag{A6}$$

$$\boldsymbol{\mu}' = \boldsymbol{\mu}_m^{\boldsymbol{p}_\mathrm{c}} + \boldsymbol{\Sigma}_m^{\boldsymbol{p}_\mathrm{c}\boldsymbol{p}_\mathrm{h}}(\boldsymbol{\Sigma}_m^{\boldsymbol{p}_\mathrm{h}\boldsymbol{p}_\mathrm{h}})^{-1}(\boldsymbol{p}_\mathrm{h} - \boldsymbol{\mu}_m^{\boldsymbol{p}_\mathrm{h}}), \tag{A7}$$

$$\boldsymbol{\Sigma}' = \boldsymbol{\Sigma}_m^{\boldsymbol{p}_\mathrm{c}\boldsymbol{p}_\mathrm{c}} - \boldsymbol{\Sigma}_m^{\boldsymbol{p}_\mathrm{c}\boldsymbol{p}_\mathrm{h}}(\boldsymbol{\Sigma}_m^{\boldsymbol{p}_\mathrm{h}\boldsymbol{p}_\mathrm{h}})^{-1}\boldsymbol{\Sigma}_m^{\boldsymbol{p}_\mathrm{h}\boldsymbol{p}_\mathrm{c}}. \tag{A8}$$

While making the prediction, the position of the worker $\boldsymbol{p}_\mathrm{c}^{(t+1)}$ at step $t+1$ is calculated as

$$\boldsymbol{p}_\mathrm{c}^{(t+1)} = \mathbb{E}[\boldsymbol{p}_\mathrm{c}^{(t+1)}|\boldsymbol{p}_\mathrm{h}^{(t+1)}],$$
$$\boldsymbol{p}_\mathrm{h}^{(t+1)} = \left(\boldsymbol{p}_\mathrm{c}^{(t)}\ \boldsymbol{p}_\mathrm{c}^{(t-1)}\ \cdots\ \boldsymbol{p}_\mathrm{c}^{(t+1-d)}\right)^\mathrm{T}. \tag{A9}$$

This calculation to predict the worker's position, shown in Equation (A9) is repeated until the length of the predicted trajectory becomes equal to the maximum prediction length T_p. The process of predicting the worker's motion is summarized in Algorithm 2. For the details of the derivation, please see [37].

References

1. Troccaz, J.; Lavallee, S.; Hellion, E. A passive arm with dynamic constraints: A solution to safety problems in medical robotics. In Proceedings of the IEEE Systems Man and Cybernetics Conference—SMC, Le Touquet, France, 17–20 October 1993; Volume 3, pp. 166–171. doi:10.1109/ICSMC.1993.385004. [CrossRef]
2. Colgate, J.; Wannasuphoprasit, W.; Peshkin, M. Cobots: Robots for collaboration with human operators. In Proceedings of the 1996 ASME International Mechanical Engineering Congress and Exposition, Atlanta, Georgia, 17–22 November 1996; pp. 433–439.
3. Yamada, Y.; Konosu, H.; Morizono, T.; Umetani, Y. Proposal of Skill-Assist: A system of assisting human workers by reflecting their skills in positioning tasks. In Proceedings of the IEEE SMC'99 Conference Proceedings 1999 IEEE International Conference on Systems, Man, and Cybernetics (Cat. No.99CH37028), Tokyo, Japan, 12–15 October 1999; Volume 4, pp. 11–16. [CrossRef]
4. ISO. *ISO 10218-1:2011—Robots and Robotic Devices—Safety Requirements for Industrial Robots—Part 1: Robots*; ISO: Geneva, Switzerland, 2011.
5. ISO. *ISO 10218-2:2011—Robots and Robotic Devices—Safety Requirements for Industrial Robots—Part 2: Robot Systems and Integration*; ISO: Geneva, Switzerland, 2011.
6. Perakovic, D.; Periša, M.; Cvitić, I.; Zorić, P. Information and communication technologies for the society 5.0 environment. In Proceedings of the XXXVIII Simpozijum o Novim Tehnologijama u Poštanskom i Telekomunikacionom Saobraćaju—PosTel 2020, Belgrade, Serbia, 1–2 December 2020; doi:10.37528/FTTE/9788673954318/POSTEL.2020.020. [CrossRef]
7. Kinugawa, J.; Kawaai, Y.; Sugahara, Y.; Kosuge, K. PaDY: Human-friendly/cooperative working support robot for production site. In Proceedings of the 2010 IEEE/RSJ International Conference on Intelligent Robots and Systems, Taipei, Taiwan, 18–22 October 2010; pp. 5472–5479. [CrossRef]
8. Hawkins, K.P.; Vo, N.; Bansal, S.; Bobick, A.F. Probabilistic human action prediction and wait-sensitive planning for responsive human-robot collaboration. In Proceedings of the 2013 13th IEEE-RAS International Conference on Humanoid Robots (Humanoids), Atlanta, GA, USA, 15–17 October 2013; pp. 499–506. [CrossRef]
9. Pérez-D'Arpino, C.; Shah, J.A. Fast target prediction of human reaching motion for cooperative human-robot manipulation tasks using time series classification. In Proceedings of the 2015 IEEE International Conference on Robotics and Automation (ICRA), Seattle, WA, USA, 26–30 May 2015; pp. 6175–6182. [CrossRef]
10. Unhelkar, V.V.; Lasota, P.A.; Tyroller, Q.; Buhai, R.D.; Marceau, L.; Deml, B.; Shah, J.A. Human-Aware Robotic Assistant for Collaborative Assembly: Integrating Human Motion Prediction With Planning in Time. *IEEE Robot. Autom. Lett.* **2018**, 3, 2394–2401. [CrossRef]
11. Bonci, A.; Cen Cheng, P.D.; Indri, M.; Nabissi, G.; Sibona, F. Human-Robot Perception in Industrial Environments: A Survey. *Sensors* **2021**, 21, 1571. [CrossRef] [PubMed]
12. Tanaka, Y.; Kinugawa, J.; Sugahara, Y.; Kosuge, K. Motion planning with worker's trajectory prediction for assembly task partner robot. In Proceedings of the 2012 IEEE/RSJ International Conference on Intelligent Robots and Systems, Vilamoura, Algarve, Portugal, 7–12 October 2012; pp. 1525–1532. [CrossRef]
13. Kanazawa, A.; Kinugawa, J.; Kosuge, K. Adaptive Motion Planning for a Collaborative Robot Based on Prediction Uncertainty to Enhance Human Safety and Work Efficiency. *IEEE Trans. Robot.* **2019**, 35, 817–832. [CrossRef]
14. Baraglia, J.; Cakmak, M.; Nagai, Y.; Rao, R.; Asada, M. Initiative in robot assistance during collaborative task execution. In Proceedings of the 2016 11th ACM/IEEE International Conference on Human-Robot Interaction (HRI), Christchurch, New Zealand, 7–10 March 2016; pp. 67–74. [CrossRef]

15. Cakmak, M.; Srinivasa, S.S.; Lee, M.K.; Forlizzi, J.; Kiesler, S. Human preferences for robot-human hand-over configurations. In Proceedings of the 2011 IEEE/RSJ International Conference on Intelligent Robots and Systems, San Francisco, CA, USA, 25–30 September 2011; pp. 1986–1993. [CrossRef]
16. Cakmak, M.; Srinivasa, S.S.; Lee, M.K.; Kiesler, S.; Forlizzi, J. Using spatial and temporal contrast for fluent robot-human hand-overs. In Proceedings of the 2011 6th ACM/IEEE International Conference on Human-Robot Interaction (HRI), Lausanne, Switzerland, 6–9 March 2011; pp. 489–496. [CrossRef]
17. Mainprice, J.; Akin Sisbot, E.; Jaillet, L.; Cortés, J.; Alami, R.; Siméon, T. Planning human-aware motions using a sampling-based costmap planner. In Proceedings of the 2011 IEEE International Conference on Robotics and Automation, Shanghai, China, 9–13 May 2011; pp. 5012–5017. [CrossRef]
18. Aleotti, J.; Micelli, V.; Caselli, S. An Affordance Sensitive System for Robot to Human Object Handover. *Int. J. Soc. Robot.* **2014**, *6*, 653–666. [CrossRef]
19. Sisbot, E.A.; Alami, R. A Human-Aware Manipulation Planner. *IEEE Trans. Robot.* **2012**, *28*, 1045–1057. [CrossRef]
20. Nagumo, Y.; Ohya, A. Human following behavior of an autonomous mobile robot using light-emitting device. In Proceedings of the 10th IEEE International Workshop on Robot and Human Interactive Communication, ROMAN 2001 (Cat. No.01TH8591), Paris, France, 18–21 September 2001; pp. 225–230. [CrossRef]
21. Hirai, N.; Mizoguchi, H. Visual tracking of human back and shoulder for person following robot. In Proceedings of the 2003 IEEE/ASME International Conference on Advanced Intelligent Mechatronics (AIM 2003), Kobe, Japan, 20–24 July 2003; Volume 1, pp. 527–532. [CrossRef]
22. Yoshimi, T.; Nishiyama, M.; Sonoura, T.; Nakamoto, H.; Tokura, S.; Sato, H.; Ozaki, F.; Matsuhira, N.; Mizoguchi, H. Development of a Person Following Robot with Vision Based Target Detection. In Proceedings of the 2006 IEEE/RSJ International Conference on Intelligent Robots and Systems, Beijing, China, 9–13 October 2006; pp. 5286–5291. [CrossRef]
23. Morioka, K.; Oinaga, Y.; Nakamura, Y. Control of human-following robot based on cooperative positioning with an intelligent space. *Electron. Commun. Jpn.* **2012**, *95*, 20–30. [CrossRef]
24. Suda, R.; Kosuge, K. Handling of object by mobile robot helper in cooperation with a human using visual information and force information. In Proceedings of the IEEE/RSJ International Conference on Intelligent Robots and Systems, Lausanne, Switzerland, 30 September–4 October 2002; Volume 2, pp. 1102–1107. [CrossRef]
25. Mainprice, J.; Berenson, D. Human-robot collaborative manipulation planning using early prediction of human motion. In Proceedings of the 2013 IEEE/RSJ International Conference on Intelligent Robots and Systems, Tokyo, Japan, 3–7 November 2013; pp. 299–306. [CrossRef]
26. Fridovich-Keil, D.; Bajcsy, A.; Fisac, J.F.; Herbert, S.L.; Wang, S.; Dragan, A.D.; Tomlin, C.J. Confidence-aware motion prediction for real-time collision avoidance. *Int. J. Robot. Res.* **2020**, *39*, 250–265. [CrossRef]
27. Park, J.S.; Park, C.; Manocha, D. I-Planner: Intention-aware motion planning using learning-based human motion prediction. *Int. J. Robot. Res.* **2019**, *38*, 23–39. [CrossRef]
28. Maeda, G.; Ewerton, M.; Neumann, G.; Lioutikov, R.; Peters, J. Phase estimation for fast action recognition and trajectory generation in human-robot collaboration. *Int. J. Robot. Res.* **2017**, *36*, 1579–1594. [CrossRef]
29. Liu, H.; Wang, L. Human motion prediction for human-robot collaboration. *J. Manuf. Syst.* **2017**, *44*, 287–294. [CrossRef]
30. Cheng, Y.; Sun, L.; Liu, C.; Tomizuka, M. Towards Efficient Human-Robot Collaboration With Robust Plan Recognition and Trajectory Prediction. *IEEE Robot. Autom. Lett.* **2020**, *5*, 2602–2609. [CrossRef]
31. Sisbot, E.A.; Marin-Urias, L.F.; Alami, R.; Simeon, T. A Human Aware Mobile Robot Motion Planner. *IEEE Trans. Robot.* **2007**, *23*, 874–883. [CrossRef]
32. Iqbal, K.F.; Kanazawa, A.; Ottaviani, S.R.; Kinugawa, J.; Kosuge, K. A real-time motion planning scheme for collaborative robots using HRI-based cost function. *Int. J. Mechatr. Autom.* **2021**, *8*, 42–52. [CrossRef]
33. Jaillet, L.; Cortes, J.; Simeon, T. Transition-based RRT for path planning in continuous cost spaces. In Proceedings of the 2008 IEEE/RSJ International Conference on Intelligent Robots and Systems, Nice, France, 22–26 September 2008; pp. 2145–2150. [CrossRef]
34. Kinugawa, J.; Kanazawa, A.; Arai, S.; Kosuge, K. Adaptive Task Scheduling for an Assembly Task Coworker Robot Based on Incremental Learning of Human's Motion Patterns. *IEEE Robot. Autom. Lett.* **2017**, *2*, 856–863. [CrossRef]
35. Calinon, S.; Bruno, D.; Caldwell, D.G. A task-parameterized probabilistic model with minimal intervention control. In Proceedings of the 2014 IEEE International Conference on Robotics and Automation (ICRA), Hong Kong, China, 31 May–7 June 2014; pp. 3339–3344. doi:10.1109/ICRA.2014.6907339. [CrossRef]
36. Du Toit, N.E.; Burdick, J.W. Robot Motion Planning in Dynamic, Uncertain Environments. *IEEE Trans. Robot.* **2012**, *28*, 101–115. [CrossRef]
37. Sung, H.G. Gaussian Mixture Regression and Classification. Ph.D. Thesis, Rice University, Houstion, TX, USA, 2004.

Article

Human–Machine Differentiation in Speed and Separation Monitoring for Improved Efficiency in Human–Robot Collaboration

Urban B. Himmelsbach *, Thomas M. Wendt, Nikolai Hangst, Philipp Gawron and Lukas Stiglmeier

Work-Life Robotics Laboratory, Department of Business and Industrial Engineering, Offenburg University of Applied Sciences, 77723 Gengenbach, Germany; thomas.wendt@hs-offenburg.de (T.M.W.); nikolai.hangst@hs-offenburg.de (N.H.); philipp.gawron@hs-offenburg.de (P.G.); lukas.stiglmeier@hs-offenburg.de (L.S.)
* Correspondence: urban.himmelsbach@hs-offenburg.de; Tel.: +49-7803-9698-4488

Abstract: Human–robot collaborative applications have been receiving increasing attention in industrial applications. The efficiency of the applications is often quite low compared to traditional robotic applications without human interaction. Especially for applications that use speed and separation monitoring, there is potential to increase the efficiency with a cost-effective and easy to implement method. In this paper, we proposed to add human–machine differentiation to the speed and separation monitoring in human–robot collaborative applications. The formula for the protective separation distance was extended with a variable for the kind of object that approaches the robot. Different sensors for differentiation of human and non-human objects are presented. Thermal cameras are used to take measurements in a proof of concept. Through differentiation of human and non-human objects, it is possible to decrease the protective separation distance between the robot and the object and therefore increase the overall efficiency of the collaborative application.

Keywords: human–robot collaboration; speed and separation monitoring; human–machine differentiation; thermal cameras; protective separation distance

Citation: Himmelsbach, U.B.; Wendt, T.M.; Hangst, N.; Gawron, P.; Stiglmeier, L. Human–Machine Differentiation in Speed and Separation Monitoring for Improved Efficiency in Human–Robot Collaboration. *Sensors* **2021**, *21*, 7144. https://doi.org/10.3390/s21217144

Academic Editor: Anne Schmitz

Received: 30 September 2021
Accepted: 26 October 2021
Published: 28 October 2021

Publisher's Note: MDPI stays neutral with regard to jurisdictional claims in published maps and institutional affiliations.

1. Introduction

Human–Robot Collaboration (HRC) is seeing an enormous growth in research interest as well as in industry applications. The highest priority in HRC applications is given to the safety of the human within the system. A human within a robotic system is called an operator. Different approaches on how to protect the operator from any harm are subject to research. There has been good progress on how to protect the operator from any harm. The efficiency of the systems suffered from most of these safety improvements. Reduced efficiency leads to a reduced acceptance of HRC. In order to increase the acceptance, it is important to examine how these methods for operator safety can become more efficient.

Operator safety does not necessarily mean preventing the operator only from any physical contact. It can also mean to prevent psychological harm through dangerous and threatening movement of the manipulator. An overview of different methods of safe human–robot interaction can be found in [1]. Lasota et al. divided their work into four major categories of safe HRC: safety through control, through motion planning, through prediction, and through psychological consideration. The category of safety through control is subdivided into pre- and post-collision methods [1].

Speed and separation monitoring (SSM), which is subject of this work, belongs to the subcategory of pre-collision methods. Other methods of this subcategory are quantitative limits and the potential field method [2].

Established methods for HRC have already been integrated into standards like the ISO/TS 15066. The Technical Specification 15066 differentiates between four different modes of collaborative operations [3]:

- safety-rated monitored stop (SRMS),
- hand guiding (HG),
- speed and separation monitoring (SSM), and
- power and force limiting (PFL).

This paper focuses on Speed and Separation Monitoring (SSM). There are different sensor systems that can detect the Separation and Speed between the robot and the operator.

There are already quite a few well working sensor systems for speed and separation monitoring on the market. One can distinguish these systems as external and internal. Internal means that the sensors are part of the robot itself, e.g., mounted somewhere on the manipulator surface. External means that the sensor is placed on the edge of the table that the robot is mounted on, or on ceiling above the robot's workspace. Examples for external sensor systems are laserscanners like [4], camera systems like the SafetyEYE [5] or pressure-sensitive floors [6]. There are only few examples for sensor systems that are mounted on the manipulator itself. A good example is the Bosch APAS system [7]. It consists of a safety skin that measures the separation distances capacitively. The main disadvantage is that it can only detect an obstacle in a distance of two to five centimeters.

All of these systems are more or less great in detecting obstacles within the workspace. It is difficult for them—if not impossible—to classify the obstacles in human and non-human objects. Non-human objects, like an automated guided vehicle (AGV), are therefore treated like an operator and safety measures are applied accordingly when they enter or pass through the workspace and its surroundings. These AGVs have fixed and well known dimensions. They have the ability to be programmed for certain behaviour and usually have a navigation system. Therefore, it should be possible to integrate an AGV with high precision into the robotic system for example in order to deliver and pick up workpieces. If the AGV is part of the entire system, it should not be treated as an operator. Instead, it should be possible to continue the robot's movement with high velocities and consequently increase the overall efficiency of the system.

A good overview on research of concepts and performance of SSM can be found in [8,9]. Lucci et al. proposed in [10] to combine speed and separation monitoring with power and force limiting. This way it is possible to continue the movement of the robot when the operator is very close to the robot. A complete halt of the robot's motion is only necessary when it comes to a contact between the operator and the robot. They showed that with their approach it is possible to increase the overall production efficiency. Kumar et al. researched on how to calculate the amount of sensors needed for a specific area as well as how SSM can be achieved from the surface of the robot [11,12]. Grushko et al. proposed the approach of giving haptic feedback to the operator through vibration on the operator's work gloves [13]. The system monitors the workspace with three RGB-D cameras. The controller calculates if the operator's hand intersects with the planned path of the robot and gives appropriate feedback to the operator. They were able to proof in user studies that the participants could finish their task more efficiently compared to the original baseline. A trajectory planning approach was taken by Palleschi et al. in [14]. Using a visual perception system to gather position data of the operator, they also used an interaction/collision model from Haddadin et al. [15] to permanently check the safety situation according to the ISO/TS 15066 standard. If the safety evaluation showed that the robot needs to slow down, their algorithm searched for an alternative path with lower risk for injury and velocities acceptable according to the safety limits. In an experimental validation the group was able to proof the effectiveness of planning safe trajectories for a task of unwrapping an object. Another approach based on dynamically scaled safety zones was proposed by Scalera et al. in [16]. Bounding Volumes around the robot links and the operator's body and extremities represent the safety zones. These safety zones vary in their size according to the velocities of the robot and operator. Information about the operator's position is gathered through a Microsoft Kinect camera. The paper proofed in a collaborative sorting operation that it was possible to shorten the task completion time by 10%.

As there is no system available that can distinguish between human and non-human objects, we see demand for such a system and the need to research methods on how to integrate the differentiation of human and non-human objects in existing safety methods.

In this paper, we propose a method on how such a differentiation is possible. Different sensor principles are presented that are capable of differentiating between human and non-human objects. Thermal cameras with two different field of views (FoV) are used to make measurements. The results show that it is possible two detect an operator in ranges up to 4 m from the sensor.

We have shown in previous publications that it is possible to do speed and separation monitoring directly from the robot arm. We showed that time-of-flight (ToF) sensors are suitable for this task. The first approaches used a camera mounted on the flange of the robot [17]. Further research investigated the use of single-pixel ToF sensors distributed over the links of the manipulator [18]. In [19], we presented a sensor solution in form of an adapter-plate that is mounted between the flange of the robot and the gripper. The previous work also showed that there is still potential for further growth of HRC applications in industrial settings. It also showed that the efficiency will play a key role in the success of HRC and that there is a need to improve the efficiency of these systems.

This paper proposes to differentiate obstacles in the vicinity of robotic systems into human and non-human objects. With this classification, it is possible to calculate object specific distance and velocity limits for the robotic system. The limits for non-human objects can be lower because there is only a financial risk associated with it instead of possible injuries to the human. As a result, it is possible to increase the efficiency of human–robot collaborative applications. The paper shows that a differentiation in certain applications is possible with thermal cameras that can be attached to manipulator or gripper. There is no need for an additional camera system surrounding the robot's workspace and there is no need for any equipment to be attached to objects that shall be differentiated. The main contributions of this work can be summarized as shown in Table 1.

Table 1. List of main contributions of this paper.

Main Contributions
Introduction of human–machine differentiation into speed and separation monitoring.
Introduction of the extended protective separation distance formula that is extended with a variable for human and non-human objects.
An algorithm for detecting an operator in distances of up to 4 m with thermal cameras directly from the manipulator.
Experimental verification of the algorithm with two thermal cameras with different field of views.

The paper is structured as follows. Section 1 introduces the topic and the state of the art. Section 2 gives an overview of sensor systems that can be used for human–machine differentiation in the context of speed and separation monitoring in human–robot collaboration. In Section 3 the protective separation distance is explained and the new object-specific protective separation distances is proposed. Furthermore, this section shows the potential efficiency improvement that can be established with this method. Section 4 explains what kind of and how the measurements were executed. Section 5 shows and discusses the results of the measurements before Section 6 concludes the paper and gives an outlook on future research.

2. Methods for Human–Machine Differentiation

There are active and passive methods to differentiate between human and non-human objects. Active methods are, for example, when camera systems are used and the AGV is marked with a sticker or QR-Code that identifies the object. Other active methods would be when the AGV sends its coordinates via wireless communication to the robotic system so that the robotic system knows exactly where the AGV is located and can therefore differentiate the AGV from other objects in the surroundings. A list of examples for active and passive methods is given in Table 2. The same is true for humans; they could wear a kind of tracker to monitor their position and send it to the robotic system. Depending on the overall situation on industrial shop floors, there might already exist a navigation system that keeps track of all machines, AGVs, and operators.

Table 2. Overview of active and passive methods for human–machine differentiation.

Active Methods	Passive Methods
Wireless Transmission of position	Temperature
Camera and QR-codes	Heart Beat
Camera and Object Recognition	Breath
Sound localization	Walking Pattern
	Dielectric constant

There are many small- and medium-sized enterprises (SMEs) that are new to automation with collaborative robots. They usually do not have any existing navigation or monitoring systems. Moreover, they require flexible solutions. Passive differentiation methods provide the most flexibility. They do not require any additional installations in the surroundings, on the operator or the AGV. This paper focuses on passive differentiation methods that will be presented in the following subsections.

These passive methods use sensors that measure properties that are characteristic for humans or machines. Here is a list of human specific properties that can be used for differentiation [20]:

- Temperature
- Size
- Weight
- Number of legs
- Heart Beat
- Breath

Depending on the specific property that needs to measured, the sensors can be placed at different locations. Three different locations are proposed that make sense to place the sensors. These are the base of the robot, the robot links itself, and the flange or gripper. For integration of the sensors in the gripper, a very flexible method is to use 3D printed grippers. Using 3D printing technology, it is possible to arrange and layout the sensors as needed. A good overview on this topic can be found in [21]. The following sections describe some possible sensor principles that can be used for passive human–machine differentiation.

2.1. Pressure-Sensitive Floor

If the mass of the automated guided vehicle (AGV) is known, and if this mass is different to the mass of the operators working around the robotic system, then it should be possible to differentiate between human and non-human objects by the difference in their mass. An improved system might be able to detect whether the object has two feet on the ground or if there are four wheels touching the ground. AGVs might have a different and more consistent footprint on the pressure sensitive floor. A human being has a variation in pressure. While walking, the human lifts up one foot and there is only one foot touching the ground with full mass.

The average weight of an adult human being is assumed to be 75 kg. The total weight of clothes, including shoes, is assumed to be an additional 3 kg. The total mass of an operator in an industrial setting is then assumed to be 78 kg. In general, an operator should be able to carry a payload of 20 kg. Considering a minimum weight of 50 kg and a maximum weight of 100 kg per operator we get a range of 50 kg of a light worker without payload and up to 120 kg for a worker with payload. Distributed on two feet, we have a range between 25 kg and 60 kg per foot.

AGVs are available in different sizes and weights. Assuming a standard AGV, we have a total mass between 200 kg up to 1000 kg. Usually the weight is distributed on four wheels. This means a weight per wheel of 50 kg up to 200 kg. As we have an overlapping range of about 50 kg to 60 kg for both, human beings on one foot and AGVs on one wheel, it is not possible to differentiate only by weight. A good overview of the research on pressure sensitive floors can be found in the work of Andries et al. [6]. Other state-of-the-art methods that use pressure-sensitive floors are in [22,23].

2.2. Capacitive Sensors

Another possible way to differentiate between human and non-human objects is to measure the change of capacity when an object is approaching a capacitive sensor. There are already sensor systems available that use a capacitive measurement to detect objects in the surroundings of the robot like [7]. The capacity of an object depends on different properties:

- size,
- material,
- humidity, and
- dielectric constant.

Depending on the kind of non-human objects that are present in the application, it could be possible to differentiate between human and non-human objects. AGVs are commonly built with materials like aluminum or steel and have motors and other metallic equipment. For such objects, a capacitive sensor system should be capable of differentiating between human and non-human objects.

Lumelsky et al. were pioneers on the topic of sensitive skin and its use on robot manipulators [24,25]. Other early work like the one from Karlsson and Järrhed [26] proposed one single huge capacitor with one plate on the floor and the second plate on the ceiling above the robot's workspace. More recent work was done by Lam et al. [27] who managed to integrate the sensors into the housing of the robot manipulator. Thus, reaching a solution where not a single part of the sensor is on the outside of the manipulator that could be damaged.

2.3. Thermal Cameras

Body temperature is a property of a human that is already used in other sensor applications. The human body temperature is usually between 36 °C and 37.8 °C. There is only a small window allowed for variations. From 37.8 °C to 41 °C, the human has a physical condition called fever. Above 41 °C, the fever can be life-threatening. Everything below 36 °C is too cold [28]. Everything above the absolute zero point irradiates infrared light or waves in the infrared spectrum. It can be detected with Bolometers or Thermopiles. In a first measurement, images were taken with a FLIR camera. Note that the human body temperature is only visible on parts of the human that are not covered with clothes or other means of protection like helmets, masks, or safety googles. For the covered parts of the body, the temperature is attenuated, as you can see in Figure 1. Even though the AGV is turned on in the picture, there is no significant heat radiation coming from the AGV next to the human.

2.4. Conclusions

There are different sensors that allow a differentiation between human and non-human objects. A differentiation in an industrial setting depends greatly on the conditions in the hall that the system is used in. The decision for a specific sensor needs to made for each individual case. In our work, we continue to focus on the differentiation with thermal cameras.

(a)

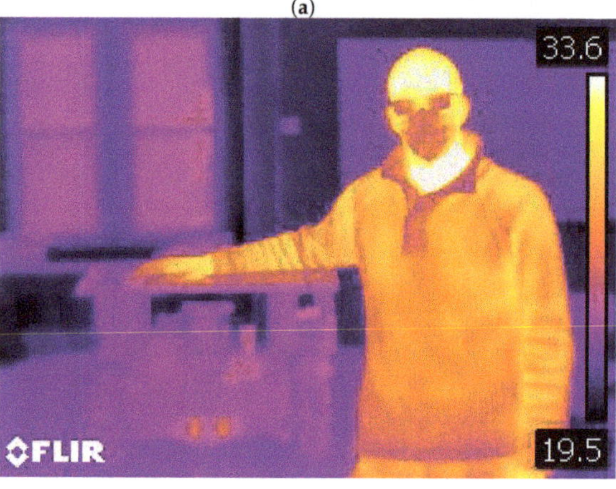

(b)

Figure 1. Comparison of visual and thermal image of a human next to an AGV. (**a**) Visual image; (**b**) Thermal image.

3. Potential Efficiency Improvement

The Cambridge Dictionary defines efficiency as follows: "the good use of time and energy in a way that does not waste any" [29]. For a standard, non-collaborative, robotic application, a common way to measure efficiency is to measure how long the robotic system needs to fulfill a sequence of tasks. With finding ways to shorten the time to fulfill the tasks, one increases the efficiency of the system.

When it comes to human–robot collaborative applications, it gets a bit more complicated. Interactions happen not only with other well-defined objects, but also with a human—and no human is like another. A human in industrial applications is called an operator. This operator might be talking to other operators, might take a break, switch with an other operator, or simply has to clean their nose.

All of these interruptions are not foreseeable for the robotic system and come along with leaving and re-entering the robot's workspace. The more of these occasions happen, the less efficient is the overall robotic system. An efficient speed and separation monitoring system is essential for these occasions and influences the overall system efficiency.

On industrial shop floors, there is usually no hard border for a transition from the walkway or driveway into the operator's or robot's workspace as shown in Figure 2. The monitored space can often reach into the walkway and driveways.

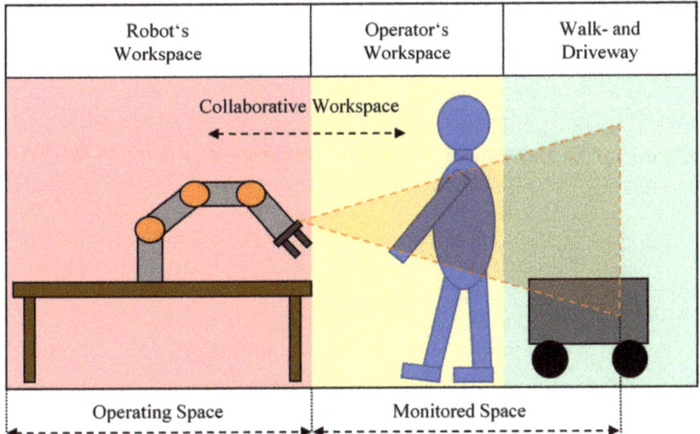

Figure 2. Different workspaces around a robotic application. Note how the monitored space ranges into the walkway and driveway area.

In human–robot collaborative applications, there is a special focus on the operator. The safety of the operator has priority over the speed and movement of the robot. This is why the robot has to slow down or come to a complete stop when an operator enters the monitored space. There are different sensor systems that can measure the operators location and speed. So far, these systems do not differentiate between an operator and another machine like an AGV. The AGV is handled like an operator and the robot has to slow down or stop when it gets closer than the protective separation distance.

This is where our work proposes to differentiate between an operator and other machines. This differentiation shall then be taken into account when calculating the protective separation distance. With smaller protective separation distances for non-human objects, we increase the time that the robot can work with higher velocities and thus increase the overall efficiency of the system.

3.1. Protective Separation Distance

The point in time when the operator enters the workspace can be variable as well as the speed of the operator while entering the workspace. Depending on the tasks, there are different types and amounts of interaction with other objects. Other objects in this context can be other robots, automated guided vehicles, or human beings. These objects can either provide workpieces, tools, or actively support the robot's task.

No matter what kind of interaction happens, the object needs to enter and exit the robot's workspace at a certain point in time. When entering the workspace, the robot has to slow down in order to prevent harm to the object. The moment when to slow down or stop depends on the speed of the robot, the robots reaction and stopping time as well as on the speed of the operator.

This moment is defined as the protective separation distance. The ISO/TS 15066 provides equations to calculate the protective separation distance. This distance depends on a large portion on the operators location and speed. Different values are needed to calculate the protective separation distance. The protective separation distance is calculated as shown in Equation (1) [3]:

$$S_p(t_0) = S_h + S_r + S_s + C + Z_d + Z_r. \tag{1}$$

The different values are defined in the ISO/TS 15066 as follows [3]:

- $S_p(t_0)$ is the protective separation distance at time t_0.
- t_0 is the present or current time.

- S_h is the contribution to the protective separation distance attributable to the operator's change in location.
- S_r is the contribution to the protective separation distance attributable to the robot system's reaction time.
- S_s is the contribution to the protective separation distance due to the robot system's stopping distance.
- C is the intrusion distance, as defined in ISO 13855. This is the distance that a part of the body can intrude into the sensing field before it is detected.
- Z_d is the position uncertainty of the operator in the collaborative workspace, as measured by the presence sensing device resulting from the sensing system measurement tolerance.
- Z_r is the position uncertainty of the robot system, resulting from the accuracy of the robot position measurement system [3].

The protective separation distance can be a fixed number if worst case values are used to calculate it. Especially the contribution by the human operator plays an important role in the equation.

It is allowed by the ISO/TS 15066 that the protective separation distance can be calculated dynamically according to the robot's and operator's speeds [3]. The operators contribution to the overall protective separation distance can be calculated as shown in Equation (2):

$$S_h = \int_{t_0}^{t_0+T_r+T_s} v_h(t)\, dt. \tag{2}$$

A constant value for S_h can be calculated with Equation (3):

$$S_h = 1.6 \cdot T_r + T_s. \tag{3}$$

Equation (4) shows how to calculate the distance that the robot moves during the reaction time of the controller of the robot:

$$S_r = \int_{t_0}^{t_0+T_r} v_r(t)\, dt. \tag{4}$$

A constant value for S_r can be calculated with Equation (5):

$$S_r = v_r(t_0) \cdot T_r. \tag{5}$$

The contribution of the stopping time can be calculated with Equation (6):

$$S_s = \int_{t_0+T_r}^{t_0+T_r+T_s} v_s(t)\, dt. \tag{6}$$

3.2. Object-Specific Protective Separation Distance

Our proposal in this paper is to introduce an additional variable in the formula for the protective separation distance for the object kind. There are two different approaches of how to handle this additional variable.

One is to treat the variable as a binary digit: the value is either 0 or 1. If the object is a human, the contribution of the operator's change in location to the protective separation distance needs to be fully accounted for and the value is set to 1. If the object is a non-human object, the variable is set to 0 and the contribution of the object to the protective separation distance is neglected.

The second approach would be to treat the value as a probability of how likely the object is a human or a non-human object. With 0 being a non-human object and 1 being a human. Equation (7) shows the formula for the extended protective separation distance:

$$S_p(t_0) = (S_h \cdot T) + S_r + S_s + C + Z_d + Z_r. \tag{7}$$

In order to get a rough estimate of values for the protective separation distance, we calculate an example for the protective separation distance. We calculate with $v_r = 2.5\,\text{m/s}$ and an operator velocity of 1.6 m/s.

The specification sheet for the KUKA LBR iiwa 7 R800 specifies a stopping distance of $5.193°$ for a category 0 Stop on axis 1 with a 100% radius and a 100% program override. With a specified radius of 800 mm for the KUKA robot, the distance traveled during stopping would be 72.47 mm according to Equation (8):

$$S_s = 2 \cdot \pi \cdot 800\,\text{mm} \cdot \frac{5.193°}{360°} = 72.47\,\text{mm}. \tag{8}$$

Neglecting the values for position uncertainties of the robot and the operator, and neglecting the intrusion distance, we can plot the protective separation distance for robot speed of 0 to 2.5 m/s with operator speeds of 0.25 m/s which is the maximum allowed speed close to the robot, 1.6 m/s as an average operator speed, and 2.5 m/s as maximum speed. Figure 3 shows the calculated protective separation distances. The protective separation distance is linearly dependent on the robot and the human velocity. If the robot moves with full speed of 2.5 m/s and the operator approaches the system with a speed of 1.6 m/s, the protective separation distance is 2.922 m.

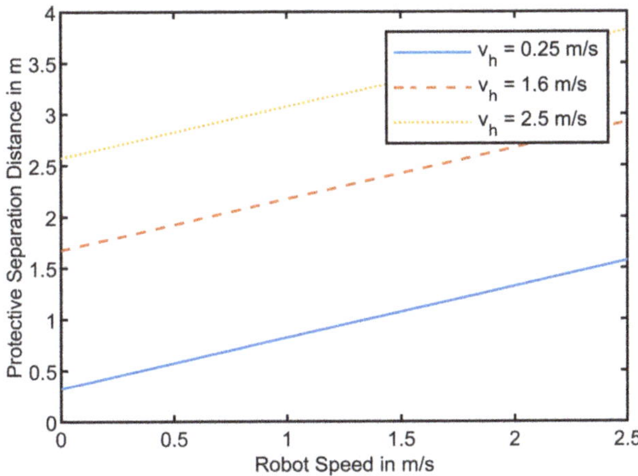

Figure 3. Protective Separation Distances for robot speeds between 0 m/s and 2.5 m/s and operator speeds of 0.25 m/s, 1.6 m/s, and 2.5 m/s.

Figure 3 shows the dependency of the robot's and the operator's speed on the protective separation distance. If it is possible to differentiate between an operator and an AGV, there would be no need for accounting for the approaching distance of the operator and S_h could be neglected. This reduces the protective separation distance for a robot's speed of $v_r = 2.5\,\text{m/s}$ from 3.5 m down to 1.5 m. This opens a range of 2 m where the AGV can drive by the robotic system without interfering with the robot's speed.

4. Measurements

4.1. Monitored Space

A difficult question is always what needs to be monitored by the sensors system. Typically, a robot's workspace is divided into two main sections, as shown in Figure 2, namely, the operating space and the collaborative workspace. The collaborative workspace is the part where the operator can work collaboratively with the robot. The operating space is the part where no human being is allowed and where the robot can work faster than in the collaborative workspace.

Considering a robot that is capable of moving $360°$ around its base, the collaborative workspace can be as small as a few degrees or as big as the full $360°$ around the base. Thus, the size of the collaborative workspace is calculated as follows:

$$Size\ of\ Collaborative\ Workspace = 360° - Size\ of\ operating\ space. \tag{9}$$

The operating space is protected by design against any access of the operator. The collaborative workspace needs to be monitored with a sensor system that is capable of measuring the separation distance to an intruding obstacle like the operator or an AGV.

A sensor for monitoring the collaborative workspace has a defined field of view (FoV). The amount of sensors needed to monitor the entire collaborative workspace is then calculated as shown in Equation (10):

$$Number\ of\ sensors\ needed = \frac{Size\ of\ Collaborative\ Workspace}{FoV}. \tag{10}$$

The collaborative workspace ends with the maximum reach of the manipulator. In order to calculate the protective separation distance we need to be able to detect obstacles before they enter the collaborative workspace.

Therefore, monitoring is necessary for the collaborative workspace and an additional extended monitoring space. This extended monitoring space usually includes walkways for other workers and AGVs. The required size of the extended monitoring space must be at least the maximum possible protective separation distance as calculated in Section 3. The sensor for differentiating between human and non-human objects must have the same range.

As seen in Figure 3, the maximum possible protective separation distance for robot speed of $v_r = 2.5$ m/s and an operator speed of $v_h = 2.5$ m/s, is $S_{p,2.5} = 3.822$ m. We round up and set the maximum separation distance to $S_{p,max} = 4$ m.

The goal is to be able to detect a temperature of a human being in an industrial surrounding in a distance of $S_{p,max} = 4$ m.

As described in Section 2, the human body temperature can usually only be measured somewhere in the head area of the operator due to clothing covering the skin of the rest of the body. Let us assume a head size of an average human being of 20 cm. We want to be able to have a minimum pixel size for measurement of 10 cm in a distance of $S_{p,max} = 4$ m. The pixel size in different distances from a sensor is calculated as shown in Equation (11):

$$x = 2\tan\left(\frac{\alpha}{2}\right)d. \tag{11}$$

With d being the distance from the sensor to the object, α the field of view of the sensor, and x the size of the viewing window in a distance d from the sensor as shown in Figure 4.

The commercially available TeraRanger Evo Thermal 33 and Evo Thermal 90 are used to make measurements. The properties of the sensors are listed in Table 3. The sensor is connected via USB to a laptop running Windows 10. Matlab is used to read the data from the sensor via a serial connection with parameters set to: Baud Rate of 115,200, 8 Data Bits, 1 Stop Bit, Parity None, and no flow control. Matlab was chosen due to its great ability to work with matrices as the data read from the sensor with its resolution of 32×32 pixels is best represented in a 32×32 matrix. Furthermore, Matlab provides a well-established set of functions for postprocessing the data. With the KUKA Sunrise Toolbox it is possible to control the KUKA LBR iiwa 7 R800 robot directly with Matlab via an Ethernet connection [30]. This allows the control of the entire measurement setup with only one laptop running Matlab.

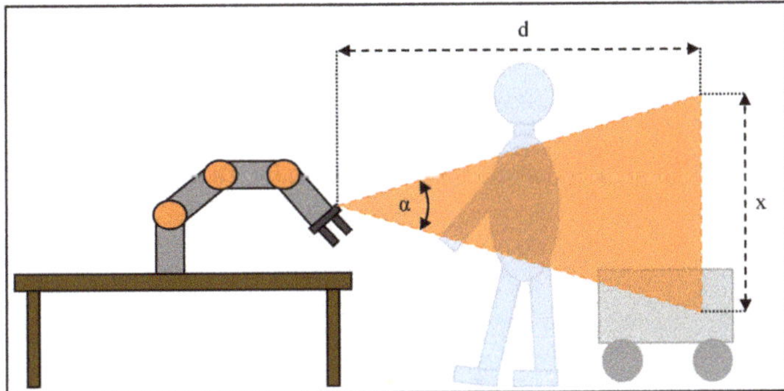

Figure 4. Schematic for the field of view of the sensor attached to the flange of the robot.

Table 3. Teraranger Evo Thermal Specifications [31].

Specification	Evo Thermal 90	Evo Thermal 33
Resolution	32 × 32 pixels	32 × 32 pixels
Field of View	90° × 90°	33° × 33°
Temperature Range	−20 °C to 670 °C	30 °C to 45 °C
Update Rate	7 Hz	7 Hz
Range	up to 5 m	up to 5 m
Size	29 × 29 × 13 mm	29 × 29 × 22 mm
Weight	10 g	12 g

The two sensors from Terabee have a field of view of 90° and 33°. The sensors are shown in Figure 5. The resolution is 32 × 32 pixels. The size of the area measured by the sensor in a distance d is calculated by dividing Equation (11) by 32 pixels as shown in Equation (12):

$$x_{Sensor} = \frac{2\tan\left(\frac{\alpha}{2}\right)d}{32}.$$

(12)

Figure 5. Teraranger Thermal 33 and 90.

The pixel sizes for both sensors for distances from 1 m up to 5 m are shown in Figure 6.

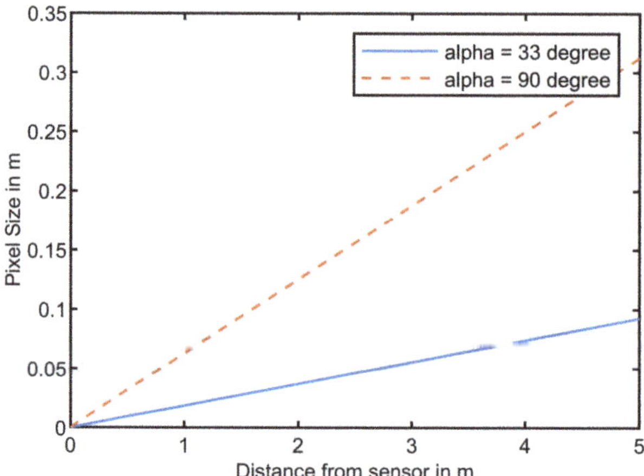

Figure 6. Size of pixel in different distances from the sensor.

The average size of a human head is assumed to be 20 cm. The pixel size of the 33° FoV sensor in a distance of 5 m is ~10 cm according to Equation (11). For the 90° FoV sensor, the pixel size would already be at 30 cm in a distance of 5 m which would not lead to good results. A pixel size of 10 cm for the 90° FoV sensor is reached at a distance of 2 m.

A first measurement was to see if it is possible to measure the human temperature in different distances of 1 m to 4 m in 1 m steps. With Matlab, the average temperature of 10 subsequent measurements was calculated and plotted in an thermal image. The room temperature during the measurement was 22.2 °C and the humidity was at 56%.

In order to find out if it is possible to detect an operator within 4 m from the robot, we make following measurement. The sensor is placed in a height of 120 cm. The sensor is connected via USB to a laptop running Windows 10 and Matlab. Matlab opens a serial connection to the sensor. The Matlab script reads the temperature values from the sensor 100 times. In a first measurement, there is no operator or other human being in the field of view of the sensor. In the next eight measurements, there is an operator with a height of 183 cm in distances of 0.5 m to 4 m in 0.5 m steps. At each distance value, 100 measurements are taken. Matlab then calculates the mean value for each measurement as well as the standard deviation. This measurement will show if it is possible to see the difference between human beings and the surroundings.

4.2. Differentiation Algorithm

In order to save computing time, the first approach is to measure the temperature and compare it to a threshold as shown in the flow chart in Figure 7.

First, the thermal data from the camera are read via a serial connection. Second, the Matlab function *max()* is used to find the maximum measured value. Third, the measured maximum temperature is compared with a threshold. If the maximum measured temperature exceeds the threshold, the variable T is set to 1, meaning that the object is treated like a human. If the measured temperature stays below the threshold, the variable T is set to 0, meaning that there is no human in the field of view of the sensor and that the object must be a machine. Fourth, the extended protective separation distance as introduced in Section 3 is calculated. In the last step, the robot's speed is adjusted according to the calculated extended protective separation distance.

The temperature threshold needs to be set depending on the application. Best results will be achieved in settings where the temperature of the surrounding equipment is significantly lower than the temperature of a human being. With typical room temperatures of less than 23 °C, a threshold for the measurements of 24 °C is chosen.

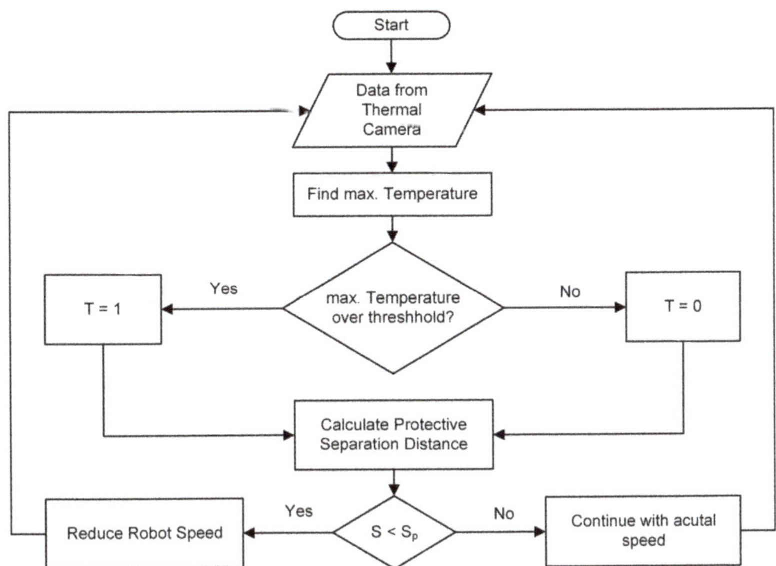

Figure 7. Flow chart of temperature decision for speed adaption.

5. Results and Discussion

Figure 8 shows the eight results for the thermal measurements of both sensors. Figure 8a,c,e,g show the results with the TeraRanger Evo Thermal 33. As seen in Figure 8a, the human temperature is measured quite well with a mean temperature over 10 measurements of 34.92 °C. In Figure 8a, one can also see that the human is wearing glasses. Glasses have poor transmission of long-wave infrared radiation and therefore we see a lower temperature on the glasses. This could be a possible solution for AGVs that show a certain heat radiation from their motors or electronics. Those parts could be covered by a glass or another material that does not transmit heat radiation. In Figure 8c,e,g, you can see that the underarms of the human being were not covered and therefore also were measured with a temperature in the range of 30 °C.

Figure 8b,d,f,h show the four results for the thermal measurement of a human-being in distances of 1 m to 4 m in 1 m steps with the Terabee Evo Thermal 90 sensor. Figure 8b shows that the bigger FoV of 90° allows to measure almost a complete standing operator in a short distance of only 1 m compared to only half the operator in Figure 8a. As calculated in Section 4, you can see that in Figure 8f,h the operator and especially the head are so far away, that one pixels measures more than just the temperature of the head. This leads to a significantly reduced average temperature. That makes it harder to differentiate the operator from its surroundings.

Figure 8 shows two main advantages and drawbacks of the sensors. For the Evo Thermal 33 sensor, the main drawback is the small field of view. Depending on the application, multiple sensors might be needed to cover the entire area that needs to monitored. The advantage is that the measured temperature is close to actual temperature for the entire distance range from 1 m up to 4 m. This is the drawback of the Evo Thermal 90 sensor, that still measures temperatures over 30 °C for distances up to 2 m. However, for distances above 2 m, the single pixels of the sensor cover areas of 12.5 cm by 12.5 cm and more, resulting in lower temperature measurements if a body part only covers a part of the pixel. Depending on the room temperature it gets more and more difficult to detect a human being in distances of more than 2 m for the Evo Thermal 90 sensor. The advantage of the Evo Thermal 90 sensor is the field of view that allows to cover a three times bigger area than the Evo Thermal 33.

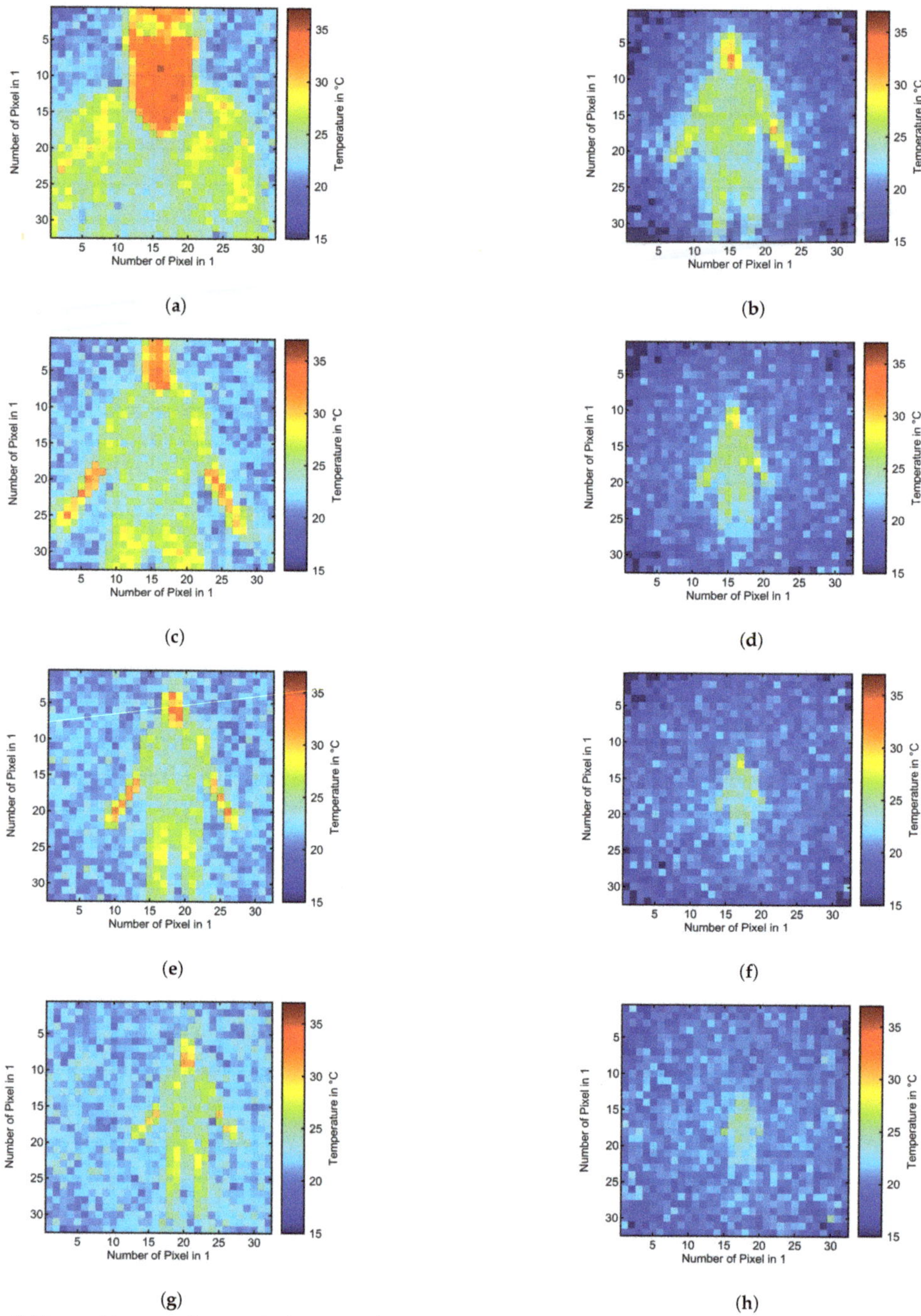

Figure 8. Thermal images of human in distances of 1 m, 2 m, 3 m, and 4 m of the two sensors TeraRanger Evo Thermal 33 and 90. (**a**) Human in 1 m distance of Evo Thermal 33; (**b**) Human in 1 m distance of Evo Thermal 90; (**c**) Human in 2 m distance of Evo Thermal 33; (**d**) Human in 2 m distance of Evo Thermal 90; (**e**) Human in 3 m distance of Evo Thermal 33; (**f**) Human in 3 m distance of Evo Thermal 90; (**g**) Human in 4 m distance of Evo Thermal 33; (**h**) Human in 4 m distance of Evo Thermal 90.

Figure 9 shows the results of the measurement where the highest temperature was measured while an operator was in distances of 0.5 m to 4 m in 0.5 m steps from the sensor. The measurement was executed once with the Evo Thermal 33 and once with the Evo Thermal 90. For each distance of the operator, 100 measurements were taken. The mean value was calculated and plotted in Figure 9 with error bars for the standard deviation. The record for a distance of 0 m represents the measurement without operator in the field of view of the sensors.

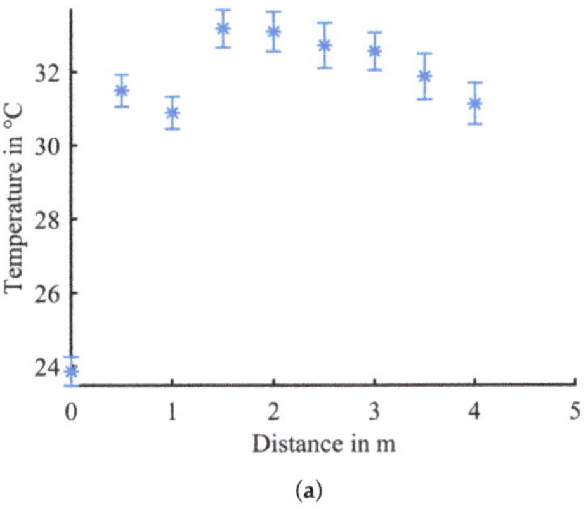

(a)

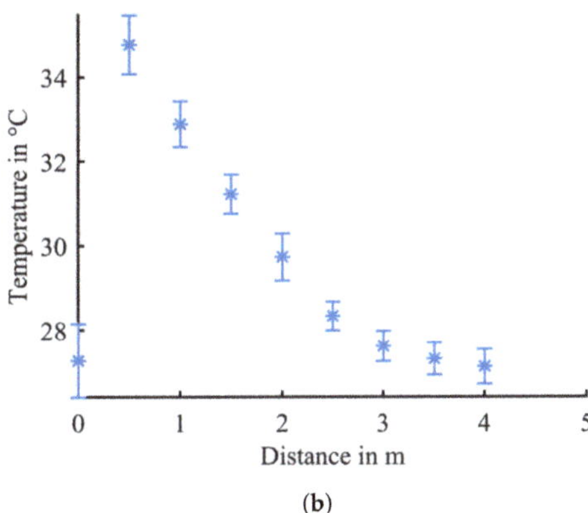

(b)

Figure 9. 100 Measurements with human-being in distances from 0.5 m to 4 m for both sensors, the TeraRanger Evo Thermal 33 and 90. (**a**) Evo Thermal 33: 100 Measurements with human in distances from 0.5 m to 4 m; (**b**) Evo Thermal 90: 100 Measurements with human in distances from 0.5 m to 4 m.

Figure 9a shows that for the Evo Thermal 33 sensor, there is a difference of more than 5 °C between the temperature measurements in all different distances compared to the temperatures measured without an operator present.

The lower mean values for distances 0.5 m and 1 m with the Evo Thermal 33 as shown in Figure 9a can be explained by the narrow field of view of the sensor. Due to the sensor being placed in a height of 120 cm and the FoV being 33 °, the sensor cannot measure the temperatures from the head of a 183 cm operator. Due to the operator wearing long-sleeved shirt, the mean values are a bit lower because the sensor does not see any naked skin that would radiate more heat. Starting at a distance of 1.5 m, the head of the operator with a lot of exposed skin is lying in the field of view of the sensor and therefore detected with a higher mean temperature than the measurements of 0.5 m and 1.0 m.

Figure 9b shows that the measurement for the scene without an operator shows a similar temperature range like the temperatures measured in distances of 2.5 m and more. Therefore, it will not be possible to make a differentiation between human and non-human objects with the Evo Thermal 90 sensor in distances above 2 m. This confirms the result of Figure 8 and is one of the main drawbacks of the Evo Thermal 90 sensor.

Regarding the proposed algorithm, these results show that for normal room temperatures below 24 °C, it is possible to make a differentiation between human beings and other machines like AGVs. One drawback is in case that an AGV exposes a heat source like a motor or an electric device that radiates heat in the same amount like a human being, the AGV could be mistakenly be treated like an operator. This might lead to a reduced efficiency, but it would not be a safety issue for the operator. A possible solution would be to cover the heat source with a material that does not allow transmission of infrared heat. The main advantage of this algorithm is its simple structure and therefore short computing time.

An interesting question arises when looking at the corona pandemic where one main indicator for human health is body temperature. Pictures on TV showed that people had their temperature measured on their forehead, a region that is also part of the measurement in our setup. Considering the entire possible temperature range of a human being between 36 °C and 41 °C, this should not affect the system performance. For setups where the human is the warmest object, it is no problem at all. The threshold will be set depending on the room temperature and the given temperatures of the surroundings. Everything above that temperature will be treated as a human being. It will become more important in setups where the system should be able to differentiate a human from objects that are warmer than the human. If the object's mean temperature is close to 41 °C, then it will be difficult to make a correct differentiation. The differentiation will be easier when the object's temperature is essentially higher than the human's core temperature.

6. Conclusions and Outlook

This paper introduced an object-specific protective separation distance for speed and separation monitoring in human–robot collaborative applications. The use case was that in small- and medium-sized enterprises the shop floor space is limited. The space that needs to be monitored for speed and separation monitoring in HRC applications overlap with the walkways and driveways for other operators and AGVs. AGVs that pass through this monitored space slow down the robotic applications because they are treated like an operator. Differentiating between operators and AGVs allows to adjust the protective separation distance and therefore let the robot move with higher speed.

The main feature that differentiates an operator from an AGV is its temperature. Using a thermal camera, it is possible to differentiate between a human and an AGV in distances of up to four meters depending on the resolution and on the field of view of the sensor. The measurements showed that the smaller FoV sensor has advantage in measuring the temperature of objects in distances of 2 m and more. The 90° FoV sensor had the advantage of being able to measure the entire height of an operator in distances as close as 1 m. A mix of both sensors will be subject for further research. A disadvantage of this method is that if the AGV exposes a heat source like a motor or an other electric device, it can mistakenly be treated as an operator. In these cases, the heat sources on the AGV must be covered.

The paper showed that there is potential of more than 50% to decrease the protective separation distance and therefore increase the efficiency of the overall collaborative robotic system. The object-specific protective separation distance differentiates between human and non-human objects in the vicinity of the robot's workspace through the use of thermal cameras.

The proposed differentiation between human and non-human objects might not only be beneficial for Speed and Separation Monitoring, but also for power- and force-limiting operations. The power- and force-limiting operation is based on maximum values for quasi-static and transient contacts [3]. The values are determined in a risk assessment for the specific application.

Similar to the situation in speed and separation monitoring, there is no need to treat non-human objects like a human object. For hon-human objects, the maximum values for quasi-static and transient contacts can be higher. The amount of how much higher these values can be set depends on the materials that the non-human objects are made of. With a sensor system that can differentiate between human and non-human objects, it is possible to adjust the maximum values for the power and force limiting operation. The robot will be able to move with higher speed when a non-human object is close by and therefor the overall efficiency will be increased. This topic will be subject for further investigation.

Furthermore, research in the future will investigate the different presented sensor systems and how well they are suitable for human–machine differentiation. Fusing the data of different sensors might lead to even better results. A first step will be to combine an infrared ToF sensor with the thermal camera in order to get a single sensor system. Another important task is to look at how the different sensor systems can be compared and how the overall system efficiency can be described to suit a broader spectrum of applications.

Author Contributions: Conceptualization, U.B.H.; Methodology, U.B.H. and T.M.W.; Software, U.B.H.; Validation, N.H., P.G. and L.S.; Formal Analysis, U.B.H.; Investigation, U.B.H., N.H. and L.S.;

Resources, U.B.H. and P.G.; Data Curation, U.B.H. and L.S.; Writing—Original Draft Preparation, U.B.H.; Writing—Review and Editing, N.H. and P.G.; Visualization, U.B.H. and L.S.; Supervision, T.M.W.; Project Administration, T.M.W. and U.B.H.; Funding Acquisition, U.B.H. and T.M.W. All authors have read and agreed to the published version of the manuscript.

Funding: This work was supported by the Federal Ministry for Economic Affairs and Energy in the Central Innovation Program for small and medium-sized enterprises (SMEs). The article processing charge was funded by the Baden-Württemberg Ministry of Science, Research and Culture and the Offenburg University of Applied Sciences in the funding program Open Access Publishing.

Institutional Review Board Statement: Not applicable.

Informed Consent Statement: Not applicable.

Data Availability Statement: Measurement data is available on request from the corresponding author.

Conflicts of Interest: The authors declare no conflict of interest.

References

1. Lasota, P.A.; Fong, T.; Shah, J.A. *A Survey of Methods for Safe Human-Robot Interaction*; Foundations and trends in robotics; Now Publishers, Inc.: Hanover, MA, USA, 2017.
2. Khatib, O. Real-Time Obstacle Avoidance for Manipulators and Mobile Robots. *Int. J. Robot. Res.* **1986**, *5*, 90–98. [CrossRef]
3. International Organization for Standardization. *Technical Specification ISO/TS 15066: Robots and Robotic Devices—Collaborative Robots*, 1st ed.; ISO: Geneva, Switzerland, 2016.
4. SICK AG. Safety Laser Scanners. Available online: https://www.sick.com/de/en/opto-electronic-protective-devices/safety-laser-scanners/c/g187225 (accessed on 30 September 2021).
5. Pilz GmbH & Co. KG. Safety EYE. Available online: https://www.pilz.com/de-DE/eshop/SafetyEYE-Tools/c/00106002257092 (accessed on 30 September 2021).
6. Andries, M.; Simonin, O.; Charpillet, F. Localization of Humans, Objects, and Robots Interacting on Load-Sensing Floors. *IEEE Sens. J.* **2016**, *16*, 1026–1037. [CrossRef]
7. Robert Bosch GmbH. Bosch APAS–Flexible Robots Collaborate in Industry 4.0. Available online: https://www.bosch.com/research/know-how/success-stories/bosch-apas-flexible-robots-collaborate-in-industry-4-0/ (accessed on 30 September 2021).
8. Byner, C.; Matthias, B.; Ding, H. Dynamic speed and separation monitoring for collaborative robot applications–Concepts and performance. *Robot. Comput.-Integr. Manuf.* **2019**, *58*, 239–252. [CrossRef]
9. Xia, F.; Bahreyni, B.; Campi, F. Multi-functional capacitive proximity sensing system for industrial safety applications. In Proceedings of the 2016 IEEE SENSORS, Orlando, FL, USA, 30 October–3 November 2016; pp. 1–3. [CrossRef]
10. Lucci, N.; Lacevic, B.; Zanchettin, A.M.; Rocco, P. Combining Speed and Separation Monitoring With Power and Force Limiting for Safe Collaborative Robotics Applications. *IEEE Robot. Autom. Lett.* **2020**, *5*, 6121–6128. [CrossRef]
11. Kumar, S.; Arora, S.; Sahin, F. Speed and Separation Monitoring using On-Robot Time-of-Flight Laser-ranging Sensor Arrays. In Proceedings of the 2019 IEEE 15th International Conference on Automation Science and Engineering (CASE), Vancouver, BC, Canada, 22–26 August 2019; pp. 1684–1691. [CrossRef]
12. Kumar, S.; Sahin, F. Sensing Volume Coverage of Robot Workspace using On-Robot Time-of-Flight Sensor Arrays for Safe Human Robot Interaction. In Proceedings of the 2019 IEEE International Conference on Systems, Man and Cybernetics (SMC), Bari, Italy, 6–9 October 2019; pp. 378–384. [CrossRef]
13. Grushko, S.; Vysocký, A.; Heczko, D.; Bobovský, Z. Intuitive Spatial Tactile Feedback for Better Awareness about Robot Trajectory during Human-Robot Collaboration. *Sensors* **2021**, *21*, 5748. [CrossRef] [PubMed]
14. Palleschi, A.; Hamad, M.; Abdolshah, S.; Garabini, M.; Haddadin, S.; Pallottino, L. Fast and Safe Trajectory Planning: Solving the Cobot Performance/Safety Trade-Off in Human-Robot Shared Environments. *IEEE Robot. Autom. Lett.* **2021**, *6*, 5445–5452. [CrossRef]
15. Haddadin, S.; Haddadin, S.; Khoury, A.; Rokahr, T.; Parusel, S.; Burgkart, R.; Bicchi, A.; Albu-Schäffer, A. On making robots understand safety: Embedding injury knowledge into control. *Int. J. Robot. Res.* **2012**, *31*, 1578–1602. [CrossRef]
16. Scalera, L.; Giusti, A.; di Cosmo, V.; Riedl, M.; Matt, D.T. Application of Dynamically Scaled Safety Zones Based on the ISO/TS 15066:2016 for Collaborative Robotics. *Int. J. Mech. Control* **2020**, *21*, 41–50.
17. Himmelsbach, U.B.; Wendt, T.M.; Lai, M. Towards Safe Speed and Separation Monitoring in Human-Robot Collaboration with 3D-Time-of-Flight Cameras. In Proceedings of the 2018 Second IEEE International Conference on Robotic Computing (IRC), Laguna Hills, CA, USA, 31 January–2 February 2018; pp. 197–200. [CrossRef]
18. Himmelsbach, U.B.; Wendt, T.M.; Hangst, N.; Gawron, P. Single Pixel Time-of-Flight Sensors for Object Detection and Self-Detection in Three-Sectional Single-Arm Robot Manipulators. In Proceedings of the 2019 Third IEEE International Conference on Robotic Computing (IRC), Naples, Italy, 25–27 February 2019; pp. 250–253. [CrossRef]

19. Himmelsbach, U.B.; Wendt, T.M. Built-In 360 Degree Separation Monitoring for Grippers on Robotic Manipulators in Human-Robot Collaboration. In Proceedings of the 2020 Fourth IEEE International Conference on Robotic Computing (IRC), Taichung, Taiwan, 9–11 November 2020; pp. 156–160. [CrossRef]
20. Fraden, J. *Handbook of Modern Sensors: Physics, Designs, and Applications*, 5th ed.; Springer International Publishing: Cham, Switzerland, 2016.
21. Hangst, N.; Junk, S.; Wendt, T. Design of an Additively Manufactured Customized Gripper System for Human Robot Collaboration. In *Industrializing Additive Manufacturing*; Meboldt, M., Klahn, C., Eds.; Springer International Publishing: Cham, Switzerland, 2021; pp. 415–425. [CrossRef]
22. Srinivasan, P.; Birchfield, D.; Qian, G.; Kidane, A. Design of a Pressure Sensitive Floor for Multimodal Sensing. In Proceedings of the Ninth International Conference on Information Visualisation (IV'05), London, UK, 6–8 July 2005; pp. 41–46. [CrossRef]
23. Sousa, M.; Techmer, A.; Steinhage, A.; Lauterbach, C.; Lukowicz, P. Human tracking and identification using a sensitive floor and wearable accelerometers. In Proceedings of the 2013 IEEE International Conference on Pervasive Computing and Communications (PerCom), San Diego, CA, USA, 18–22 March 2013; pp. 166–171. [CrossRef]
24. Lumelsky, V.J.; Shur, M.S.; Wagner, S. Sensitive skin. *IEEE Sens. J.* **2001**, *1*, 41–51. [CrossRef]
25. Lumelsky, V.J.; Cheung, E. Real-time collision avoidance in teleoperated whole-sensitive robot arm manipulators. *IEEE Trans. Syst. Man Cybern.* **1993**, *23*, 194–203. [CrossRef]
26. Karlsson, N.; Jarrhed, J.O. A capacitive sensor for the detection of humans in a robot cell. In Proceedings of the 1993 IEEE Instrumentation and Measurement Technology Conference, Irvine, CA, USA, 18–20 May 1993; pp. 164–166. [CrossRef]
27. Lam, T.L.; Yip, H.W.; Qian, H.; Xu, Y. Collision avoidance of industrial robot arms using an invisible sensitive skin. In Proceedings of the 2012 IEEE/RSJ International Conference on Intelligent Robots and Systems (IROS 2012), Vilamoura-Algarve, Portugal, 7–12 October 2012; IEEE: Piscataway, NJ, USA, 2012; pp. 4542–4543. [CrossRef]
28. Hasday, J.D.; Singh, I.S. Fever and the heat shock response: Distinct, partially overlapping processes. *Cell Stress Chaperones* **2000**, *5*, 471. [CrossRef]
29. Cambridge University Press. Efficiency. Available online: https://dictionary.cambridge.org/de/worter-buch/englisch/efficiency (accessed on 30 September 2021).
30. Safeea, M.; Neto, P. KUKA Sunrise Toolbox: Interfacing Collaborative Robots with MATLAB. *IEEE Robot. Autom. Mag.* **2018**, *26*, 91–96. [CrossRef]
31. Terabee SAS. TeraRanger Evo Thermal Specifications. Available online: https://www.terabee.com/sensors-modules/thermal-cameras/ (accessed on 30 September 2021).

Article

Uncertainty-Aware Knowledge Distillation for Collision Identification of Collaborative Robots

Wookyong Kwon [1], Yongsik Jin [1] and Sang Jun Lee [2,*]

[1] Daegu-Gyeongbuk Research Center, Electronics and Telecommunications Research Institute (ETRI), Daegu 42994, Korea; wkwon@etri.re.kr (W.K.); yongsik@etri.re.kr (Y.J.)

[2] Division of Electronic Engineering, Jeonbuk National University, 567 Baekje-daero, Deokjin-gu, Jeonju 54896, Korea

* Correspondence: sj.lee@jbnu.ac.kr; Tel.: +82-63-270-2463

Abstract: Human-robot interaction has received a lot of attention as collaborative robots became widely utilized in many industrial fields. Among techniques for human-robot interaction, collision identification is an indispensable element in collaborative robots to prevent fatal accidents. This paper proposes a deep learning method for identifying external collisions in 6-DoF articulated robots. The proposed method expands the idea of CollisionNet, which was previously proposed for collision detection, to identify the locations of external forces. The key contribution of this paper is uncertainty-aware knowledge distillation for improving the accuracy of a deep neural network. Sample-level uncertainties are estimated from a teacher network, and larger penalties are imposed for uncertain samples during the training of a student network. Experiments demonstrate that the proposed method is effective for improving the performance of collision identification.

Keywords: collision identification; collaborative robot; deep learning; uncertainty estimation; knowledge distillation

Citation: Kwon, W.; Jin, Y.; Lee, S.J. Uncertainty-Aware Knowledge Distillation for Collision Identification of Collaborative Robots. *Sensors* **2021**, *21*, 6674. https://doi.org/10.3390/s21196674

Academic Editor: Anne Schmitz

Received: 22 August 2021
Accepted: 5 October 2021
Published: 8 October 2021

Publisher's Note: MDPI stays neutral with regard to jurisdictional claims in published maps and institutional affiliations.

1. Introduction

With the increasing demands of collaborative tasks between humans and robots, the research on human–robot interaction has received great attention from researchers and engineers in the field of robotics [1]. Robots that can collaborate with humans are called collaborative robots (cobots), and cobots differ from conventional industrial robots in that they do not require a fence to prevent access. Previously, the application of robots is limited to performing simple and repetitive tasks in well-structured and standardized environments such as factories and warehouses. However, the development of sensing and control technologies has significantly expanded the area of application of cobots [2], and they are beginning to be applied to several tasks around us. More specifically, their applications have been diversified from traditional automated manufacturing and logistics industries to more general tasks such as medical [3], service [4,5], food and beverage industries [6], and these tasks require more elaborate sensing and complicated control techniques. Furthermore, with the development of intelligent algorithms including intention estimation [7] and gesture recognition [8], cobots can be utilized in wider application areas.

In general, robots have advantages over humans in repetitive tasks, and humans are better at making comprehensive decisions and judgments. Therefore, human–robot collaboration possibly increases the efficiency of intelligent systems through complementary synergies. As the scope of robotics applications gradually expands through collaborative work, interaction with humans or unstructured environments has become an important technical issue, which requires the implementation of advanced perception and control algorithms. Especially, collision detection and identification techniques are indispensable elements to improve the safety and reliability of collaborative robots [9,10].

To perform cooperative tasks with the aid of human–robot interactions, several studies have been carried out to detect and identify robot collisions for the safety of workers [11]. Previous work can be categorized into two approaches: the first category is the study on the control of collaborative robots by predicting possible collisions and the other is the study of responses after impacts. While collision avoidance is more advantageous in terms of safety [12], this approach inevitably requires additional camera sensors for action recognition of coworkers or 3D reconstruction of surrounding environments [13]. Furthermore, it is difficult to completely avoid abrupt and unpredictable collisions. Therefore, techniques for collision identification are essential to improve the safety and reliability of collaborative robots.

Collision detection algorithms investigate external forces [14] or currents [15] to determine whether a true collision has occurred on an articulated robot. A key element in the procedure of collision detection is the estimatation of external torques. A major approach to estimating external torques is utilizing torque sensor signals to compute internal joint torques based on the physical dynamics of robots, and several other methods to construct momentum observers to estimate external torques without the use of torque sensors. The method that does not use torque sensors is called sensorless external force estimation, and an elaborate modeling of the observer and filter is essential for the precise estimation of external forces [16–19]. External forces are further processed by a thresholding method [20] or classification algorithm [21], to determine whether a collision has occurred. Recently, deep-learning-based methods have outperformed traditional model-based methods in detecting collisions [22]. Beyond collision detection, the identification of collision locations is beneficial for the construction of more reliable collaborate robots, by making them react appropriately in collision situations.

To ensure the proper responses of collaborative robots in cases of collisions, it is necessary to identify collision locations. The collision identification technique can be defined as a multiclass classification of time series sensor data according to collision locations. In early studies, collision identification was mainly based on the elaborate modeling of filters [23] and observers [24], and a frequency domain analysis was conducted to improve the accuracy of collision identification [25]. To address the classification problem, machine learning techniques, which were employed to analyze time series data, have also been applied to collision identification [26]. Recently, support vector machines [27] and probabilistic methods [28] were applied to improve the reliability of collision identification systems. In [29], the collision identification performance was improved by utilizing additional, sensors such as inertial measurement units, and analyzing their vibration features.

In this paper, we propose a method that can identify collisions on articulated robots by utilizing deep neural networks for joint sensor signals. Collision identification refers to a technique that not only detects the occurrence of a collision, but also determines its location. Recently, a collision detection method was proposed by Heo et al. [22]; we extend this existing method for collision identification and improve the robustness of the deep neural network. To improve the performance of the collision identification system, we construct a deeper network, which is called a teacher network, to distill its probabilistic knowledge to a student network. In the process of distilling knowledge, we employ the uncertainties of the teacher network to focus on learning difficult examples, mostly collision samples. This paper is organized as follows. Section 2 presents related work, Section 3 explains collision modeling and data collection, and Section 4 presents the proposed method. Section 5 and Section 6 presents the experimental results and conclusion, respectively.

2. Related Work

2.1. Deep Learning Methods for Collision Identification of Collaborative Robots

Collision detection is a key technology to ensure the safety and reliability of collaborative robots. Although most previous methods were based on the mathematical modeling of robots [30–32], recently, deep learning methods have shown promising results for this goal. Min et al. [33] estimated vibration features based on the physical modeling of robots and utilized neural networks for collision identification. Xu et al. [34] combined neural networks and nonlinear disturbance observer for collision detection. Park et al. [35] combined a convolutional neural network and support vector machine to detect collisions, and Heo et al. [22] employed causal convolutions, which were previously utilized for auto-regressive models in WaveNet [36] to detect collisions based on joint sensor signals including torque, position, and velocity. Maceira et al. [37] employed recurrent neural networks to infer the intentions of external forces in collaborative tasks, and Czubenko et al. [38] proposed an MC-LSTM, which combines convolutions and recurrent layers for collision detection. Mohammadi et al. [13] utilized external vision sensors to further recognize human actions and collisions.

2.2. Knowledge Distillation

Knowledge distillation was proposed by Hinton et al. [39] to train a student network with the aid of a deeper network, which is called a teacher network. Probabilistic responses of the teacher network are beneficial to improve the accuracy of the student network because the probabilities of false categories were also utilized during knowledge distillation. Although most early methods directly distill the logits of a teacher network, Park et al. [40] utilized the logits' relations, and Meng et al. [41] proposed a conditional teacher–student learning framework. Furthermore, knowledge from intermediate feature maps was distilled for network minimization [42] and performance improvement [43,44]. Knowledge distillation has been employed in various applications such as object detection [45], semantic segmentation [46], domain adaptation [47], and defense for adversarial examples [48]. Recently, the teacher–student learning framework has been applied with other advanced learning methodologies such as adversarial learning [49] and semi-supervised learning [50].

2.3. Uncertainty Estimation

Uncertainty plays an important role in interpreting the reliability of machine learning models and their predictions. Probabilistic approaches and Bayesian methods have been regarded as useful mathematical tools to quantify predictive uncertainties [51]. Recently, Gal and Ghahramani proposed Monte Carlo dropout (MC-dropout) [52], which can be interpreted as an approximate Bayesian inference of deep Gaussian processes, by utilizing dropout [53] at test time. Lakshminarayanan et al. [54] proposed deep ensembles for the better quantification of uncertainties, and Amersfoort et al. [55] proposed deterministic uncertainty quantification, which is based on a single model to address the problem of computational cost of MC-dropout and deep ensembles. Uncertainties have been utilized to quantify network confidences [56], selecting out-of-distribution samples [57], and improving the performance of deep neural networks [58,59], in various application areas such as medical image analysis [60] and autonomous driving [61].

3. Collision Modeling and Data Collection

3.1. Mathematical Modelling of Collisions

This section explains the mathematical modeling of dynamic equations for 6 Degrees of Freedom (DoF) articulated robots. In order to operate a robot through a desired trajectory and move it to a target position, precise control torque is required for each joint motor, and the control torque can be represented as the following dynamic equation:

$$\tau = M(q)\ddot{q} + C(q,\dot{q})\dot{q} + g(q), \tag{1}$$

where $\tau \in \mathbb{R}^n$ is the control torque, $M(q) \in \mathbb{R}^{n \times n}$ is the inertia matrix of the articulated robot, $C(q,\dot{q}) \in \mathbb{R}^{n \times n}$ is the matrix of Coriolis and Centrifugal torques, $g(q) \in \mathbb{R}^n$ is the vector of gravitational torques, and q, $\dot{q}$, $\ddot{q}$ are the angular position, velocity, and acceleration of each joint, respectively. The dynamic equation can be obtained through the Newton–Euler method or the Euler–Lagrange equation using the mechanical and physical information of the robot. Since the dynamic equation of the robot is given as (1), in the absence of external force, external torques can be computed by subtracting the control torques from measured torques.

When a joint torque sensor is installed onto each joint, the torque generated on each joint due to external force is given as follows:

$$\tau_{ext} = \tau_s - \tau, \tag{2}$$

where τ_{ext} is the external torques generated onto each joint due to a collision, and τ_s is torque values measured by joint torque sensors. The external torque can be precisely estimated under an accurate estimation of robot dynamics and physical parameters of the articulated robot such as the mass and center of mass of each link.

In robots that are not equipped with a joint torque sensor, sensorless methods are utilized to estimate external torques. Sensorless methods are basically based on the current signal of each joint motor, and an additional state variable $p = M(q)\dot{q}$ is defined to reformulate the dynamic equation as follows:

$$\dot{p} = C(q,\dot{q})^{\top}\dot{q} - g(q) - f(q,\dot{q}) + \tau_m, \tag{3}$$

where $f(q,\dot{q})$ is the friction matrix, and τ_m is the motor torque. In the case of the sensorless method, it is necessary to obtain the transmitted torque from the motor to the link to estimate the collision torque. Therefore, the friction must additionally be considered in the existing robot dynamics equation. A main issue in sensorless external torque estimation is the elaborate design of observer and filter under the dynamic Equation (3), and the effect of disturbance can be reduced using momentum state variables. Due to the effect of noise and nonlinear frictional force, sensorless methods are generally less precise in the estimation of external torques compared to methods that utilize joint torque sensors. Through the methods mentioned above, it is possible to obtain the torques generated in each joint due to the collision of the robot. Then, the collision identification algorithm can determine collision locations from joint torques obtained through sensor or sensorless methods.

3.2. Data Collection and Labeling

Figure 1a presents the 6-DoF articulated robot to collect sensor data, which include the information of joint torque, current, angular position, and angular velocity. The Denavit–Hartenberg parameters of the articulated robot are presented in [62]. From the 6-DoF articulated robot, joint sensor signals were obtained with the sampling rate of 1 kHz, and a data sample collected at time t can be expressed as

$$\mathbf{x}_t = [\boldsymbol{\tau}_t^{\top}, \mathbf{i}_t^{\top}, \boldsymbol{\theta}_t^{\top}, \mathbf{w}_t^{\top}]^{\top} \in \mathbb{R}^{24}, \tag{4}$$

where $\boldsymbol{\tau}_t, \mathbf{i}_t, \boldsymbol{\theta}_t, \mathbf{w}_t$ are six-dimensional vectors corresponding to torque, current, angular position, and angular velocity, respectively; the i-th components of these vectors indicate the sensor signals obtained at the i-th joint. Figure 1b shows the definition of collision categories according to collision locations. Collisions were generated at six locations, and in the case of no collision, which refers to the normal state, a label of 0 was assigned. In the case of a collision, a categorical label corresponding to the location was assigned to generate ground truth data.

Joint sensor data were collected, along with collision time and category, by applying intentional collisions at different locations. The collision time and category were converted into ground truth data which have an identical length to the corresponding sensor signals, as shown in Figure 2. For a collision occurrence, the corresponding category was assigned to 0.2 s of data samples from the collision time; each collision is represented as 200 collision samples in the ground truth data. We collected joint sensor signals for 5586 intentional collisions along with their ground truth data; the number of collisions, which were applied to different locations, is equal. This dataset was divided into training, validation, and test sets with the ratio of 70%, 10%, and 20%, as presented in Table 1.

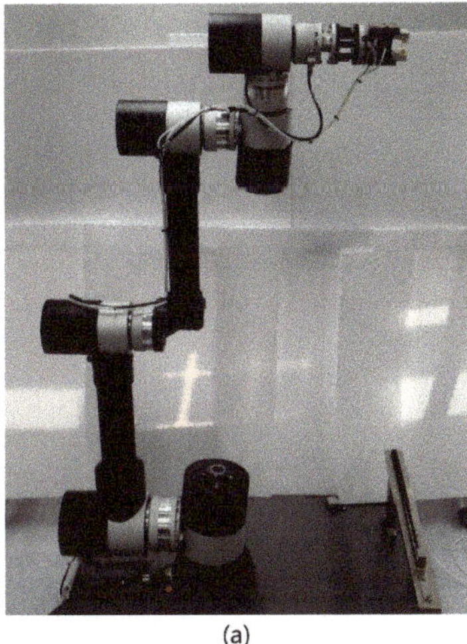
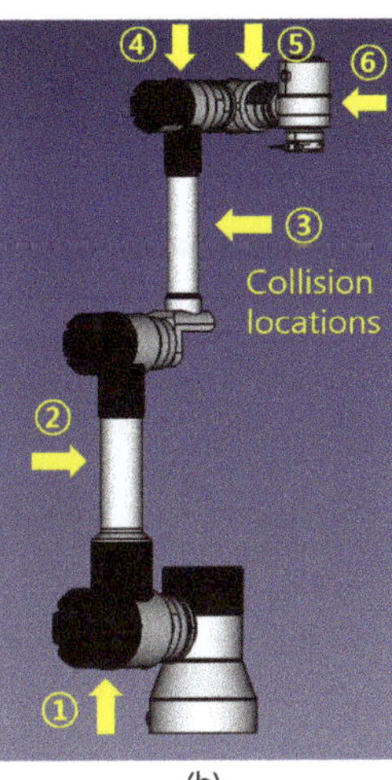

(a)   (b)

Figure 1. The definition of labels. (**a**) presents 6-DoF articulated robot, and (**b**) presents the definition of categories; yellow arrows in (**b**) indicate categorical labels according to collision locations.

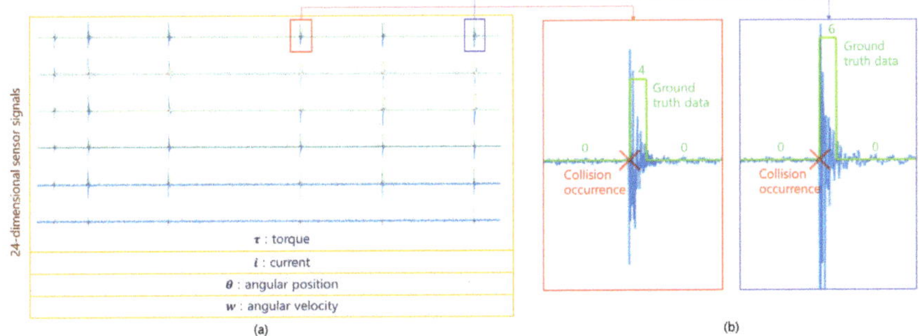

Figure 2. Examples of sensor signals and ground truth data. (**a**) shows a part of the acquired sensor signals, and (**b**) presents examples of generated ground truth data around collision occurrences. Green lines with numbers in (**b**) indicate labeled categories in the ground truth data.

Table 1. The number of collisions and data samples. *Total* indicates the number of data samples, which were collected with a sampling rate of 1 kHz, and *Collision* indicates the number of collision samples.

	Training Set		Validation Set		Test Set	
Collisions	3906		558		1122	
	Total	*Collision*	*Total*	*Collision*	*Total*	*Collision*
Samples	19,563,048	781,200	2,778,777	111,600	5,798,685	224,400

4. Proposed Method

This section presents the proposed method for the collision identification of articulated robots. Firstly, two neural network architectures are presented; one of them is a student network and the other architecture is a teacher network for knowledge distillation. The second part explains the proposed knowledge distillation method, which considers the predictive uncertainties of the teacher network. Lastly, a post-processing is utilized to improve the robustness of the proposed algorithm by reducing noise in network predictions.

4.1. Network Architectures

This paper employs the network architecture presented by Heo et al. [22] as a base network model. Heo et al. [22] proposed a deep neural network, called CollisionNet, to detect collisions in articulated robots. Its architecture is composed of causal convolutions to reduce detection delay and dilated convolutions to achieve large receptive fields. We modeled the base network by modifying the last fully connected layer in CollisionNet to conduct multiclass classification and identify collision locations. The base network is composed of seven convolution layers and three fully connected layers, and its details are identical to CollisionNet except the last layer; convolution filters with the size of 3 are utilized for both regular and dilated convolutions, the depth of the intermediate features is increased from 128 to 512, and the dilation ratio is increased by a factor of two. The architecture of the base network is identically utilized as a student network in the process of knowledge distillation.

Figure 3 shows the architecture of the teacher network. To construct the teacher network, three regular convolutions in the base network are replaced into convolution blocks. A convolution block contains four convolution layers with a skip connection, and therefore, the number of parametric layers in the teacher network increases to 19. The number of channels in the second and third convolution layers in a convolution block are identical to the number of output channels of the corresponding regular convolution layers. The number of trainable parameters in the teacher network is 6.63 million; therefore, it has more capacity to learn the training data compared to the base network, which has 2.79 million parameters. Dropout layers with a dropout ratio of 0.5 are added to the fully connected layers in the teacher network, and Monte Carlo samples from the teacher network are acquired by applying dropout at the test time.

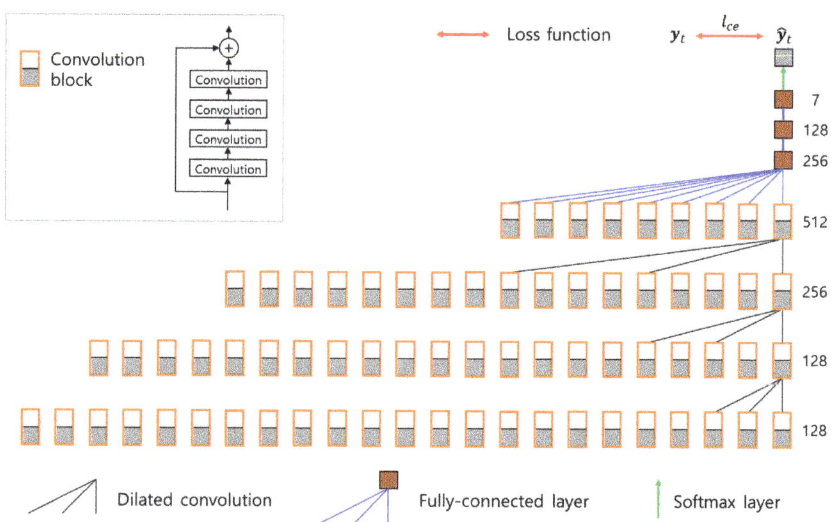

Figure 3. The architecture of the teacher network.

4.2. Uncertainty-Aware Knowledge Distillation

The teacher network is trained with the cross-entropy loss between the softmax prediction $\hat{y}_T$ and its one hot encoded label **y**. The i-th component of $\hat{y}_T$ indicates the predicted probability that the input sample belongs to the i-th category. In our case, seven categories exist, which contain the normal state and six possible collision locations. The loss function for the training of the teacher network is defined as

$$l_{ce}(\mathbf{y}, \hat{\mathbf{y}}_T) = -\sum_i y_i \log(\hat{y}_{T,i}), \tag{5}$$

where y_i and $\hat{y}_{T,i}$ are the i-th components of $\mathbf{y}$ and $\hat{\mathbf{y}}_T$, respectively.

After training the teacher network, K logits, $\hat{\mathbf{z}}_T^1, \cdots, \hat{\mathbf{z}}_T^K$ are obtained from an input sample by utilizing MC-dropout [52]. These logits are computed by randomly ignoring 50% of neurons in the fully connected layers in the teacher network. Based on the K logits of the teacher network, the i-th component of the uncertainty vector is computed by

$$u_i = \frac{1}{K} \sum_k (\hat{z}_{T,i}^k - \bar{z}_{T,i})^2, \tag{6}$$

where $\bar{z}_{T,i}$ is the i-th component of the averaged logit $\bar{\mathbf{z}}_T$, which is computed by

$$\bar{\mathbf{z}}_T = \frac{1}{K} \sum_k \hat{\mathbf{z}}_T^k. \tag{7}$$

The uncertainty u_i is the variance of logits; therefore, the value of the uncertainty increases as distances between the logits increase.

The total loss $\mathcal{L}$ for the training of the student network is composed of two loss functions, as follows:

$$\mathcal{L} = l_{ce}(\mathbf{y}, \hat{\mathbf{y}}_S) + l_{kd}(\bar{\mathbf{z}}_T, \hat{\mathbf{z}}_S, \mathbf{u}), \tag{8}$$

where $l_{ce}(\mathbf{y}, \hat{\mathbf{y}}_S)$ is the cross-entropy loss between the softmax prediction of the student network and its corresponding label, $\mathbf{u}$ is the uncertainty vector whose i-th component is u_i, and $l_{kd}(\bar{\mathbf{z}}_T, \hat{\mathbf{z}}_S, \mathbf{u})$ is the uncertainty-aware knowledge distillation loss. The knowledge distillation loss os obtained by computing uncertainty-weighted Kullback–Leibler divergence (KL divergence) between $\sigma(\hat{\mathbf{z}}_S, T)$ and $\sigma(\bar{\mathbf{z}}_T, T)$, as follows:

$$l_{kd}(\bar{\mathbf{z}}_T, \hat{\mathbf{z}}_S, \mathbf{u}) = -\sum_i u_i \sigma(\bar{\mathbf{z}}_T, T)_i \{\log(\sigma(\hat{\mathbf{z}}_S, T)_i) - \log(\sigma(\bar{\mathbf{z}}_T, T)_i)\}, \tag{9}$$

where $\sigma(\mathbf{z}, T)$ is the softmax function with the temperature T, and $\sigma(\mathbf{z}, T)_i$ is the i-th component of $\sigma(\mathbf{z}, T)$. In (9), $\sigma(\mathbf{z}, T)_i$ can be computed as

$$\sigma(\mathbf{z}, T)_i = \frac{\exp(z_i/T)}{\sum_j \exp(z_j/T)}. \tag{10}$$

The overall procedure for the training of the student network is presented in Figure 4.

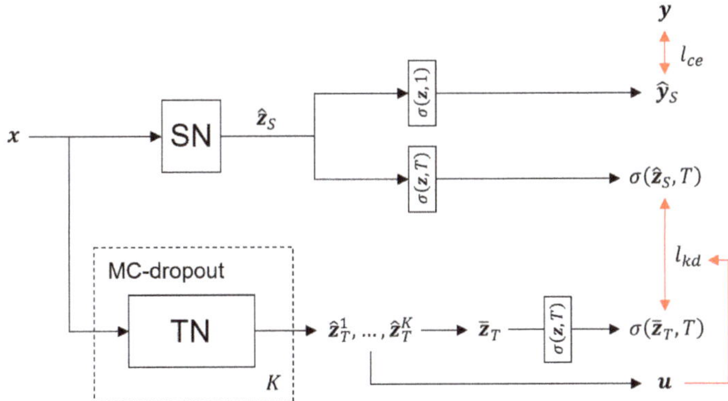

Figure 4. The procedure of uncertainty-aware knowledge distillation for the training of the student network; SN and TN indicate the student and teacher networks, respectively, and $\sigma(\mathbf{z}, T)$ is the softmax function with the temperature T.

4.3. Post-Processing

The post-processing to reduce errors in network predictions is inspired by a connected component analysis in image-processing techniques. In the labeled data, a collision is represented by connected samples, with a non-zero number corresponding to its location. However, a few predictions may differ from their adjacent predictions, because a neural network independently infers predictions for different data samples. Based on the collision properties in the labeled data, incorrect predictions are reduced by the post-processing presented in Figure 5.

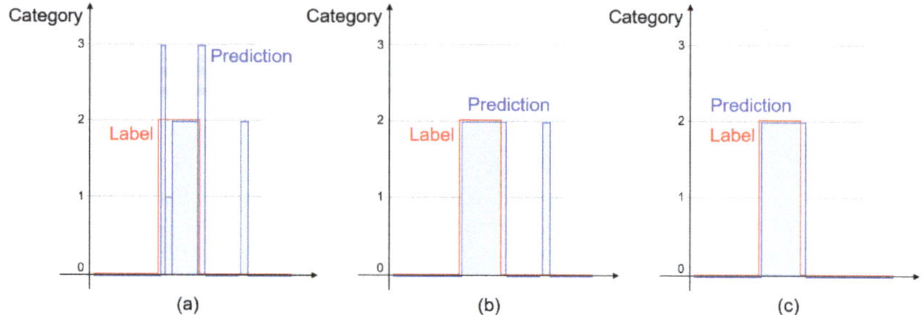

Figure 5. The procedure for the post-processing. (**a**) presents the predictions from the student network, and (**b**) presents the result of grouping non-zero connected samples and assigning an identical category of the maximum frequent. (**c**) presents the result of a thresholding method.

The post-processing is composed of two steps; in Figure 5, (a) shows predictions from the student network, and (b) and (c) present the results after the first and second post-processing steps, respectively. In the first step, non-zero connected samples are grouped, and the number of samples for each category are counted. Predictions in a group are replaced into the category which corresponds to the maximum frequency, as presented in Figure 5b. In the second step, if the number of non-zero connected samples is less than a threshold value, then these samples are regarded as the normal state. The threshold value of 10 samples is utilized in experiments, and Figure 6 presents examples of the results of the post-processing.

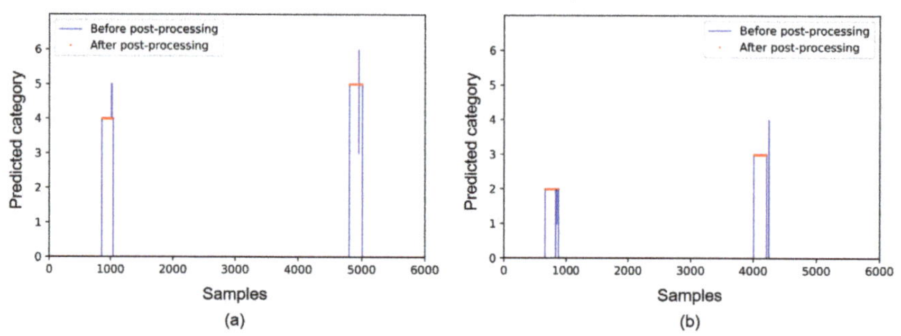

Figure 6. Examples of predictions before and after the post-processing. (**a**) presents predictions for the collision categories of 4 and 5, and (**b**) presents predictions for the collision categories of 2 and 3.

5. Experiments

5.1. Experimental Environment and Evaluation Measures

The proposed algorithm is developed within a hardware environment including Intel core i7-10700 CPU, 32GB DDR4 RAM, and RTX 3080 GPU. In experiments, Python and Pytorch are mainly utilized to implement the proposed algorithm and to conduct an ablation study. To demonstrate the proposed method, the dataset is gathered from a collaborative robot, which consists of six rotating joints. The cobot weighs 47 kg, has a maximum payload of 10 kg, and reaches up to 1300 mm. The actuator consists of motors manufactured from Parker, motor drivers from Welcon, and embedded joint torque sensors in each joint. The hardware of the cobot contains a custom embedded controller, based on real-time linux kernel, and it communicates with drivers through EtherCAT with a cycle time of 1 ms.

To demonstrate the effectiveness of the proposed method, we evaluate the algorithm in three ways: (1) sample-level accuracy, (2) collision-level accuracy, and (3) time delay. In the process of collision identification, deep neural networks perform sample-level multiclass classification, which classifies each sample, composed of a 24-dimensional sequence of sensor data, into the normal state or one of six collision locations. To evaluate the sample-level accuracy of deep neural networks, we measure *Recall*, *Precision*, and *F1-score* for each sample, which are defined as follows:

$$Recall = TP/(TP + FN),$$
$$Precision = TP/(TP + FP),$$
$$F1\text{-}score = 2 \times \frac{precision \times recall}{precision + recall}, \tag{11}$$

where TP, FP, FN are the numbers of true positives, false positives, and false negatives, respectively. True positive is a correctly identified collision sample, false positive is an incorrect prediction, which is classified into a collision, and false negative is an incorrect prediction which is classified into the normal state.

Collision-level accuracy is another important measure for evaluating a collision identification system. Because collaborative robots respond to each collision, reducing the number of false positive collisions is an important issue. *Recall*, *Precision*, and *F1-score* are computed as (11) with different definitions of TP, FP, and FN to measure the collision-level accuracy. A group of connected samples that are classified into a collision is regarded as a true positive if the intersection over union (IoU) between the connected predictions and its corresponding true collision samples is greater than 0.5. A group of predictions that are classified into a false category of collisions is regarded as a false positive, and a false negative is a missed collision. Figure 7 shows several cases of TP, FP, and FN for measuring the collision-level accuracy.

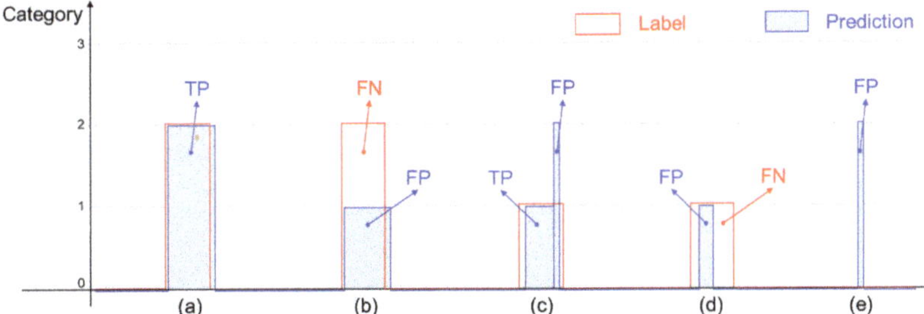

Figure 7. Examples of true-positive, false-positive, and false-negative collisions for computing collision-level accuracies. (**a**) presents a TP collision, (**b**,**d**) present FP and FN cases, (**c**) presents TP and FP cases, and (**e**) presents a FP collision.

Finally, the time delay is measured to evaluate the processing time of the collision identification system. For the safe and reliable collaborations of human and robots, the processing time is required to be reduced as possible. The total processing time is composed of the inference time of a neural network, detection delay for collisions, and post-processing time. Based on these three types of evaluation measure, the effectiveness of the proposed method is demonstrated in experiments.

5.2. Training of Neural Networks

To train the neural networks, Adam optimizer [63] is utilized with a learning rate of 10^{-4}. The learning rate is decreased to 10^{-5} after training 200 epochs. Figure 8 presents f1-scores for the training and validation datasets during the training of 500 epochs. As shown in Figure 8, after training a sufficientl number of epochs, the validation accuracy was not further decreased. Therefore, in the following experiments, the accuracies of deep neural networks are evaluated for the test set after training 300 epochs.

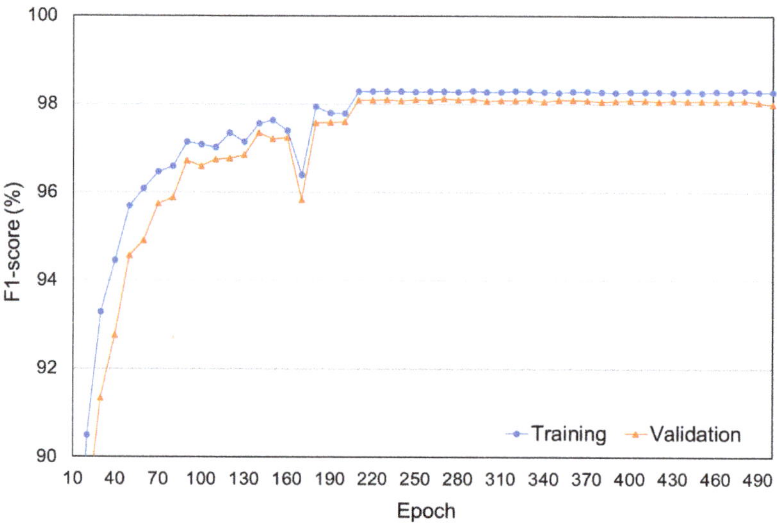

Figure 8. F1-scores for the training and validation datasets.

To train the student network, the temperature of the softmax function is set to 5 during the process of knowledge distillation. The temperature value has to be greater than 1 to soften probabilistic predictions of neural network, and temperature values between 2 and 5 are usually used for knowledge distillation in the previous literature [39]. In our experiments, modifications to the temperature value glead to insignificant changes in the experimental results. In Figure 9, (a) shows the first dimension of 24-dimensional sensor data, which corresponds to the torque signal at the first joint, and (b) presents uncertainties measured by MC-dropout with the value of $K = 4$. As shown in Figure 9, the uncertainties of collision samples are high compared to normal state samples. By weighting the uncertainties on the KL-divergence between probabilistic predictions of the student and teacher network, the student network is able to focus on learning difficult data samples.

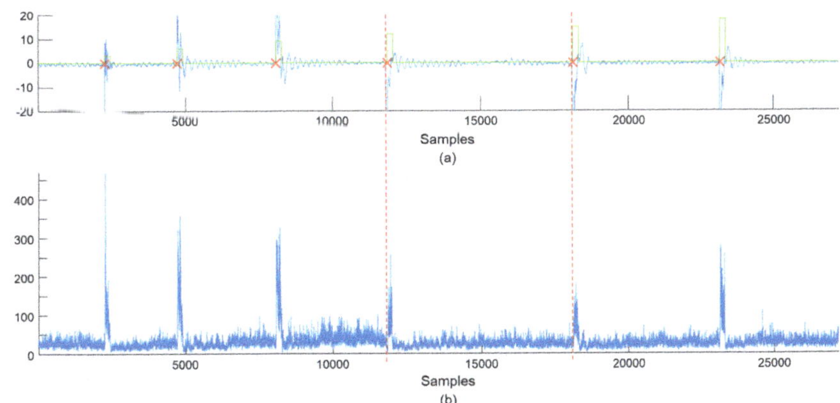

Figure 9. Uncertainties measured by MC-dropout of the teacher network. (**a**) shows the first dimension of 24-dimensional sensor data, and (**b**) presents uncertainties measured by MC-dropout. In (**a**), red × marks indicate collision moments, and green lines represent labels for the normal state and locations of collisions.

5.3. Sample-Level Accuracy

The first measure to evaluate the performance of deep neural networks is the sample-level accuracy. As explained in Section 4.1, the architecture of the deep neural network proposed in [22] is employed to construct the base model. To demonstrate the effectiveness of uncertainty-aware knowledge distillation for the problem of collision identification, we compare the accuracies of the proposed method with those of the base model and a student network. The student network has an identical architecture to the base model, and is trained by distilling knowledge in the teacher network without employing uncertainty information. Table 2 presents the sample-level recall, precision, and f1-score of four neural network models; the proposed method means another student network, which is trained by uncertainty-aware knowledge distillation. The last row of Table 2 presents the sample-level accuracies of the teacher network. As presented in Table 2, the f1-scores of the proposed method are comparable to those of the teacher network; it is worth noting that the proposed method employs a lightweight network compared to the teacher network.

Table 2. Sample-level accuracies of the four different neural network models before and after the post-processing.

	Before Post-Processing			After Post-Processing		
	Recall	*Precision*	*F1-Score*	*Recall*	*Precision*	*F1-Score*
Base model	98.1611	98.3985	98.2796	98.5473	99.0617	98.8038
Student network	98.2015	98.3458	98.2736	98.5992	99.0198	98.8091
Proposed method	98.3110	98.4516	98.3812	98.7119	99.0465	98.8789
Teacher network	98.2729	98.5337	98.4031	98.5629	99.1011	98.8313

5.4. Collision-Level Accuracy

This section presents the collision-level accuracies. As collaborative robots react to each collision, reducing the number of false-positive collisions is a very important issue in reliable collision identification systems. In the labeled data, a collision is represented by 200 non-zero samples; therefore, false-positive collisions, which are composed of a few fals- positive samples, are not effectively reflected in the sample-level accuracies. Although the sample-level accuracies of the four neural network models are above 98%, there are a considerable number of false-positive collisions. To compute the collision-level accuracies, a group of non-zero predictions is regarded as a collision, and Table 3 presents the numbers of true-positive, false-positive, and false-negative collisions of the four neural network models. In Table 3, the base model, student network, and proposed method have an identical network architecture to CollisionNet [22]; the student network is trained by regular knowledge distillation, and the proposed method employs uncertainties during knowledge distillation. As shown in Table 3, the number of false positives is significantly reduced after the post-processing. Table 4 presents the collision-level recall, precision, and f1-score of the four neural networks. By utilizing probabilistic labels and uncertainties from the teacher network, the proposed method produces better accuracies, despite its lightweight network architecture compared to the teacher network.

Table 3. The numbers of true-positive (TP), false-positive (FP), and false-negative (FN) collisions of the four neural network models before and after post-processing.

	Before Post-Processing			After Post-Processing		
	TP	*FP*	*FN*	*TP*	*FP*	*FN*
Base model	1119	229	3	1119	121	3
Student network	1118	295	4	1118	109	4
Proposed method	1120	205	2	1120	76	2
Teacher network	1119	267	3	1119	77	3

Table 4. Collision-level accuracies of the four different neural network models before and after the post-processing.

	Before Post-Processing			After Post-Processing		
	Recall	*Precision*	*F1-Score*	*Recall*	*Precision*	*F1-Score*
Base model	99.7326	78.9139	88.1102	99.7326	90.2419	94.7502
Student network	99.6436	79.1224	88.2052	99.6435	91.1165	95.1894
Proposed method	99.8217	84.5283	91.5406	99.8217	93.6454	96.6350
Teacher network	99.7326	80.7359	89.2344	99.7326	93.5619	96.5487

5.5. Analysis for the Processing Time

The processing time is another important factor for responding to external forces within an acceptable timeframe. In the collision identification system, the total processing time is composed of the inference time of a neural network, time delay for detecting a collision, and post-processing time. Table 5 presents the averaged processing time for each step. The teacher network requires an 83% longer inference time compared to the base model, student network, and proposed method. The detection delay is measured by averaging the time intervals between collision occurrences and their corresponding first true-positive samples. As presented in Table 5, the proposed method requires 2.6350 ms to identify a collision occurrence, and this satisfies the requirement for the safety of collaborative robots.

Table 5. The averaged processing time in ms for the collision identification.

	Inference Time	Detection Delay	Post-Processing	Total
Base model	1.7641	0.8239	0.2057	2.7938
Student network	1.7641	0.6198	0.2057	2.5897
Proposed method	1.7641	0.6651	0.2057	2.6350
Teacher network	3.2348	0.7006	0.2057	4.1412

6. Conclusions

This paper proposes a collision identification method for collaborative robots. To identify the locations of external forces, the propose method employs a deep neural network, which is composed of causal convolutions and dilated convolutions. The key contribution is the method of capturing sample-level uncertainties and distilling the knowledge of a teacher network to train a student network, with consideration of predictive uncertainties. In the knowledge distillation, KL-divergence between the predictions of the student and teacher networks are weighted by the predictive uncertainties to focus on data samples that are difficult to train. Furthermore, we also propose a post-processing to reduce the number of false-positive collisions. Experiments were conducted with a 6-DoF-articulated robot, and we demonstrated that the uncertainty is beneficial to improving the accuracy of the collision identification method.

Author Contributions: Conceptualization, W.K. and S.J.L.; methodology, S.J.L.; software, S.J.L.; validation, S.J.L.; formal analysis, S.J.L.; data curation, W.K.; writing—original draft preparation, W.K. and S.J.L.; writing—review and editing, W.K., Y.J. and S.J.L.; visualization, S.J.L. All authors have read and agreed to the published version of the manuscript.

Funding: This work was supported by the National Research Foundation of Korea (NRF) grant funded by the Korea government (MSIT) (No. 2021R1G1A1009792). This research was supported by "Research Base Construction Fund Support Program" funded by Jeonbuk National University in 2021, and partially supported by Electronics and Telecommunications Research Institute (ETRI) grant funded by the Korean government. (21ZD1130, Development of ICT Convergence Technology for Daegu-Gyeongbuk Regional Industry).

Institutional Review Board Statement: Not applicable.

Informed Consent Statement: Not applicable.

Conflicts of Interest: The authors declare no conflict of interest.

References

1. Goodrich, M.A.; Schultz, A.C. *Human-Robot Interaction: A Survey*; Now Publishers Inc.: Boston, MA, USA, 2008.
2. Ajoudani, A.; Zanchettin, A.M.; Ivaldi, S.; Albu-Schäffer, A.; Kosuge, K.; Khatib, O. Progress and prospects of the human–robot collaboration. *Auton. Rob.* **2018**, *42*, 957–975. [CrossRef]
3. Haidegger, T. Autonomy for surgical robots: Concepts and paradigms. *IEEE Trans. Med. Rob. Bionics* **2019**, *1*, 65–76. [CrossRef]
4. Berezina, K.; Ciftci, O.; Cobanoglu, C. *Robots, Artificial Intelligence, and Service Automation in Restaurants*; Emerald Group Publishing: Bingley, UK, 2020.
5. Wilson, G.; Pereyda, C.; Raghunath, N.; de la Cruz, G.; Goel, S.; Nesaei, S.; Minor, B.; Schmitter-Edgecombe, M.; Taylor, M.E.; Cook, D.J. Robot-enabled support of daily activities in smart home environments. *Cognitive Syst. Res.* **2019**, *54*, 258–272. [CrossRef] [PubMed]
6. Iqbal, J.; Khan, Z.H.; Khalid, A. Prospects of robotics in food industry. *Food Sci. Technol.* **2017**, *37*, 159–165. [CrossRef]
7. Petković, T.; Puljiz, D.; Marković, I.; Hein, B. Human intention estimation based on hidden Markov model motion validation for safe flexible robotized warehouses. *Rob. Comput. Integr. Manuf.* **2019**, *57*, 182–196. [CrossRef]
8. Oudah, M.; Al-Naji, A.; Chahl, J. Hand gesture recognition based on computer vision: A review of techniques. *J. Imaging* **2020**, *6*, 73. [CrossRef] [PubMed]
9. Vicentini, F. Collaborative robotics: A survey. *J. Mech. Des.* **2021**, *143*, 040802. [CrossRef]
10. Zhang, S.; Wang, S.; Jing, F.; Tan, M. A sensorless hand guiding scheme based on model identification and control for industrial robot. *IEEE Trans. Ind. Inf.* **2019**, *15*, 5204–5213. [CrossRef]
11. Haddadin, S.; De Luca, A.; Albu-Schäffer, A. Robot collisions: A survey on detection, isolation, and identification. *IEEE Trans. Rob.* **2017**, *33*, 1292–1312. [CrossRef]
12. Morikawa, S.; Senoo, T.; Namiki, A.; Ishikawa, M. Realtime collision avoidance using a robot manipulator with light-weight small high-speed vision systems. In Proceedings of the 2007 IEEE International Conference on Robotics and Automation (ICRA), Roma, Italy, 10–14 April 2007; pp. 794–799.
13. Mohammadi Amin, F.; Rezayati, M.; van de Venn, H.W.; Karimpour, H. A mixed-perception approach for safe human–robot collaboration in industrial automation. *Sensors* **2020**, *20*, 6347. [CrossRef]
14. Lu, S.; Chung, J.H.; Velinsky, S.A. Human-robot collision detection and identification based on wrist and base force/torque sensors. In Proceedings of the 2005 IEEE international Conference on Robotics and Automation (ICRA), Barcelona, Spain, 18–22 April 2005; pp. 3796–3801.

15. Lee, S.D.; Kim, M.C.; Song, J.B. Sensorless collision detection for safe human-robot collaboration. In Proceedings of the 2015 IEEE/RSJ International Conference on Intelligent Robots and Systems (IROS), Hamburg, Germany, 28 September–2 October 2015; pp. 2392–2397.

16. Han, L.; Xu, W.; Li, B.; Kang, P. Collision detection and coordinated compliance control for a dual-arm robot without force/torque sensing based on momentum observer. *IEEE/ASME Trans. Mechatron.* **2019**, *24*, 2261–2272. [CrossRef]

17. De Luca, A.; Albu-Schaffer, A.; Haddadin, S.; Hirzinger, G. Collision detection and safe reaction with the DLR-III lightweight manipulator arm. In Proceedings of the 2006 IEEE/RSJ International Conference on Intelligent Robots and Systems (IROS), Beijing, China, 9–15 October 2006; pp. 1623–1630.

18. Briquet-Kerestedjian, N.; Makarov, M.; Grossard, M.; Rodriguez-Ayerbe, P. Generalized momentum based-observer for robot impact detection—Insights and guidelines under characterized uncertainties. In Proceedings of the 2017 IEEE Conference on Control Technology and Applications (CCTA), Maui, HI, USA, 27–30 August 2017; pp. 1282–1287.

19. Mamedov, S.; Mikhel, S. Practical aspects of model-based collision detection. *Front. Rob. AI* **2020**, *7*, 162.

20. Geravand, M.; Flacco, F.; De Luca, A. Human-robot physical interaction and collaboration using an industrial robot with a closed control architecture. In Proceedings of the 2013 IEEE International Conference on Robotics and Automation (ICRA), Karlsruhe, Germany, 6–10 May 2013; pp. 4000–4007.

21. Sharkawy, A.N.; Koustoumpardis, P.N.; Aspragathos, N. Neural network design for manipulator collision detection based only on the joint position sensors. *Robotica* **2020**, *38*, 1737–1755. [CrossRef]

22. Heo, Y.J.; Kim, D.; Lee, W.; Kim, H.; Park, J.; Chung, W.K. Collision detection for industrial collaborative robots: A deep learning approach. *IEEE Rob. Autom. Lett.* **2019**, *4*, 740–746. [CrossRef]

23. Hu, J.; Xiong, R. Contact force estimation for robot manipulator using semiparametric model and disturbance Kalman filter. *IEEE Trans. Ind. Electron.* **2017**, *65*, 3365–3375. [CrossRef]

24. Ren, T.; Dong, Y.; Wu, D.; Chen, K. Collision detection and identification for robot manipulators based on extended state observer. *Control Eng. Pract.* **2018**, *79*, 144–153. [CrossRef]

25. Kouris, A.; Dimeas, F.; Aspragathos, N. A frequency domain approach for contact type distinction in human–robot collaboration. *IEEE Rob. Autom. Lett.* **2018**, *3*, 720–727. [CrossRef]

26. Jo, S.; Kwon, W. A Comparative Study on Collision Detection Algorithms based on Joint Torque Sensor using Machine Learning. *J. Korea Robot. Soc.* **2020**, *15*, 169–176. [CrossRef]

27. Pan, J.; Manocha, D. Efficient configuration space construction and optimization for motion planning. *Engineering* **2015**, *1*, 46–57. [CrossRef]

28. Zhang, Z.; Qian, K.; Schuller, B.W.; Wollherr, D. An online robot collision detection and identification scheme by supervised learning and bayesian decision theory. *IEEE Trans. Autom. Sci. Eng.* **2020**, *18*, 1144–1156. [CrossRef]

29. Birjandi, S.A.B.; Kühn, J.; Haddadin, S. Observer-extended direct method for collision monitoring in robot manipulators using proprioception and imu sensing. *IEEE Rob. Autom. Lett.* **2020**, *5*, 954–961. [CrossRef]

30. Caldas, A.; Makarov, M.; Grossard, M.; Rodriguez-Ayerbe, P.; Dumur, D. Adaptive residual filtering for safe human-robot collision detection under modeling uncertainties. In Proceedings of the 2013 IEEE/ASME International Conference on Advanced Intelligent Mechatronics, Wollongong, Australia, 9–12 July 2013; pp. 722–727.

31. Makarov, M.; Caldas, A.; Grossard, M.; Rodriguez-Ayerbe, P.; Dumur, D. Adaptive filtering for robust proprioceptive robot impact detection under model uncertainties. *IEEE/ASME Trans. Mechatron.* **2014**, *19*, 1917–1928. [CrossRef]

32. Birjandi, S.A.B.; Haddadin, S. Model-Adaptive High-Speed Collision Detection for Serial-Chain Robot Manipulators. *IEEE Rob. Autom. Lett.* **2020**, *5*, 6544–6551. [CrossRef]

33. Min, F.; Wang, G.; Liu, N. Collision detection and identification on robot manipulators based on vibration analysis. *Sensors* **2019**, *19*, 1080. [CrossRef] [PubMed]

34. Xu, T.; Fan, J.; Fang, Q.; Zhu, Y.; Zhao, J. A new robot collision detection method: A modified nonlinear disturbance observer based-on neural networks. *J. Intell. Fuzzy Syst.* **2020**, *38*, 175–186. [CrossRef]

35. Park, K.M.; Kim, J.; Park, J.; Park, F.C. Learning-based real-time detection of robot collisions without joint torque sensors. *IEEE Rob. Autom. Lett.* **2020**, *6*, 103–110. [CrossRef]

36. Oord, A.V.D.; Dieleman, S.; Zen, H.; Simonyan, K.; Vinyals, O.; Graves, A.; Kalchbrenner, N.; Senior, A.; Kavukcuoglu, K. Wavenet: A generative model for raw audio. *arXiv* **2016**, arXiv:1609.03499.

37. Maceira, M.; Olivares-Alarcos, A.; Alenya, G. Recurrent neural networks for inferring intentions in shared tasks for industrial collaborative robots. In Proceedings of the 2020 29th IEEE International Conference on Robot and Human Interactive Communication (RO-MAN), Naples, Italy, 31 August–4 September 2020; pp. 665–670.

38. Czubenko, M.; Kowalczuk, Z. A Simple Neural Network for Collision Detection of Collaborative Robots. *Sensors* **2021**, *21*, 4235. [CrossRef]

39. Hinton, G.; Vinyals, O.; Dean, J. Distilling the knowledge in a neural network. *arXiv* **2015**, arXiv:1503.02531.

40. Park, W.; Kim, D.; Lu, Y.; Cho, M. Relational knowledge distillation. In Proceedings of the IEEE/CVF Conference on Computer Vision and Pattern Recognition, Long Beach, CA, USA, 15–20 June 2019; pp. 3967–3976.

41. Meng, Z.; Li, J.; Zhao, Y.; Gong, Y. Conditional teacher-student learning. In Proceedings of the IEEE International Conference on Acoustics, Speech and Signal Processing (ICASSP), Brighton, UK, 12–17 May 2019; pp. 6445–6449.

42. Yim, J.; Joo, D.; Bae, J.; Kim, J. A gift from knowledge distillation: Fast optimization, network minimization and transfer learning. In Proceedings of the IEEE Conference on Computer Vision and Pattern Recognition, Honolulu, HI, USA, 21–26 July 2017; pp. 4133–4141.

43. Romero, A.; Ballas, N.; Kahou, S.E.; Chassang, A.; Gatta, C.; Bengio, Y. Fitnets: Hints for thin deep nets. *arXiv* **2014**, arXiv:1412.6550.

44. Zagoruyko, S.; Komodakis, N. Paying more attention to attention: Improving the performance of convolutional neural networks via attention transfer. *arXiv* **2016**, arXiv:1612.03928.

45. Yim, J.; Joo, D.; Bae, J.; Kim, J. Learning efficient object detection models with knowledge distillation. In Proceedings of the 31st Conference on Neural Information Processing Systems (NIPS 2017), Long Beach, CA, USA, 4–7 December 2017; pp. 742–751.

46. Hou, Y.; Ma, Z.; Liu, C.; Hui, T.W.; Loy, C.C. Inter-region affinity distillation for road marking segmentation. In Proceedings of the IEEE/CVF Conference on Computer Vision and Pattern Recognition, Seattle, WA, USA, 13–19 June 2020; pp. 12486–12495.

47. Gupta, S.; Hoffman, J.; Malik, J. Cross modal distillation for supervision transfer. In Proceedings of the IEEE conference on computer vision and pattern recognition, Las Vegas, NV, USA, 27–30 June 2016; pp. 2827–2836.

48. Papernot, N.; McDaniel, P.; Wu, X.; Jha, S.; Swami, A. Distillation as a defense to adversarial perturbations against deep neural networks. In Proceedings of the 2016 IEEE symposium on security and privacy (SP), San Jose, CA, USA, 22–26 May 2016; pp. 582–597.

49. Liu, Y.; Chen, K.; Liu, C.; Qin, Z.; Luo, Z.; Wang, J. Structured knowledge distillation for semantic segmentation. In Proceedings of the IEEE/CVF Conference on Computer Vision and Pattern Recognition, Long Beach, CA, USA, 15–20 June 2019; pp. 2604–2613.

50. Tarvainen, A.; Valpola, H. Mean teachers are better role models: Weight-averaged consistency targets improve semi-supervised deep learning results. *arXiv* **2017**, arXiv:1703.01780.

51. Ghahramani, Z. Probabilistic machine learning and artificial intelligence. *Nature* **2015**, *521*, 452–459. [CrossRef]

52. Gal, Y.; Ghahramani, Z. Dropout as a bayesian approximation: Representing model uncertainty in deep learning. In Proceedings of the 33rd international conference on machine learning, New York, NY, USA, 19–24 June 2016; pp. 1050–1059.

53. Srivastava, N.; Hinton, G.; Krizhevsky, A.; Sutskever, I.; Salakhutdinov, R. Dropout: A simple way to prevent neural networks from overfitting. *J. Mach. Learn. Res.* **2014**, *15*, 1929–1958.

54. Lakshminarayanan, B.; Pritzel, A.; Blundell, C. Simple and scalable predictive uncertainty estimation using deep ensembles. *arXiv* **2016**, arXiv:1612.01474.

55. Van Amersfoort, J.; Smith, L.; Teh, Y.W.; Gal, Y. Uncertainty estimation using a single deep deterministic neural network. In Proceedings of the 37rd International Conference on Machine Learning, Online, 12–18 July 2020; pp. 9690–9700.

56. Zhang, Z.; Dalca, A.V.; Sabuncu, M.R. Confidence calibration for convolutional neural networks using structured dropout. *arXiv* **2019**, arXiv:1906.09551.

57. Tagasovska, N.; Lopez-Paz, D. Single-model uncertainties for deep learning. *arXiv* **2018**, arXiv:1811.00908.

58. Shen, Y.; Zhang, Z.; Sabuncu, M.R.; Sun, L. Real-time uncertainty estimation in computer vision via uncertainty-aware distribution distillation. In Proceedings of the IEEE/CVF Winter Conference on Applications of Computer Vision, Online, 5–9 January 2021; pp. 707–716.

59. Jin, X.; Lan, C.; Zeng, W.; Chen, Z. Uncertainty-aware multi-shot knowledge distillation for image-based object re-identification. In Proceedings of the AAAI Conference on Artificial Intelligence, New York, NY, USA, 7–12 February 2020; pp. 11165–11172.

60. Mehrtash, A.; Wells, W.M.; Tempany, C.M.; Abolmaesumi, P.; Kapur, T. Confidence calibration and predictive uncertainty estimation for deep medical image segmentation. *IEEE Trans. Med. Imaging* **2020**, *39*, 3868–3878. [CrossRef]

61. Oh, D.; Ji, D.; Jang, C.; Hyunv, Y.; Bae, H.S.; Hwang, S. Segmenting 2k-videos at 36.5 fps with 24.3 gflops: Accurate and lightweight realtime semantic segmentation network. In Proceedings of the 2020 IEEE International Conference on Robotics and Automation (ICRA), Online, 31 May–31 August 2020; pp. 3153–3160.

62. Kwon, W.; Jin, Y.; Lee, S.J. Collision Identification of Collaborative Robots Using a Deep Neural Network. *IEMEK J. Embed. Syst. Appl.* **2021**, *16*, 35–41.

63. Kingma, D.P.; Ba, J. Adam: A method for stochastic optimization. *arXiv* **2014**, arXiv:1412.6980.

Article

Improved Mutual Understanding for Human-Robot Collaboration: Combining Human-Aware Motion Planning with Haptic Feedback Devices for Communicating Planned Trajectory

Stefan Grushko *, Aleš Vysocký, Petr Oščádal, Michal Vocetka, Petr Novák and Zdenko Bobovský

Department of Robotics, Faculty of Mechanical Engineering, VSB-TU Ostrava, 70800 Ostrava, Czech Republic; ales.vysocky@vsb.cz (A.V.); petr.oscadal@vsb.cz (P.O.); michal.vocetka@vsb.cz (M.V.); petr.novak@vsb.cz (P.N.); zdenko.bobovsky@vsb.cz (Z.B.)
* Correspondence: stefan.grushko@vsb.cz

Citation: Grushko, S.; Vysocký, A.; Oščádal, P.; Vocetka, M.; Novák, P.; Bobovský, Z. Improved Mutual Understanding for Human-Robot Collaboration: Combining Human-Aware Motion Planning with Haptic Feedback Devices for Communicating Planned Trajectory. *Sensors* **2021**, *21*, 3673. https://doi.org/10.3390/s21113673

Academic Editor: Anne Schmitz

Received: 5 May 2021
Accepted: 23 May 2021
Published: 25 May 2021

Publisher's Note: MDPI stays neutral with regard to jurisdictional claims in published maps and institutional affiliations.

Abstract: In a collaborative scenario, the communication between humans and robots is a fundamental aspect to achieve good efficiency and ergonomics in the task execution. A lot of research has been made related to enabling a robot system to understand and predict human behaviour, allowing the robot to adapt its motion to avoid collisions with human workers. Assuming the production task has a high degree of variability, the robot's movements can be difficult to predict, leading to a feeling of anxiety in the worker when the robot changes its trajectory and approaches since the worker has no information about the planned movement of the robot. Additionally, without information about the robot's movement, the human worker cannot effectively plan own activity without forcing the robot to constantly replan its movement. We propose a novel approach to communicating the robot's intentions to a human worker. The improvement to the collaboration is presented by introducing haptic feedback devices, whose task is to notify the human worker about the currently planned robot's trajectory and changes in its status. In order to verify the effectiveness of the developed human-machine interface in the conditions of a shared collaborative workspace, a user study was designed and conducted among 16 participants, whose objective was to accurately recognise the goal position of the robot during its movement. Data collected during the experiment included both objective and subjective parameters. Statistically significant results of the experiment indicated that all the participants could improve their task completion time by over 45% and generally were more subjectively satisfied when completing the task with equipped haptic feedback devices. The results also suggest the usefulness of the developed notification system since it improved users' awareness about the motion plan of the robot.

Keywords: human robot collaboration; human robot interaction; path planning; bidirectional awareness; haptic feedback device; human machine interface

1. Introduction

Human-robot collaboration (HRC) is a promising trend in the field of industrial and service robotics. Collaborative robots created new opportunities in the field of human-robot cooperation by enabling the robot to share the workspace with the personnel where it helps with non-ergonomic, repetitive, uncomfortable, or even dangerous operations. With the growing level of cooperation, there is a tendency to increase the intertwining of human and robot workspaces in the future, potentially leading to complete unification [1,2]. By allowing a fully shared workspace between humans and robots, it is possible to utilise the advantages of both, maximise their efficiency and minimise the time needed to complete a task. Shared workspaces are examples of a dynamic environment, as the human operators represent moving obstacles, which motions are difficult to predict accurately. Moreover, in a typically shared workplace, the collaborative robot does not have any perception about the position of the operator and can only react to the collisions by detecting the contacts with the tool or robot body (by measuring joint torques or monitoring the predicted joint position deviation [1,3]). These approaches have apparent limitations defined by the fact that in the case of adaptive tasks with high variability, the operator may be unaware of the actions planned by the robot, and the robot cannot predictively avoid a collision with the operator. In such adaptable HRC scenarios, the understanding between the human operator and the robot is crucial. On one side, the robot must be aware of a human operator, and on the other side, the operator needs to be aware of the current status of the robot system. Advanced workplaces may include monitoring systems [4–6] enabling the robot to react to (and potentially predict [7–11]) the operator's movements by immediately stopping the activity or by replanning the trajectory [12–14]. However, a workspace monitoring system enabling the robot to avoid collisions can be considered as one-side awareness, but the other side of communication remains unresolved. An extensive review of the existing challenges in the field of human-robot interaction is available in the work of P. Tsarouchi et al. [15]. Despite many efforts made to make robots understand and predict human actions, there is still a lack of communication from the robot to the human operator. Difficulties in understanding the robot's intent (planned trajectory, current task, internal status) during demanding cooperation can lead to dangerous situations, reduced work efficiency, and a general feeling of anxiety when working close to the robot (even if it is a collaborative robot). Better awareness can be achieved by providing the operator with intuitive communication channels that allow them to understand the motion plans and status of the robot. These communication channels may be realised with the help of feedback devices—Human-Machine Interfaces (HMIs). To convey information, these systems may utilise the primary sensory modalities of a human: vision, hearing, touch.

Many existing methods for communicating robot motion intent use graphical clues which notify the human worker about the status of the robot. Typical information for visualisation may include the internal status of the robot, command acknowledgement, planned trajectory, and current workspace borders. In the simplest example of such an approach, the data visualisations are displayed on 2D monitors [16] (static or hand-held tablets), which, however, require the operator to interrupt the current task and check the visualisation on display. Light projectors represent a straightforward solution for representing additional graphical clues and notifications to the operator, possibly directly in the operator's vicinity, making it easier for the clues to be noticed [17,18]. Projector-based systems have a number of limitations, the main one being that various obstacles and the operator himself can block the system from both displaying the graphical clues and tracking the work objects, leading to an increased risk of misinterpretation of the visualised information by the operator.

Multiple Augmented Reality (AR) and Mixed Reality (MR) approaches have been developed as a subsequent improvement of the projector-based solutions. It allows a more intuitive overlay of the visual notifications with the real environment and objects in the workspace. Augmented reality headsets allow 3D graphics to be displayed directly in the user's field of view without completely overlapping visual information from the real world [19–21]. One of the problems associated with the visualisation of the planned movement of the robot and other contextual information is that it cannot be guaranteed that this information is always in the field of view of the operator (the operator may be watching in another direction). It is also worth mentioning that MR/AR devices themselves present an interfering component that may distract the operator during the task. Experimental user studies performed by A. Hietanen [21] performed a comprehensive comparison and evaluation of HRC in a realistic manufacturing task in two conditions: a projector-mirror setup and a wearable AR headset. The results indicated that HoloLens was experienced less safe (comparing to the projector-based notification system) due to the intrusiveness of the device. Even though it was used as an augmented reality display, it blocked, to some extent, the view for the operator. An extensive review of the collaborative aspects of graphical interfaces for supporting workers during HRC is covered in the work of L. Wang et al. [2].

Another option of improving the awareness of a human worker during HRC is by utilising audio feedback. Auditory cues provide a wide range of contextual information that promotes awareness of a person about its surrounding. While vision feedback is traditionally preferred in applications that require a high level of accuracy, audio information is important in scenarios when other modalities are limited or blocked. An example of an application of this approach was demonstrated by A. Clair et al. [22], where the efficiency of the collaborative task was enhanced by enabling the robot to use synthesised speech to give a human teammate acoustic feedback about the currently performed action. G. Bolano et al. [14] focused on a multimodal feedback system for HRC by combining graphical and acoustic feedback channels. It is worth noting that due to potentially noisy manufacturing conditions, the operator may be unable to hear the audio signals.

Sense of touch represents a robust and direct way of transferring information to the user, making it suitable to convey information to workers in industrial environments, where visual and auditory modalities might be busy or blocked. P. Barros et al. [23] performed a set of tests using the simulation model of the teleoperated robot and enhancing the users with tactile feedback that could notify them about the actual collisions of the robot with the surroundings. Vibration devices can also be used during the control of an industrial robot, notifying the user about, for instance, approaching singularities and joint limits [24,25] or commencing the next phase of the manufacturing process [5].

It was demonstrated that the ability to communicate the robot's motion to the worker in advance has an influence on the human propensity to accept the robot [2,26,27]. In this work, we propose a novel wearable haptic notification system for informing the human operator about the robot's status, its currently planned trajectory, and the space that will be occupied by the robot during the movement. The haptic notification system consists of compact devices placed on both hands of the user, which provided vibrational alerts depending on the distance from the robot trajectory. Our approach combines the notifications with active collision avoidance [2], which enables the robot to continue on task execution even if the worker, despite the alerts, has precluded the initial trajectory.

Compared with existing interfaces, the proposed haptic notification system has the advantage of reliability alerting the user compared with graphical and acoustic feedback devices, whose efficiency can be limited or blocked during engagement in the task. The effectiveness of the robot's trajectory intent communication to the user was verified in a user study. The results of the performed user study showed that the users had a better understanding of the robot's motion. The developed haptic feedback system for collaborative workspaces can improve the efficiency and safety of human-robot cooperation in industrial conditions. The system may be able to reduce the time required for unskilled operators to get used to the manufacturing process and the movement of the robot in the near vicinity, along with helping to avoid the discomfort caused by unawareness about the intentions of the robot.

2. Materials and Methods

2.1. Concept

The proposed system is based on the concept of a shared collaborative workspace where the robot can adapt its movement to avoid collision with human workers. The workspace is monitored by multiple RGB-D sensors, and data provided by these sensors is used to construct a map of the robot's surroundings and obstacles. At each step of the task execution, the robot creates a collision-free motion plan according to the currently available free space. If during the execution of the planned movement there is a change in the environment (for example, existing obstacles change their location) and the movement can no longer be completed due to possible collisions with obstacles, the robot can create a new motion plan (active collision avoidance). The improvement to the collaboration is presented by introducing haptic feedback devices (hereinafter, Human-Machine Interfaces, HMIs), whose task is to reliably notify the human worker about the currently planned robot's trajectory and changes in its status. A wearable device is used to improve the operator's awareness during the human-robot collaborative assembly task through vibrotactile feedback.

With regard to involvement in the work process, the hands are the parts of the body that are most often present in the shared workspace (especially when the work process is taking place at a table). For this reason, it was decided to develop a haptic HMI in the form of a compact device attached to the dorsal side of a work glove providing vibrational feedback to the user's palm. Besides, it is a common practice in many industrial domains that the workers wear work gloves. Still, this placement requires a compact, wireless, and lightweight design that ensures the comfort of use (HMI must not limit the user's capabilities during manual work). The human worker is equipped with two haptic feedback devices placed at each hand. The system utilises three types of notifications to inform the operator about the status of the robot: distance notification, replanning notification, and inaccessible goal notification.

The distance notification provides a continuous vibration alert to the user about the proximity to the currently planned trajectory of the robot (see Figure 1a). The future trajectory segment is defined as the part of the feasible trajectory that yet has not been executed (see Figure 1b).

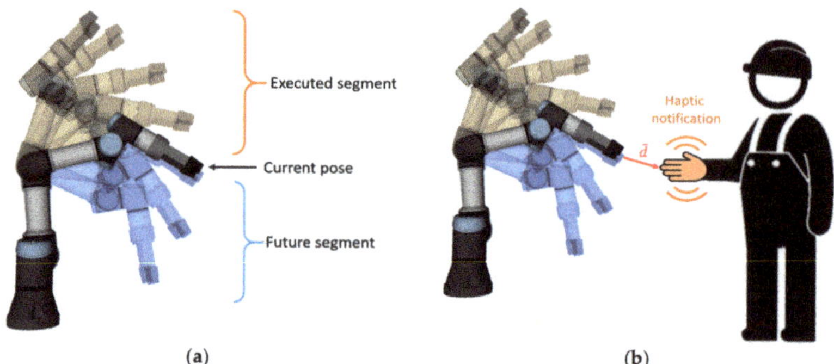

Figure 1. (a) Trajectory execution: already executed segment of trajectory, current state, and future segment (planned trajectory segment yet to be executed); (b) human worker equipped with haptic feedback device, which provides vibration alert about the proximity to the future trajectory of the robot; collision vector is denoted as $\vec{d}$. Note: collision vector is calculated considering all points of the robot body, including all links and joints.

The closer the worker's hand (equipped with HMI) approaches the future segment of the trajectory, the stronger the vibration provided by the device (see Figure 2a,b). The length of the vector between the nearest points of HMI and the robot body in all timesteps of the future trajectory (hereinafter "collision vector", $\vec{d}$) is considered to be the distance, which is used to calculate the vibration intensity. Calculation of collision vector considers all the links and joints of the robot. The user receives vibration notifications while approaching any part of the robot body. There is also an upper limit of the distance at which HMIs provide feedback (reaction distance d_r, Figure 2c). This ensures that the worker receives an alert only if his/her current actions may interfere with the robot trajectory.

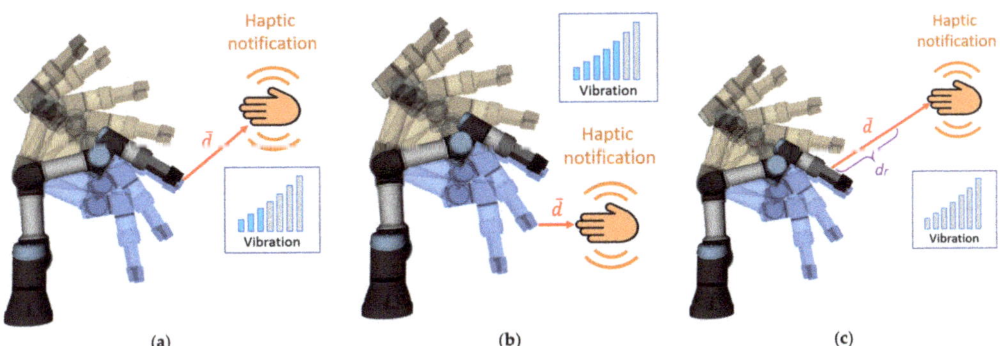

Figure 2. Illustration depicting the principle of distance notification: (**a,b**) vibration intensity is proportional to the distance to the closes point of the robot body in any timestep of the future trajectory; (**c**) distance notification is not active when the hand is further than the safe distance (reaction distance d_r).

The vibration intensity of the distance notification is calculated according to (1).

$$v_d = \frac{(v_{p,max} - v_{p,min}) * \left(d_r - \left| \vec{d} \right| \right)}{d_r + v_{p,min}} \tag{1}$$

where

v_d—calculated vibration intensity for distance notification,

$v_{p,max}$—maximum vibration intensity for distance notification,

$v_{p,min}$—minimum vibration intensity for distance notification,

d_r—reaction distance—distance threshold at which the distance notification is activated,

$\vec{d}$—current collision vector (see Figure 1).

The reaction distance d_r was set to 15 cm as it was found the most convenient for the users. The calculated intensity v_d is applied to all tactors of the corresponding haptic device.

When a person, despite a warning, interferes with the currently planned trajectory, the robot attempts to find a new feasible path to the goal position and continue the activity. Every time a new trajectory is planned, as a result of an environment change, both feedback devices use strong vibration notification (hereinafter "replanning notification"; this notification has a duration of 0.3 s) to draw the attention of the human worker and to indicate that the robot has detected an environment change and has replanned its movement (Figure 3a). If no feasible path to the target is found, both feedback devices also provide a strong vibration alert (hereinafter "inaccessible goal notification" Figure 3b), but this alert will last until the robot is able to continue its activity (movement to the goal). Both notifications may be utilised not only due to the obstacle presented by the hands of the user but also because of any obstacles (work tools, miscellaneous personal items) present in the workspace since the monitoring system maps all obstacles in the environment (using three depth sensors placed at the workplace). This way, even if the hands of the user will be completely hidden from the upper camera (which provides data for HMI localisation), the robot will be able to avoid collision with the user.

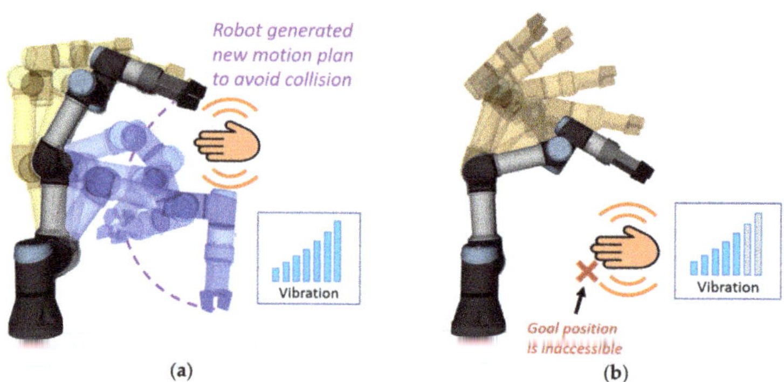

Figure 3. Status-related notifications: (**a**) replanning notification; (**b**) inaccessible goal notification.

Vibration intensities were handpicked to make the haptic feedback unobtrusive when active. Vibration intensities for individual notifications are shown in Table 1, where the intensity of vibration is proportional to the provided PWM duty cycle represented in the range 0–100%: 0% represents no vibration, 100% represents the maximum attainable vibration. Preliminary tests have demonstrated that the vibration motors have high inertia and start spinning only at intensities above 30% PWM duty cycle. However, it was also noticed that for most users, only vibration above 50% of maximum intensity was noticeable.

Table 1. Notification vibration intensities.

Notification	Vibration Intensity (PWM Duty Cycle)	Duration	Description
Inaccessible goal notification	85%	Continuous	The robot was not able to find a feasible path to the goal
Replanning notification	95%	0.3 s	The robot has replanned its motion in order to avoid a collision
Distance notification (maximum)	80%	Continuous	The user's hand is about to block the currently planned robot trajectory
Distance notification (minimum)	60%	Continuous	The user's hand is approaching the future segment of the robot's trajectory

Taken together (see illustration in Figure 4), it is anticipated that the introduced system would minimise the interference between the robot and worker, leading to less frequent replanning, improve the comfort and acceptance for the human worker while working close to the robot and allow more fluent collaboration. The transparent behaviour of the robot can also lead to increased efficiency in performing the task since the worker will be informed when hindering the robot from continuing its activity.

2.2. Experimental Workspace

The proposed system was tested on an experimental workspace with Universal Robots UR3e collaborative robot (Universal Robots, Odense, Denmark, see Figure 5a). UR3e robot has 6 DoFs, a working radius of 500 mm, and a maximum payload of 3 kg. In order to correctly map the obstacles within the workspace, three RealSense D435 RGB-D cameras (Santa Clara, CA, USA, see Figure 5b) are mounted on the workplace frame at different locations. Due to the principle of operation of the sensors, there is no limit on the number of cameras observing the same object simultaneously since the cameras do not interfere with each other. The streaming resolution was set to 424 × 240 at 15 FPS for both RGB and depth data streams, which is considered sufficient for the application.

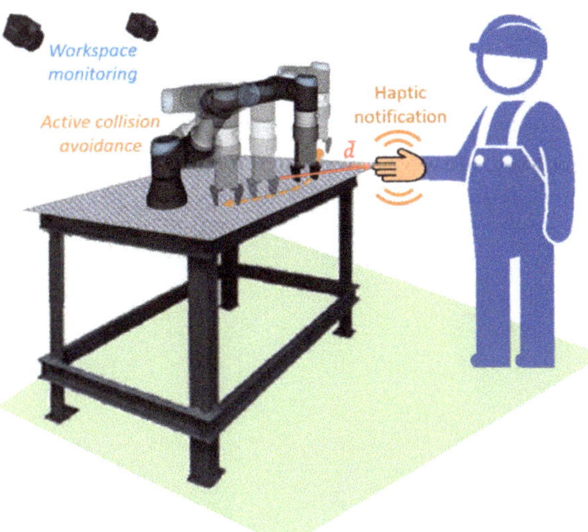

Figure 4. The concept of improved HRC combines principles of active collision avoidance with an increased awareness provided by the wearable haptic feedback device.

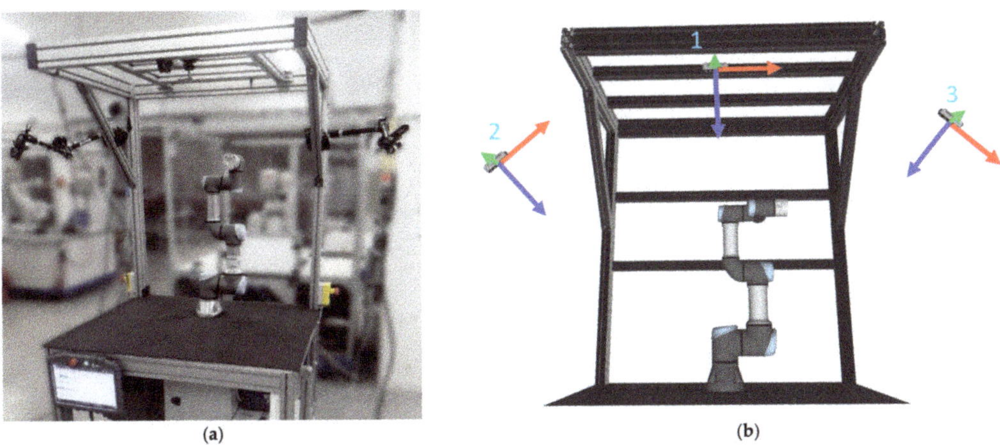

Figure 5. Experimental workspace: (**a**) overview; (**b**) Locations of three RGB-D sensors. View directions (z-axes) are depicted by the blue vectors.

It was decided to concentrate the control over the system into a single PC (Intel i7 2.80 GHz processor and 16 GB of RAM, Santa Clara, CA, USA) and to use the modular architecture offered by ROS (Melodic, Ubuntu 18.04, Linux, San Francisco, CA, USA) and to divide the software implementation into several separate components. The internal data flow is shown in Figure 6.

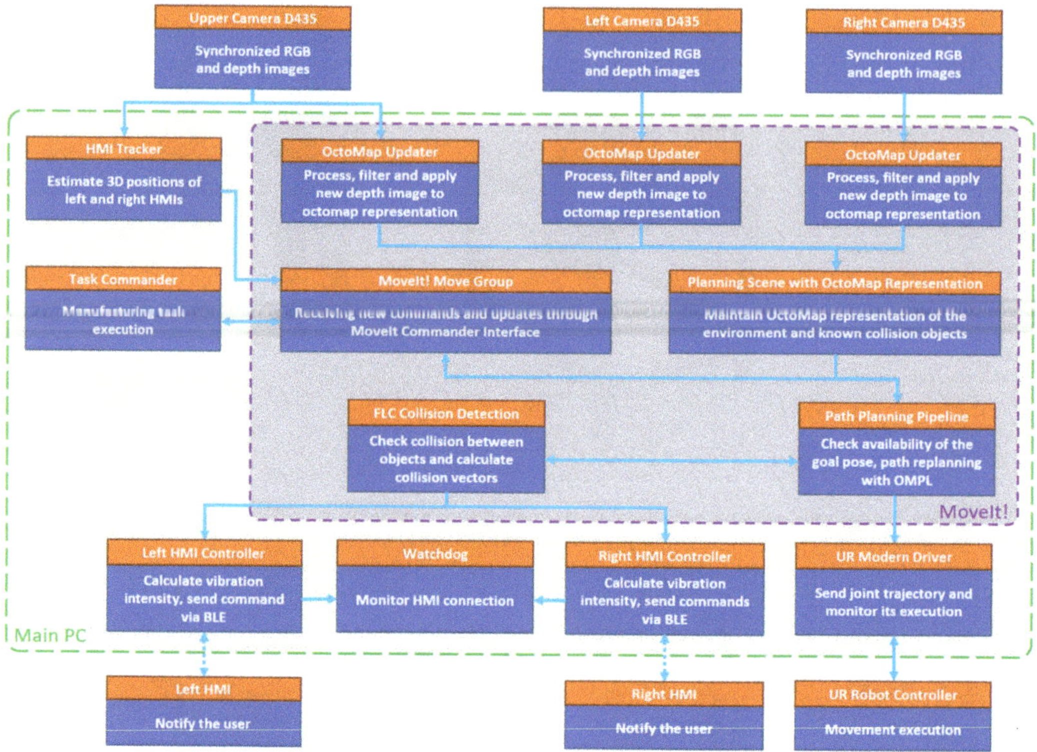

Figure 6. Data flow diagram.

The task of HMI localisation is executed by the HMI Tracker node and will be explained in more detail further in the text. Task Commander node has the primary goal of abstracting high-level commands to the robot's movement control and performing the production task. MoveIt! provides motion planning capabilities and calculations of collision vector during robot movement. Each HMI communicates with its software counterpart (HMI Controller) responsible for sustaining wireless Bluetooth data transfer, which is additionally monitored by a watchdog node. Communication with the real robot controller is implemented using a standard ROS driver.

The primary purpose of the developed haptic feedback devices is to provide the user with a reliable notification about the changes in the robot's status and its currently planned trajectory. Both HMIs were implemented in the form of compact devices attached to the dorsal side of the working gloves (see Figure 7). This design ensures that the cover of the HMI control unit does not limit the user during manual work, as it does not restrict finger movements. The battery is placed on the user's wrist, which also improves the overall ergonomics of the device. The total weight of a single HMI is 132 g. Each glove is equipped with six vibration motors, which provide haptic feedback. The motors are glued to the glove around the whole palm. The vibrational motors are controlled using DC motor drives (PWM control) by a single Arduino-compatible microcontroller that has an inbuilt battery management system and a Bluetooth Low Energy (BLE) chip, which enables communication with the main PC. During system operation, the PC constantly updates the vibration intensities of all HMI motors by sending the speed change requests via BLE. In the current version, the cover of the control unit is glued to the work gloves and cannot be detached.

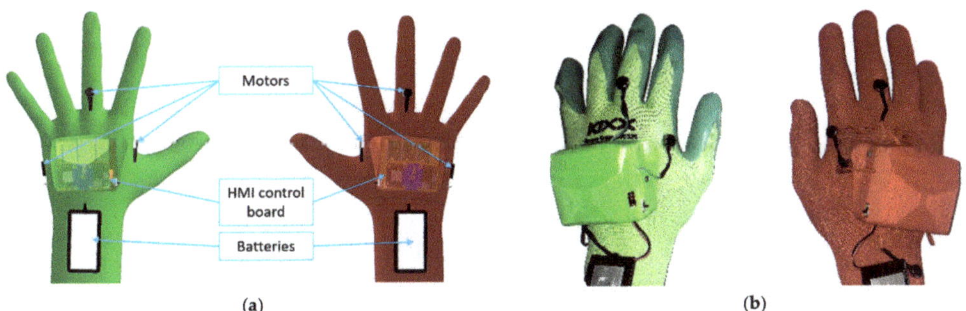

Figure 7. Prototype of HMI: (**a**) 3D model of HMI prototype, HMI cover is set as translucent in order to visualise the internal placement of the components; (**b**) implemented prototype.

It was decided to use distinctive colours for HMIs: the left HMI is green, and the right HMI is red. These colours will further simplify the task of hand tracking and determining the side of the hand.

2.3. Hand Tracking

The task of the HMI Tracker node is to determine the relative position of HMIs using data from the top depth camera located at the workplace. HMI segmentation from a 2D RGB camera image is simplified by the fact that the left and right HMIs have distinctive colours (it additionally greatly simplifies the correct determination of the side of the hand at different view angles). Colour-based segmentation is used as a simple alternative to the complex task of segmenting a 3D object in space, which otherwise would require the application of machine learning approaches such as SVM [28], deep learning-based object recognition [29], and image segmentation [30]. An extensive review of object localisation methods is demonstrated in the work of Y. Tang et al. [31].

The utilised colour-based HMI localisation approach has several limitations imposed by the nature of the detection process, the main one being that the arbitrary items with similar to the target (HMI) colours may interfere with localisation if present in the view. However, for our case, the localisation approach is considered sufficient, as it allows to reliably recognise the side of the hand regardless of the hand surface being partially occluded (providing that at least some part of the hand can be observed by the camera). In the future, the localisation system may integrate data from multiple cameras [13,32] available in the workplace in order to increase the stability and reliability of the tracking.

The position tracking of both HMIs is performed by an implemented ROS node called HMI Tracker. A schematic representation of the data processing at this node is displayed in Figure 8. Found HMI positions are published as transformations (ROS TF).

HMI Tracker node processes synchronised data messages from the upper depth camera ROS node. In order to mitigate the effect of environment luminance on the recognition accuracy, the 2D image from the camera is converted to the HVS colour space. The converted image is further processed by a 2D segmentation procedure (separately for each HMI). First, a colour range filter is used, and the resulting mask is further enhanced by the use of morphological operations (dilation and erosion), which allows for a reduction in noise in the computed mask. The largest contour is then located in the mask. The filtered HMI point cloud is then analysed to obtain the radius and the centre (centroid) of the HMI bounding sphere. The HMI bounding sphere is defined as the smallest sphere that envelopes all the HMI points, and its centre is defined as the geometrical centre of these points.

HMIs are internally represented as spheres, which allows a fast collision validation. A snapshot of the HMI tracking process (in the simulation model of the workspace) is shown in Figure 9.

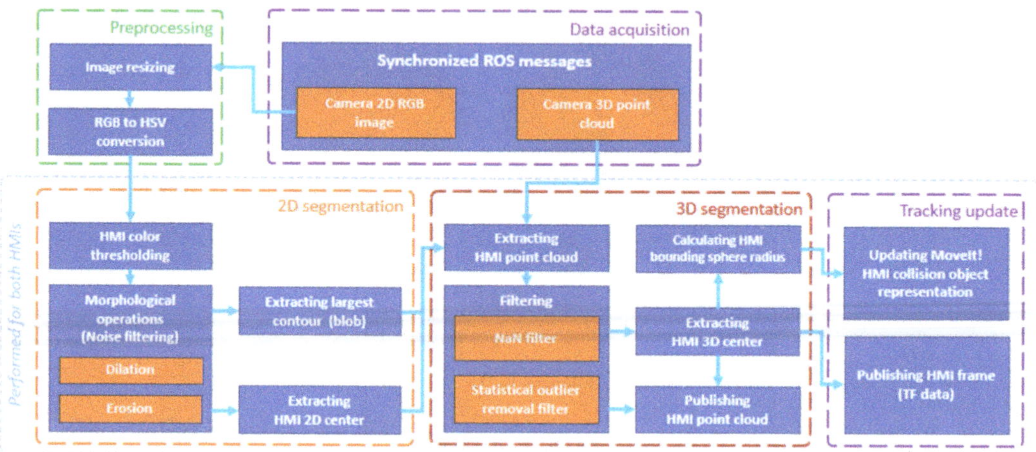

Figure 8. HMI Tracker data flow chart.

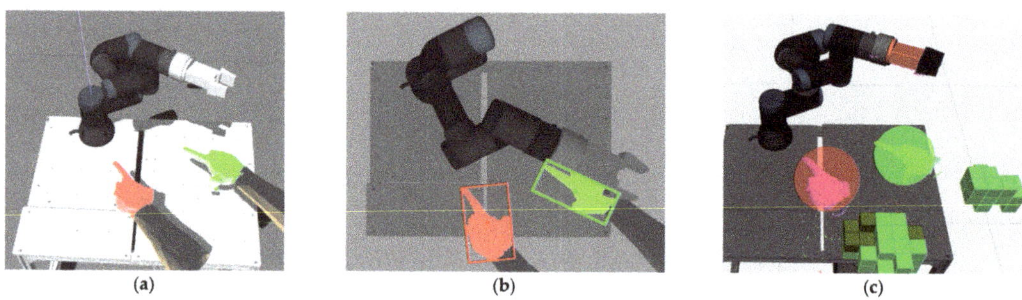

Figure 9. Snapshot of HMI tracking process: (**a**) Gazebo simulation; (**b**) upper camera image; (**c**) visualisation of the recognised HMI positions and segmented point cloud in RViz (HMIs are represented by spheres).

2.4. Motion Planning

The motion replanning subsystem is based on a modified version of ROS MoveIt! [33], which provides dynamic planning of the robot's trajectory with respect to the current position of obstacles in surrounding space: if the operator precludes the currently planned movement of the robot, the robot is able to replan this movement to avoid collisions with the human operator and continue on the performed activity. However, it is assumed that the developed haptic feedback device will minimise the probability of interference between the robot and humans. Another task of MoveIt!, in addition to trajectory planning, is to calculate the distance between the HMI and the future trajectory of the robot. MoveIt! internally uses FCL (Fast Collision Library [34]) library to perform collision validation between scene object pairs. FCL also allows to perform a distance query that returns the nearest points between an object pair. We utilise this functionality to calculate the collision vector to both HMIs in each moment of the trajectory execution and publish it to other ROS nodes. An example of how calculated collision vector changes during the execution of a motion is illustrated in Figure 10.

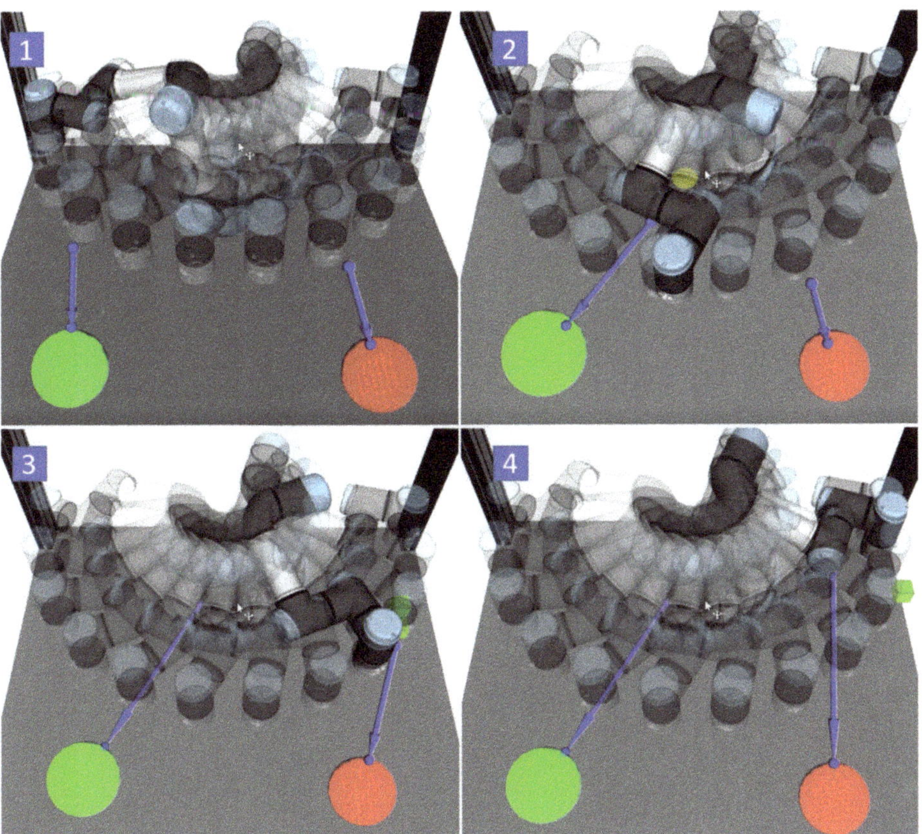

Figure 10. Example of how collision vector changing during motion sequence: robot moving from left to the right (**1–4**), collision vector is depicted by a blue arrow, HMIs are represented by green and red spheres.

3. Testing and User Study

Before starting the user studies, the system was tested in a simple task in both the real workspace and its simulated model (the parameters of the Gazebo simulation closely match those of the real workspace). During the testing, the task of the robot was to repeatedly move between two positions (from left to right and back, see Figure 10).

During the test, the individual system parameters were recorded and subsequently plotted as a timeline presented in Figure 11. The graph displays distances to both HMIs (magnitudes of the computed collision vectors), momentary notification intensities, goal accessibility status, and status of replanning routine.

Graphs in Figure 11 show that hands' movement triggered trajectory replanning several times, causing activation of replanning notification. From the timeline, it can also be observed that notification intensity is in inverse proportion to the distance to HMIs. In general, it was observed that the reaction time of the system is mainly limited by the update rate of the collision object representation of the HMIs in MoveIt!: even though the HMI tracking rate is high enough to enable smooth processing at up to 50 Hz (currently intentionally limited to 15 Hz), the update time for MoveIt! collision objects are in a range from 7 to 10 Hz (which causes the step-like changes of the distance as shown in Figure 11). This set the limit to the overall reaction speed of the system, so fast movements of the user may remain undetected by the system; however, in the future, it can be addressed by distributing the process to multiple computation units.

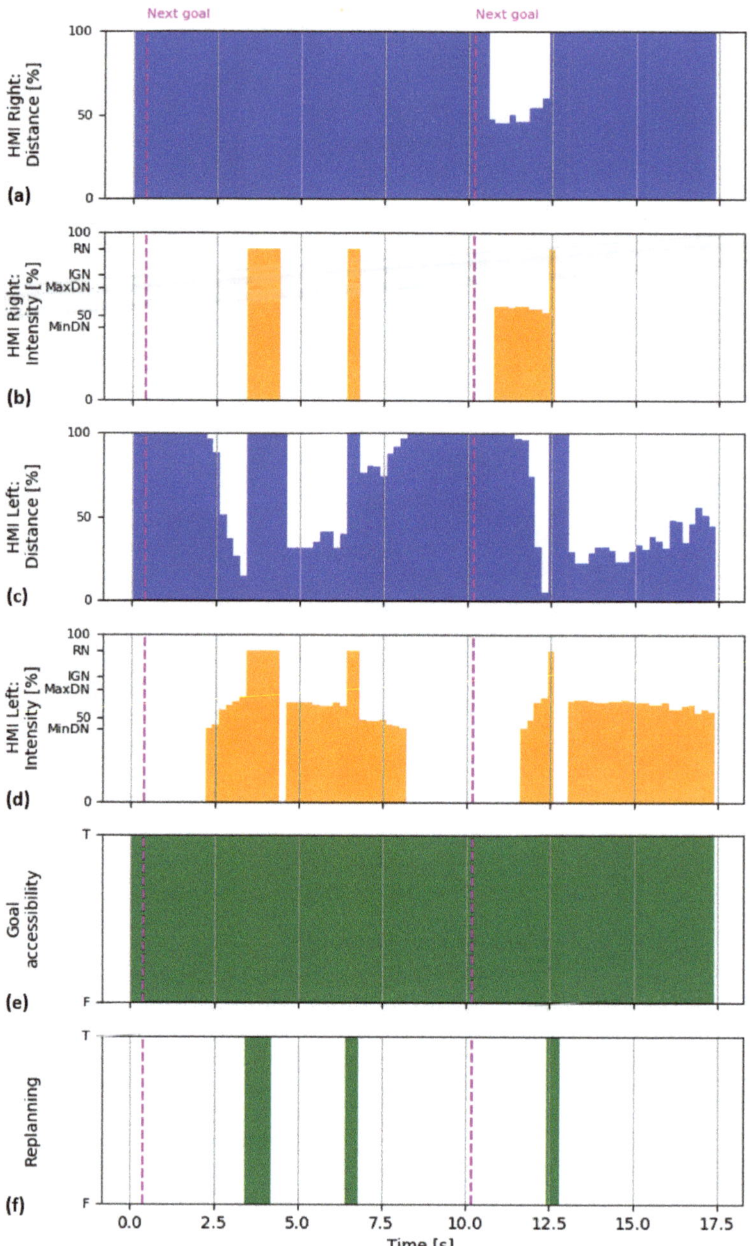

Figure 11. Timeline of the testing on the real workspace: (**a**,**c**) distances to HMIs displayed as a percentage of reaction distance; (**b**,**d**) momentary notification intensities displayed as a percentage from the maximum intensity, where RN—replanning notification, IGN—inaccessible goal notification, MaxDN—maximum distance notification, MinDN—minimum distance notification; (**e**) goal accessibility status where F—false, the planner was not able to find a feasible path to the goal); (**f**) status of replanning routine where T—true, path replanning was required; change of the goal is depicted by the purple dash line.

3.1. User Study

In order to verify the effectiveness of the developed HMI (Human-Machine Interface) in the conditions of a shared workspace, a user study was developed (see Figure 12). The conducted user study included a testing system among 16 participants.

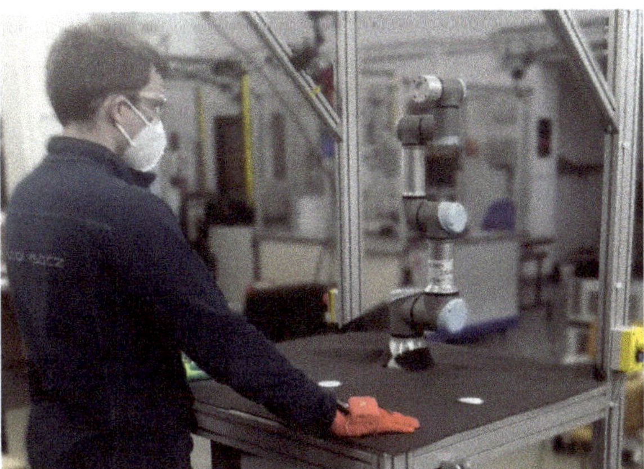

Figure 12. Performed user study.

Data collected during the experiment included both objective (measured parameters of each testing cycle) and subjective (survey-based) parameters. Data analysis further allowed us to compare the interfaces and evaluate the research hypotheses.

3.2. Experiment Description

In order to evaluate the usability of the developed interface, the following experiment assessed the responses of the group of volunteers (test subjects, test participants), whose goal was to accurately recognise the goal position of the robot during its movement. The quantitative parameters measured during trials (time required to recognise the goal position and the percentage of the successfully recognised goals) were statistically evaluated as objective parameters in order to decide the status of the research hypotheses. Each volunteer additionally filled in the task-specific questionary consisting of questions related to the usability of each interface and the perceived naturalness during the task execution. The experiment was conducted with 16 volunteered university members, which included five unexperienced participants with the background different from the field of robotics. During the experiment run no harm was done to the volunteers.

The experimental task was based on the idea of sorting parts into different containers, where it is not known in advance in which container the robot will have to place the next part. In each round of the experiment, the robot planned its trajectory from its starting position (this position did not change over the trials) to one of 5 possible TCP goal positions (see Figure 13). Each goal position was marked and numbered on the worktable.

During the experiment, each volunteer tested the following variations of completing the task:

- V1—Without HMI: the volunteer is not equipped with HMIs and has no feedback on approaching the robot's future trajectory. The volunteer must determine the robot's target position by visually examining the initial movement the robot makes to reach its goal.
- V2—Equipped HMIs: the volunteer is equipped with HMIs on both hands. The volunteer can use the HMI feedback (when moving hands) to determine the goal position of the robot. The volunteer is also instructed to watch their hands instead of watching the robot.

In both cases, the volunteers did not see a visualisation of the trajectory on the monitor.

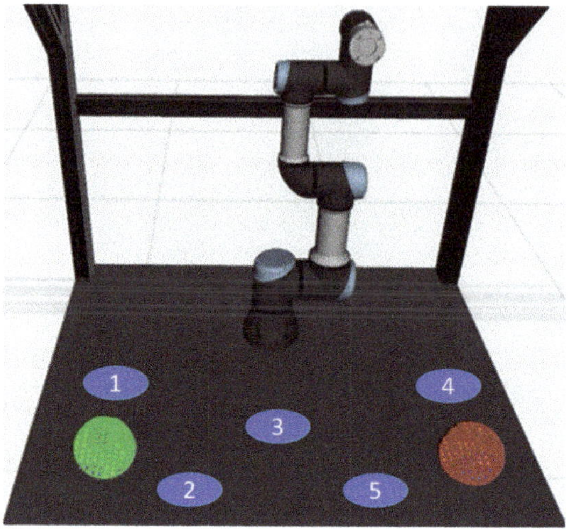

Figure 13. Preview of the workplace and the robot's goal positions (marked with numbers 1–5); hand positions are marked with red (**right hand**) and green (**left hand**); the robot is in the initial position.

The necessary safety precautions were taken during all the pilot experiments, and all the test subjects were informed about the potential risks and behaviour in safety-critical situations. Participation was not mandatory, and participants could leave any time they chose. All participants were required to review and sign a consent form and the definition of the experiment before beginning the experiment. After they agreed to participate, the experimenter clarified the ambiguities regarding the task and the principles of each interface. Additionally, before starting the experiment, each volunteer was allowed to observe the standard trajectory the robot takes while moving to all five predefined targets.

Each round started with a 3-2-1 countdown ensuring the volunteer knew the moment when the robots started to move. During movement to the goal position, the robot TCP speed was limited to the maximum value of 80 mm/s. The task of the volunteers was to determine the target position to which the robot is currently heading and, at the same time, avoid hindering the robot's future path. The following results were possible:

- If the number reported by the volunteer was correct, the task completion time was recorded, and the attempt was counted as successful.
- In case that the goal position number stated by the volunteer was incorrect, the final result of the attempt was recorded as unsuccessful, and the measured time was discarded.
- If the volunteer intervened in the planned trajectory of the robot (or collided with the robot itself), the attempt was counted as unsuccessful, and the measured time was discarded.
- Each attempt was limited in time by the duration of the robot's movement to the goal position. If the volunteer failed to determine the robot's goal by this time, the attempt was counted as unsuccessful, and the measured time was discarded.

The objective parameters (time, percentage of success) of each attempt were measured and evaluated. At the end of the round, the robot automatically returned to the starting position at high speed (150 mm/s). If the user had reported the guessed goal position before the robot reached it, the person responsible for carrying out the experiment might have interrupted the movement of the robot and move it back to the initial position to spare the time of the experiment.

Testing of each interface option was performed in five rounds, i.e., a total of 10 rounds for each volunteer. Before testing a specific interface, the volunteer had three trial rounds. The order of the tested interfaces (V1, V2) was selected at random for each volunteer to mitigate the order effect on the measured parameters. In each round, the robot started its movement from the same starting position (arm vertically straightened above the worktable, see Figure 13), and the goal position of the robot was selected randomly (i.e., the randomly selected goal positions may have repeated multiple times). Before the start of each round, the volunteer placed hands on the starting positions marked in Figure 13; these positions were selected so that the user in the starting position did not preclude the robot in any of the target positions. At the end of all rounds, the volunteer filled in the questionnaire.

3.3. Hypotheses

The initial hypotheses are based on the suggestion is that the developed feedback system should improve the awareness and comfort of the human operator while working close to the robot. The transparent robot behaviour should also lead to an increase in efficiency in the completion of the task. The dependent measures (objective dependent variables) were defined as task completion time and task success rate. The within-subjects independent variable was defined as a robot intent-communication interface:

- V1: Without HMI—baseline.
- V2: Equipped HMIs.

Overall, it is expected that the test subjects will perform better (lower task completion times, higher task completion rate) and will have higher subjective ratings when equipped with HMIs. The experimental hypotheses were defined as follows:

Hypothesis 1.1. (H1.1): *The efficiency of the test subjects will be greater with equipped HMIs than without HMIs.*

Higher efficiency is categorised as lower task completion time. This hypothesis is based on the suggestion that HMI contributes to task performance.

Hypothesis 1.2 (H1.2): *Time taken by each test subject to correctly determine the robot's goal position will be similar during all rounds when equipped with HMIs (V2). In contrast, there will be high variation in task completion times between the rounds when the determination will be based solely on the available visual information.*

To test this hypothesis, task completion time will be measured for each subject in each task condition, and the standard deviation of the measurements will be compared. This hypothesis is based on the suggestion that users' awareness enhanced by the haptic feedback is not influenced by the differences in the trajectories, whereas in the case of visual feedback, the awareness is dependent on the actual trajectory shape.

Hypothesis 2 (H2): *Volunteers will subjectively percept the task as simpler when performing the tasks while equipped with HMIs (V2) than by relying solely on the available visual information (V1).*

This hypothesis is based on the suggestion that the haptic feedback significantly contributes to the user awareness about the future trajectory of the robot, thus making the task cognitively easier.

Efficiency was defined as the time taken for the human subjects to complete the task and effectiveness as the percentage of successful task completion. Efficiency and effectiveness were evaluated objectively by measuring these parameters for each test subject during rounds of the experiment.

Apart from objective parameters, multiple subjective aspects of interacting with different types of interfaces were mapped. The analysis of subjective findings was based on responses to 17 questions. The main points of interest focused on the understanding of the robot's goals and motions, the feel of security when working closely with the robot, ergonomics, and the overall task difficulty. For both collaboration approaches, the participants were asked to indicate on a 1–7 Likert scale (scaling from 1—"totally disagree" to 5—"totally agree") the extent to which they agreed with the defined statements. Questionnaire items were inspired by works by R. Ganesan et al. [35], G. Bolano et al. [14], and A. Hietanen [21]; however, due to the differences in the tested interfaces, the questions have been significantly changed.

The first four questions (Q1–Q4, see Table 2) were task- and awareness-related and aimed to map the comparative aspects of the collaboration during the task execution with all tested interface variants V1, V2 and to test the general clarity of the provided instructions.

Table 2. General questions for all the tested interfaces.

General Questions
Q1. The task was clear for me
Q2. The task was demanding
Q3. It was simple to determine the goal position of the robot
Q4. More information was needed to accurately determine goal position of the robot

For HMI interface (V2) were additionally stated the questions defined in Table 3.

Table 3. HMI-related questions.

HMI-Related Questions
QH1. HMI improved my awareness of the robot's future trajectory
QH2. Work with HMI required long training
QH3. Work with HMI improved my confidence in safety during the task
QH4. Haptic feedback (vibration) from HMI was misleading
QH5. Haptic feedback (vibration) from HMI was too strong
QH6. Haptic feedback (vibration) from HMI was too weak
QH7. Haptic feedback (vibration) from HMI was sufficient
QH8. Haptic feedback (vibration) from HMI overwhelmed my perceptions
QH9. Use of HMI was inconvenient or caused unpleasant sensations during activation

The test subjects also could choose an interface of their own preference. The volunteer could additionally leave a free comment about any topic related to each interface option. To minimise the effect of bias (practice effect [36]) caused by the order in which participants interacted with each interface, the order was chosen randomly for each participant during the experiment.

4. Results

To determine whether the differences between objective measures in the conditions (V1, V2) were significant at the 95% confidence level, t-test was applied.

Hypothesis H1.1 states that the efficiency of the human-robot collaborative team will be higher in the case of the equipped HMIs (V2). The total time taken for completing the tasks was measured and compared between the conditions. t-tests revealed statistically significant difference (average improvement over 45%) in mean task completion times between V1 (M = 11.28, SE = 0.71) and V2 (M = 6.15, SE = 0.38) interfaces, $t(15) = 6.34$, $p < 0.00001$, Figure 14. Thus, hypothesis H1.1 was supported.

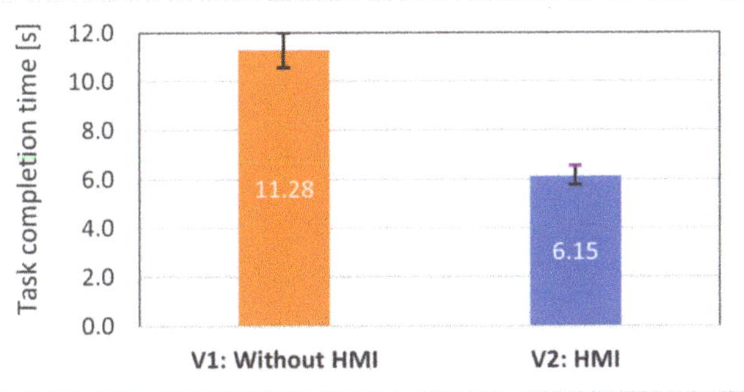

Figure 14. Average task completion time with standard errors for all 16 participants: lower is better.

Hypothesis H1.2 stated that the time taken by test subjects to correctly determine the goal position would be similar in all rounds when equipped with HMIs (V2). To investigate the hypothesis, the task completion time for all test subjects was analysed using the standard deviations. It was observed that the standard deviation for each subject for HMI modes (M = 2.17, SE = 0.37) was significantly lower than for visual inspection (V1: M = 3.30, SE = 0.29), implying that V2 allowed the participants to perform the task within a similar amount of time, whereas the time need with V1 interface was highly distinct in each round—see Figure 15. The deviations were compared using t-test, and statistically significant differences were found, $t(15) = 2.27$, $p < 0.019$. This suggests that the provided haptic feedback was intuitive and took approximately the same amount of time for the participants in each round to determine the goal position, thus supporting the H1.2 hypothesis.

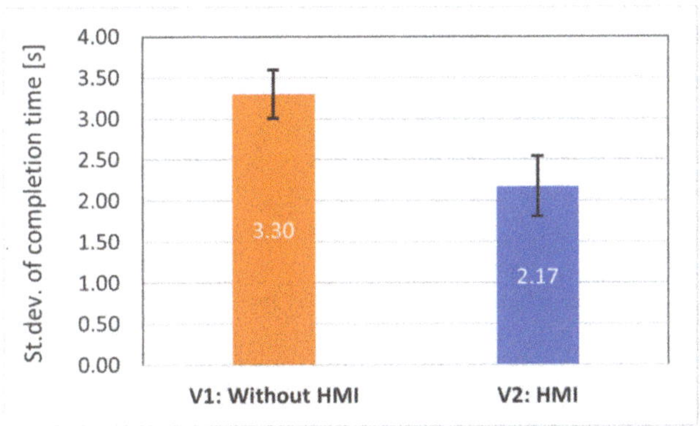

Figure 15. Standard deviations of the task completion time: lower is better.

It is also anticipated that the effectiveness of the V2 and the baseline will be comparable; however, proving this hypothesis (basically null-hypothesis) with a high confidence level requires a large number of test subjects. The average task success rates with standard errors are shown in Figure 16. One of the factors that may have led to a lower success rate with the V2 interface may be that the haptic feedback (V2) did not provide the test subjects with enough spatial information, which led to an ambiguous determination of the goal positions.

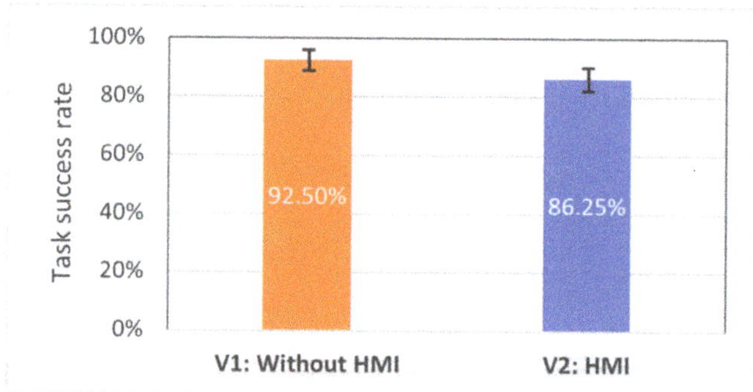

Figure 16. Average task success rate for all 16 participants: higher is better.

All 16 participants answered a total of 17 template questions, and the results were analysed. The first four questions (Q1–Q4) were common for both interfaces and intended to capture the differences in subjective usability between them. The average scores with the standard errors are shown in Figure 17. To determine whether the differences between ratings for each questionnaire item between test conditions (V1, V2) were significant at the 95% confidence level, *t*-test was applied.

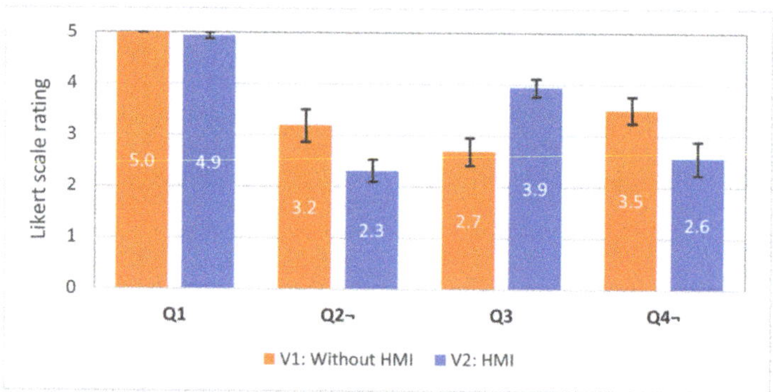

Figure 17. Average scores with standard errors for the questions Q1–Q4 used in the user studies (16 participants). Score 5 denotes "totally agree" and 1—"totally disagree". Questions Q1, Q3 —higher is better; Q2, Q4—lower is better.

Q1 was intended as an indicator of clarity of the task between the test subjects. The results show that the test subjects understood the provided instructions (there are no statistically significant differences in the responses in all three conditions). Q2 was related to the subjective perception of the task difficulty in each condition. The *t*-test revealed a statistically significant difference in responses between V1 (M = 3.19, SE = 0.32) and V2 (M = 2.31, SE = 0.22) interfaces, *t*(15) = 2.67, *p* < 0.01, the test subjects perceived the task as more simple while equipped with HMIs. Q3 intended to compare the subjectively perceived improvement in the awareness related to the robot's trajectory. *t*-test revealed a statistically significant difference in responses between V1 (M = 2.69, SE = 0.27) and V2 (M = 3.94, SE = 0.17) interfaces, *t*(15) = 3.873, *p* < 0.001, the participants subjectively perceived an improvement of their awareness about the robot trajectory when equipped with HMI. Q4 was designed to find out if the subjects subjectively felt the need for additional information in order to reliably recognise the trajectory of the robot. Statistically significant difference in responses was found between V1 (M = 3.50, SE = 0.26) and V2 (M = 2.56, SE = 0.33) interfaces, *t*(15) = 3.34, *p* < 0.01. Thus, hypothesis H2 was supported by the responses to Q2–Q3. The subjective findings are also supported by the previous evaluation of the objective performance of the test subjects.

The additional QH1–QH9 questions were intended to capture the subjective usability of the HMI (V2) interface. According to the high ranking (see Figure 18), it can be concluded that the users were satisfied with the interface and the provided feedback. There were few participants who reported that the vibration feedback was too strong and participants who reported feedback as too weak, thus leading to a conclusion that the optimal intensity of vibration feedback should be further investigated or should allow personal adjustments.

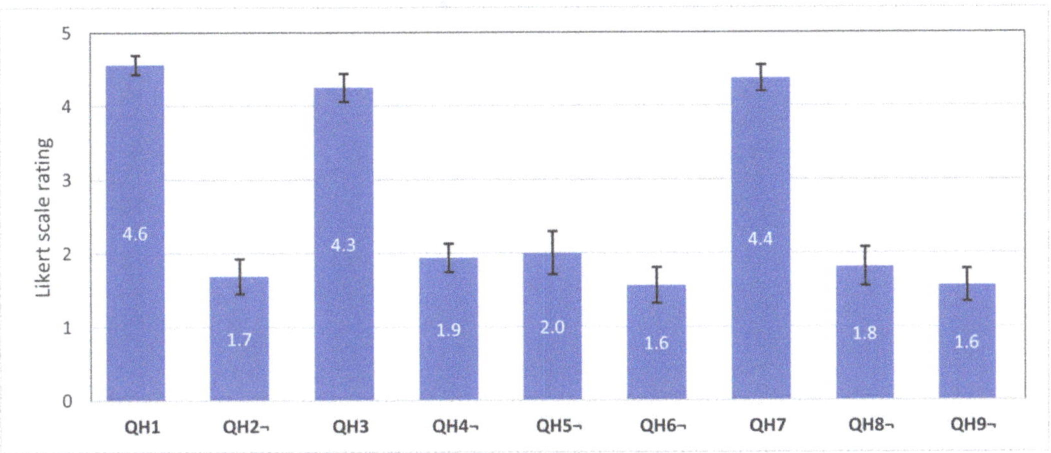

Figure 18. Average scores with standard errors for the questions QH1–QH9 used for evaluation of the HMI usability. Score 5 denotes "totally agree", and 1—"totally disagree". Questions QH1, QH3, QH7—higher is better; QH2, QH4, QH5, QH6, QH8, QH9—lower is better.

The users could also choose the one interface of their personal preference: all users favoured the HMI interface (V2). The user survey additionally offered to leave a free comment for each of the tested interfaces. Major themes mentioned in the comments included user perceptions of the haptic feedback. Some subjects noted that the vibration intensity was perceived differently in different places around the arm (for example, the vibration of the tractor +Z was perceived less). This suggests that additional research is needed to find the optimal arrangement of vibration motors. Participants also noted that it was hard for them to recognise the movement of the robot to certain positions, since, for example, in the case of goal 1 and 2, there sometimes was an ambiguity in the provided feedback, which lead to incorrectly reported number (which manifested in the task success rate). Multiple test subjects mentioned that the gloves' size was not optimal for their hands. This issue may be solved by implementing HMI in the form of modules that will be locked onto the universal work gloves.

5. Discussion

Evaluation of the objective parameters has shown significant evidence supporting hypotheses H1.1, H1.2. The participants took less time to recognise the goal position when equipped with HMIs. The time required by test subjects to successfully complete the task with HMIs equipped had low variation (unlike with visual feedback), potentially indicating that user awareness was independent of the shape of the robot movement.

Subjective findings from structured and free response questions supported hypothesis 2, which stated that participants would be more satisfied with the HMI interface compared to baseline. Overall, participants favoured the HMI with regard to human-robot fluency, clarity, and feedback.

Taken together, the results of the experiments indicate the usefulness of the developed system since it improves user's awareness about the motion plan of the robot. While not conclusive, these results indicate a potential of a haptic feedback-based approach in improving the interaction quality in human-robot collaboration.

In order to assess the developed system, we critically analyse its features and limitations. To the best of our knowledge, the presented system is the first implementation of haptic notifications, which provide the user with the information about the space that will be occupied by the robot during the movement execution while simultaneously allowing the robot to adapt the trajectory to avoid collisions with the worker. Previous studies have only demonstrated utilisation of the haptic feedback for informing the user about robot reaching some predefined positions [5,23]. The chosen modality (haptic feedback) has the advantage of more reliable data transfer compared to the graphical and acoustic feedback, whose efficiency can be limited or blocked during the engagement of the worker in the task. The utilisation of the haptic feedback as an information channel may also off-load the corresponding senses (vision and hearing). Yet, it is worth noting that in certain cases, the executed task activity (such as handling an electric screwdriver) may block the perception of the provided vibration alerts in the worker's hands. According to the results of the analysis of the subjective rating of the system, more investigation is needed in order to find the optimal placement of the vibration motors and vibration intensity levels, which will be considered equally sufficient by the majority of the users. This is necessary given that tactile sensitivity varies from person to person. However, this was beyond the scope of this study.

The HMI localisation system currently relies on data from a single depth camera, potentially enabling situations where the HMI would be occluded by an obstacle, leading to incorrect localisation. This, however, can be solved in the future by integrating data from multiple cameras into the HMI localisation pipeline. It should be emphasised that even if the hands of the user will be completely hidden from the upper camera, the robot will be able to avoid collision with the user since the obstacles in the workspace are mapped by all three cameras observing it from different perspectives. Since the overall reaction speed of the system is limited by 10 Hz updates of the workspace representation, swift movements of the user may remain undetected by the system; however, in the future, it can be addressed by distributing the process to multiple computation units and further optimising the system performance.

During the test, it was noted that there is a possibility to further improve the notification devices by enabling proportional activation of the tactors depending on the direction to the closest point of the robot's trajectory and HMI's orientation. This would allow avoiding ambiguity in the determination of the goal positions by providing the user with additional information. A similar solution was implemented in the work of M. Aggravi et al. [37], where they implemented a solution for guiding the hand of a human user using a vibrotactile haptic device placed on the user's forearm. Another example was implemented in the study of S. Scheggi et al. [38,39], in which a mobile robot had the task of steering a human (possibly sightless) from an initial to the desired target position through a cluttered corridor by only interacting with the human via HMI bracelet with three embedded vibration motors.

6. Conclusions

In this work, we propose an approach to improve human-robot collaboration. The system is based on the concept of a shared collaborative workspace where the robot can adapt its movement to avoid collision with human workers. The workspace is monitored by multiple RGB-D sensors, and data provided by these sensors allow the construction of a map of the robot's surroundings and obstacles. At each step of the task execution, the robot creates a collision-free motion plan according to the currently available free space. If during the execution of the planned movement there is a change in the environment and the movement can no longer be completed due to possible collisions with obstacles, the robot can create a new motion plan. The improvement of the collaboration is presented by introducing haptic feedback devices, whose task is to notify the human worker about the currently planned robot's trajectory and changes in its status. The human worker is equipped with two haptic feedback devices placed at each hand. These feedback devices provide continuous vibration alerts to the user about the proximity to the currently planned trajectory of the robot: the closer the worker's hand (equipped with the feedback device) approaches the future segment of the trajectory, the stronger the vibration provided by the device. A prototype of the proposed notification system was implemented and tested on an experimental collaborative workspace.

An experimental study was performed to evaluate the effect of the developed haptic feedback system for improving human awareness during HRC. The experiment data analysis included both quantitative (objective and subjective) and qualitative (subjective) findings from the experiment. Haptic HMI was found superior to the baseline and allowed the test subjects to complete the tasks faster.

Future research will focus on the improvement of the developed notification system to allow more information to be transferred to the user. The described concept of the haptic distance notifications may communicate not only the distance to the robot's trajectory but also the direction of the nearest point of the trajectory (relatively to the corresponding HMI). By organising the vibration motors into a spatial structure resembling an orthogonal coordinate system, the vibration of each motor may alert the user with the direction of the possible impact. This improved approach additionally requires measuring the relative orientation of each HMI, which can be implemented by incorporating IMU sensors into the design of HMIs. The HMI localisation system may be further improved by combining and processing the data from all the available cameras in the workplace, which would ensure that the HMI can be localised even if it is occluded or is outside the field of view of multiple cameras as long as it is visible to one of them. More types of notifications may represent more states of the robot and the surrounding machines, allowing the user to quickly react in a situation when human intervention is needed. However, due to the computational and overall complexity of the system and the supporting device infrastructure, industrial implementation of the system would require a standalone processing unit in addition to the robot controller. In general, the usability of the system can be extended for use in large production line conditions, including several robots and automatic production systems. As a human worker moves along the production line to various machines and multiple robots, the system can provide notifications of any danger that is at a reaction distance from the human.

Author Contributions: The initial idea, conceptualisation, S.G.; methodology, S.G., A.V., Z.B. and P.N.; software, S.G., P.O. and P.N.; validation, M.V. and P.O.; formal analysis, M.V. and P.N.; investigation, M.V. and P.O.; resources, P.O.; data curation, Z.B. and P.N.; writing—original draft preparation, S.G. and A.V.; writing—review and editing, A.V., M.V. and P.O.; visualisation, M.V. and P.O.; supervision, Z.B. and P.N.; project administration, Z.B. and P.N.; funding acquisition, Z.B. and P.N. All authors have read and agreed to the published version of the manuscript.

Funding: This work was supported by the Research Platform focused on Industry 4.0 and Robotics in Ostrava Agglomeration project, project number CZ.02.1.01/0.0/0.0/17_049/0008425 within the Operational Program Research, Development and Education. This article has also been supported by specific research project SP2021/47 and financed by the state budget of the Czech Republic.

Institutional Review Board Statement: All subjects gave their informed consent for inclusion before they participated in the study. The study was conducted according to the guidelines of the Declaration of Helsinki and approved by the Ethics Committee of VSB—Technical University of Ostrava (on 3 May 2021).

Informed Consent Statement: Informed consent was obtained from all subjects involved in the study.

Data Availability Statement: The data presented in this study are available on request from the corresponding author. The data are not publicly available due to project restrictions.

Conflicts of Interest: The authors declare no conflict of interest.

References

1. Vysocky, A.; Novak, P. Human-robot collaboration in industry. *MM Sci. J.* **2016**, *2016*, 903–906. [CrossRef]
2. Wang, L.; Gao, R.; Váncza, J.; Krüger, J.; Wang, X.V.; Makris, S.; Chryssolouris, G. Symbiotic Human-robot collaborative assembly. *CIRP Ann.* **2019**, *68*, 701–726. [CrossRef]
3. Villani, V.; Pini, F.; Leali, F.; Secchi, C. Survey on human–robot collaboration in industrial settings: Safety, intuitive interfaces and applications. *Mechatronics* **2018**, *55*, 248–266. [CrossRef]
4. Mohammadi Amin, F.; Rezayati, M.; van de Venn, H.W.; Karimpour, H. A mixed-perception approach for safe human–robot collaboration in industrial automation. *Sensors* **2020**, *20*, 6347. [CrossRef] [PubMed]
5. Casalino, A.; Messeri, C.; Pozzi, M.; Zanchettin, A.M.; Rocco, P.; Prattichizzo, D. Operator awareness in human–robot collaboration through wearable vibrotactile feedback. *IEEE Robot. Autom. Lett.* **2018**, *3*, 4289–4296. [CrossRef]
6. Bonci, A.; Cen Cheng, P.D.; Indri, M.; Nabissi, G.; Sibona, F. Human-robot perception in industrial environments: A survey. *Sensors* **2021**, *21*, 1571. [CrossRef] [PubMed]
7. Tang, K.-H.; Ho, C.-F.; Mehlich, J.; Chen, S.-T. Assessment of handover prediction models in estimation of cycle times for manual assembly tasks in a human–robot collaborative environment. *Appl. Sci.* **2020**, *10*, 556. [CrossRef]
8. Mainprice, J.; Berenson, D. Human-robot collaborative manipulation planning using early prediction of human motion. In Proceedings of the 2013 IEEE/RSJ International Conference on Intelligent Robots and Systems, Tokyo, Japan, 3–7 November 2013; pp. 299–306.
9. Hermann, A.; Mauch, F.; Fischnaller, K.; Klemm, S.; Roennau, A.; Dillmann, R. Anticipate your surroundings: Predictive collision detection between dynamic obstacles and planned robot trajectories on the GPU. In Proceedings of the 2015 European Conference on Mobile Robots (ECMR), Lincoln, UK, 2–4 September 2015; pp. 1–8.
10. Li, G.; Liu, Z.; Cai, L.; Yan, J. Standing-posture recognition in human–robot collaboration based on deep learning and the dempster–shafer evidence theory. *Sensors* **2020**, *20*, 1158. [CrossRef]
11. Feleke, A.G.; Bi, L.; Fei, W. EMG-based 3D hand motor intention prediction for information transfer from human to robot. *Sensors* **2021**, *21*, 1316. [CrossRef]
12. Scimmi, L.S.; Melchiorre, M.; Troise, M.; Mauro, S.; Pastorelli, S. A practical and effective layout for a safe human-robot collaborative assembly task. *Appl. Sci.* **2021**, *11*, 1763. [CrossRef]
13. Mišeikis, J.; Glette, K.; Elle, O.J.; Torresen, J. Multi 3D camera mapping for predictive and reflexive robot manipulator trajectory estimation. In Proceedings of the 2016 IEEE Symposium Series on Computational Intelligence (SSCI), Athens, Greece, 6–9 December 2016; pp. 1–8.
14. Bolano, G.; Roennau, A.; Dillmann, R. Transparent robot behavior by adding intuitive visual and acoustic feedback to motion replanning. In Proceedings of the 2018 27th IEEE International Symposium on Robot and Human Interactive Communication (RO-MAN), Nanjing, China, 27–31 August 2018; pp. 1075–1080.
15. Tsarouchi, P.; Makris, S.; Chryssolouris, G. Human–Robot Interaction review and challenges on task planning and programming. *Int. J. Comput. Integr. Manuf.* **2016**, *29*, 916–931. [CrossRef]
16. Lee, W.; Park, C.H.; Jang, S.; Cho, H.-K. Design of effective robotic gaze-based social cueing for users in task-oriented situations: How to overcome in-attentional blindness? *Appl. Sci.* **2020**, *10*, 5413. [CrossRef]
17. Kalpagam Ganesan, R.; Rathore, Y.K.; Ross, H.M.; Ben Amor, H. Better teaming through visual cues: how projecting imagery in a workspace can improve human-robot collaboration. *IEEE Robot. Autom. Mag.* **2018**, *25*, 59–71. [CrossRef]
18. Andersen, R.S.; Madsen, O.; Moeslund, T.B.; Amor, H.B. Projecting robot intentions into human environments. In Proceedings of the 2016 25th IEEE International Symposium on Robot and Human Interactive Communication (RO-MAN), New York, NY, USA, 26–31 August 2016; pp. 294–301.
19. Bambušek, D.; Materna, Z.; Kapinus, M.; Beran, V.; Smrž, P. Combining interactive spatial augmented reality with head-mounted display for end-user collaborative robot programming. In Proceedings of the 2019 28th IEEE International Conference on Robot and Human Interactive Communication (RO-MAN), New Delhi, India, 14–18 October 2019; pp. 1–8.
20. Fang, H.C.; Ong, S.K.; Nee, A.Y.C. A novel augmented reality-based interface for robot path planning. *Int. J. Interact. Des. Manuf.* **2014**, *8*, 33–42. [CrossRef]
21. Hietanen, A.; Pieters, R.; Lanz, M.; Latokartano, J.; Kämäräinen, J.-K. AR-based interaction for human-robot collaborative manufacturing. *Robot. Comput. Integr. Manuf.* **2020**, *63*, 101891. [CrossRef]
22. Clair, A.S.; Matarić, M. How robot verbal feedback can improve team performance in human-robot task collaborations. In Proceedings of the 2015 10th ACM/IEEE International Conference on Human-Robot Interaction (HRI), Portland, OR, USA, 2–5 March 2015; pp. 213–220.
23. Barros, P.G.; de Lindeman, R.W.; Ward, M.O. Enhancing robot teleoperator situation awareness and performance using vibro-tactile and graphical feedback. In Proceedings of the 2011 IEEE Symposium on 3D User Interfaces (3DUI), Singapore, 19–20 March 2011; pp. 47–54.
24. Li, H.; Sarter, N.B.; Sebok, A.; Wickens, C.D. The design and evaluation of visual and tactile warnings in support of space teleoperation. *Proc. Hum. Factors Ergon. Soc. Annu. Meet.* **2012**. [CrossRef]

25. Sziebig, G.; Korondi, P. Remote operation and assistance in human robot interactions with vibrotactile feedback. In Proceedings of the 2017 IEEE 26th International Symposium on Industrial Electronics (ISIE), Edinburgh, UK, 19–21 June 2017; pp. 1753–1758.
26. Lasota, P.; Shah, J. Analyzing the effects of human-aware motion planning on close-proximity human-robot collaboration. *Hum. Factors J. Hum. Factors Ergon. Soc.* **2015**, *57*, 21–33. [CrossRef]
27. Unhelkar, V.V.; Lasota, P.A.; Tyroller, Q.; Buhai, R.-D.; Marceau, L.; Deml, B.; Shah, J.A. Human-aware robotic assistant for collaborative assembly: Integrating human motion prediction with planning in time. *IEEE Robot. Autom. Lett.* **2018**, *3*, 2394–2401. [CrossRef]
28. Fu, L.; Duan, J.; Zou, X.; Lin, G.; Song, S.; Ji, B.; Yang, Z. Banana detection based on color and texture features in the natural environment. *Comput. Electron. Agric.* **2019**, *167*, 105057. [CrossRef]
29. Song, Z.; Fu, L.; Wu, J.; Liu, Z.; Li, R.; Cui, Y. Kiwifruit detection in field images using faster R-CNN with VGG16. *IFAC Pap. OnLine* **2019**, *52*, 76–81. [CrossRef]
30. Lim, G.M.; Jatesiktat, P.; Keong Kuah, C.W.; Tech Ang, W. Hand and object segmentation from depth image using fully convolutional network. In Proceedings of the 2019 41st Annual International Conference of the IEEE Engineering in Medicine and Biology Society (EMBC), Berlin, Germany, 23–27 July 2019; pp. 2082–2086.
31. Tang, Y.; Chen, M.; Wang, C.; Luo, L.; Li, J.; Lian, G.; Zou, X. Recognition and localization methods for vision-based fruit picking robots: A review. *Front. Plant Sci.* **2020**, *11*. [CrossRef] [PubMed]
32. Chen, M.; Tang, Y.; Zou, X.; Huang, K.; Li, L.; He, Y. High-accuracy multi-camera reconstruction enhanced by adaptive point cloud correction algorithm. *Opt. Lasers Eng.* **2019**, *122*, 170–183. [CrossRef]
33. Ioan Sucan, S.C. MoveIt! Available online: http://moveit.ros.org (accessed on 4 March 2021).
34. Pan, J.; Chitta, S.; Manocha, D. FCL: A general purpose library for collision and proximity queries. In Proceedings of the 2012 IEEE International Conference on Robotics and Automation, Saint Paul, MN, USA, 14–18 May 2012; pp. 3859–3866.
35. Ganesan, R.K. Mediating Human-Robot Collaboration Through Mixed Reality Cues. Available online: https://www.semanticscholar.org/paper/Mediating-Human-Robot-Collaboration-through-Mixed-Ganesan/de797205f4359044639071fa8935cd23aa3fa5c9 (accessed on 2 March 2021).
36. Practice Effect—APA Dictionary of Psychology. Available online: https://dictionary.apa.org/practice-effect (accessed on 4 March 2021).
37. Aggravi, M.; Salvietti, G.; Prattichizzo, D. Haptic wrist guidance using vibrations for human-robot teams. In Proceedings of the 2016 25th IEEE International Symposium on Robot and Human Interactive Communication (RO-MAN), New York, NY, USA, 26–31 August 2016; pp. 113–118.
38. Scheggi, S.; Chinello, F.; Prattichizzo, D. Vibrotactile Haptic Feedback for Human-Robot. Interaction in Leader-Follower Tasks. In *ACM PETRA '12: Proceedings of the 5th International Conference on PErvasive Technologies Related to Assistive Environments, Heraklion, Crete, Greece, 6–8 June 2012*; pp. 1–4.
39. Scheggi, S.; Aggravi, M.; Morbidi, F.; Prattichizzo, D. Cooperative Human-Robot. Haptic Navigation. In Proceedings of the 2014 IEEE International Conference on Robotics and Automation (ICRA), Hong Kong, China, 31 May–7 June 2014; pp. 2693–2698.

Article

A Mixed-Perception Approach for Safe Human–Robot Collaboration in Industrial Automation

Fatemeh Mohammadi Amin [1], **Maryam Rezayati** [1], **Hans Wernher van de Venn** [1,*] and **Hossein Karimpour** [2]

[1] Institute of Mechatronics System, Zurich University of Applied Science, 8400 Winterthur, Switzerland; mohm@zhaw.ch (F.M.A.); rzma@zhaw.ch (M.R.)

[2] Mechanical Engineering Department, University of Isfahan, Isfahan 81746-73441, Iran; h.karimpour@eng.ui.ac.ir

* Correspondence: vhns@zhaw.ch; Tel.: +41-58-934-77-89

Received: 4 September 2020; Accepted: 2 November 2020; Published: 7 November 2020

Abstract: Digital-enabled manufacturing systems require a high level of automation for fast and low-cost production but should also present flexibility and adaptiveness to varying and dynamic conditions in their environment, including the presence of human beings; however, this presence of workers in the shared workspace with robots decreases the productivity, as the robot is not aware about the human position and intention, which leads to concerns about human safety. This issue is addressed in this work by designing a reliable safety monitoring system for collaborative robots (cobots). The main idea here is to significantly enhance safety using a combination of recognition of human actions using visual perception and at the same time interpreting physical human–robot contact by tactile perception. Two datasets containing contact and vision data are collected by using different volunteers. The action recognition system classifies human actions using the skeleton representation of the latter when entering the shared workspace and the contact detection system distinguishes between intentional and incidental interactions if physical contact between human and cobot takes place. Two different deep learning networks are used for human action recognition and contact detection, which in combination, are expected to lead to the enhancement of human safety and an increase in the level of cobot perception about human intentions. The results show a promising path for future AI-driven solutions in safe and productive human–robot collaboration (HRC) in industrial automation.

Keywords: safe physical human–robot collaboration; collision detection; human action recognition; artificial intelligence; industrial automation

---

1. Introduction

As the manufacturing industry evolves from rigid conventional procedures of production to a much more flexible and intelligent way of automation within the frame of the Industry 4.0 paradigm, human–robot collaboration (HRC) has gained rising attention [1,2]. To increase manufacturing flexibility, the present industrial need is to develop a new generation of robots that are able to interact with humans and support operators by leveraging tasks in terms of cognitive skills requirements [1]. Consequently, the robot becomes a companion or so-called collaborative robot (cobot) for flexible task accomplishment rather than a preprogrammed slave for repetitive, rigid automation. It is expected that cobots actively assist operators in performing complex tasks, with highest priority on human safety in cases humans and cobots need to physically cooperate and/or share their workspace [3]. This is problematic because the current settings of cobots do not provide an adequate perception of human presence in the shared workspace. Although there are some safety monitoring systems [4–7], they can only provide a real or virtual fence for the cobot to stop or slow down when an object, including a human being, enters the defined safety zone. However, this reduces productivity in two ways as follows:

1. It is not possible to differentiate between people and other objects that enter the workspace of the cobot. Therefore, the speed is always reduced regardless of the object.
2. It is also not possible to differentiate whether an interaction with the robot should really take place or not; this also always forces a maximum reduction in speed.

This issue can only be tackled by implementing a cascaded, multi-objective safety system, which primarily recognizes human actions and detects human–robot contact [8] to percept human intention in order to avoid collisions. Therefore, the primary goal of this work is to conduct a step-change in safety for HRC in enhancing the perception of cobots by providing visual and tactile feedback to the robot from which it is able to interpret the human intention. The task is divided into two parts, human action recognition (HAR) using visual perception and contact type detection using tactile perception, which will be subsequently investigated. Finally, by combining these subsystems, it is considered to attain a more reliable and intelligent safety system, which takes advantage of considerably enhanced robot perceptional abilities.

1.1. Human Action Recognition (HAR)

Based on the existing safety regulation related to HRC applications and by inspiring from human perception and cognition ability in different situations, adding the visual perception to the cobot can enhance HRC performance. Nevertheless, the main challenge is how cobots are able to adapt to human behavior. HAR as part of visual perception plays a crucial role in overcoming this challenge and increasing productivity and safety. HAR can be used to allow the cobot keeping a safe distance with its human counterpart or the environment, ensuring an essential requirement for fulfilling safety in shared workspaces. Recent studies have been concentrated on visual and non-visual perception systems to recognize human actions [9]. One method amongst non-visual approaches consists of using wearable devices [10–15]. Nevertheless, applying this technology as a possible solution for an industrial situation seems at present neither feasible nor comfortable in industrial environments because of restrictions that it will impose on the operator's movements. On the other hand, active vision-based systems are widely used in such applications for recognizing human gestures and actions. In general, vision-based HAR approaches consist of two main steps: proper human detection and action classification.

As alluded by recent researches, machine learning methods are essential in recognizing human actions and interpreting them. Some traditional machine learning methods such as support vector machine (SVM) [16–19], hidden Markov model (HMM) [20,21], neural networks [22], and Gaussian mixture models (GMM) [23,24] have been used for human action detection with a reported accuracy of about 70 to 90 percent. On the other hand, deep learning (DL) algorithms prevail as a new generation of machine learning algorithms with significant capabilities in discovering and learning complex underlying patterns from a large amount of data [25]. This algorithm provides a new approach to improve the recognition accuracy of human actions by using depth data provided by time-of-flight, depth, or stereo cameras, extracting human location and skeleton pose. DL researchers either use video stream data [26–28], RGB-D images [29–31], or 3D skeleton tracking and joints extraction [32–35] for classification of arbitrary actions. Among different types of deep networks, convolutional neural networks (CNN) stand for a popular approach, which can be represented as 2D or 3D network in action recognition but still needs a large set of labeled data for training and contains many layers. The first 3D-CNN for HAR has been introduced by [36–38] providing an average accuracy of 91 percent. Recent researches based on 3D-CNN techniques [39–42] have obtained a high performance on the KTH dataset [43] in comparison to 2D-CNN networks [44–47]. Yet, the maximum accuracy of this research is reported to be at 98.5 percent but is not capable of classifying in real-time. In addition, most of these articles mainly focus on action classification based on domestic scenarios, only few have an approach for industrial scenarios [19,48,49] and a restricted number works on unsupervised human activities in the presence of mobile robots [50,51]. Thus, there is a need to introduce a fast and more precise network for HAR in industrial applications, which can be presented as a new 3D network architecture by considering an outperforming result in action classification.

In this work, we use a deep learning approach for real-time human action recognition in an industrial automation scenario. A convolutional analysis is applied on RGB images of the scene in order to model the human motion over the frames by skeleton-based action recognition. The artificial-intelligence-based human action recognition algorithm provides the core part for distinguishing between collision and intentional contact.

1.2. Contact Type Detection

In more and more HRC applications, there is a need of having direct, physical collaboration between human and cobot, physical human–robot collaboration (pHRC) due to an unmatched degree of flexibility in the execution of various tasks. Indeed, when a cobot is performing its task, it should be aware of its contact with the human. In addition, from a cobot's point of view, the type of this contact is not immediately obvious, due to the fact that the cobot cannot distinguish whether a human gets in contact with the robot incidentally or intentionally, when a collaborative task is executed. Therefore, it is important that the cobot needs to percept human contact with deeper understanding. Towards this goal, it is imperative firstly, to detect the human–robot contact and secondly, distinguish between intentional and incidental contacts, a process called collision detection. Some researchers propose sensor-less procedures for detecting a collision based on the robot dynamics model [52,53], but through momentum observers [52,54–57], using extended state observers [58], vibration analysis models [59], finite-time disturbance observers [56], energy observers [57], or joint velocity observers [60]. Among these methods, the momentum observer is the most common method of collision detection, because it provides better performance compared to the other methods, although the disadvantage is that it requires precise dynamic parameters of the robot [61]. For this reason, machine learning approaches such as artificial neural networks [62–64] and deep learning [65,66] have recently been applied for collision detection based on robot sensors' stream data due to their performance in modeling the uncertain systems with lower analytics effort.

Deep neural networks are extremely effective in feature extraction and learning complex patterns [67]. Recurrent neural networks (RNN) such as long short-term memory network approaches (LSTM) are frequently used in research for processing time series and sequential data [68–71]. However, the main drawback of this network is the difficulty and time consumption for training in comparison to convolutional neural networks (CNN) [65]. Additionally, current researches showed that CNN has a great performance for image processing in real time situations [26,65,72–74], where the input data are much more complicated than 1D time series signals. As proposed in [65], a 1D-CNN, named CollisionNet, has a proper potential in detecting collision, although only incidental contacts have been considered. Moreover, depending on whether the contact is intentional or incidental, the cobot should provide an adequate response, which in every case, ensures the safety of the human operator. At this step, identifying the cobot link where the collision occurred is important information for anticipating proper robot reaction, which needs to be considered in contact perception [61].

With this background and considering the fact that contact properties' patterns of incidental and intentional states are different according to the contacted link, we aim to use supervised learning, convolutional neural network, to have a model-free contact detection. Indeed, not only does the proposed system detect the contact, which in other papers [61,65,75,76] is named collision detection, it can also recognize the types of contact, incidental or intentional, provide information about a contacted link and consequently increase the robot awareness and perception about human intention during physical contact.

2. Material and Methods

2.1. Mixed Perception Terminology and Design

We hypothesize that combining two types of perception, visual and tactile, in a mixed perception approach can enhance the safety of human during collaborating with a robot by additional information to the robot's perception spectrum. It is easy to imagine that a robot then would be able to see and feel a human in its immediate vicinity at the same time. Using visual perception, a robot can notice:

1. A human entering the shared workspace (Passing)
2. A human observing its tasks when he/she is near to the robot and wants to supervise the robot task (Observation)
3. A dangerous situation when the human is not in a proper situation to do collaboration or observation, which can threaten human safety (Dangerous Observation)
4. A human interaction when the human is close to robot and doing the collaboration (Interaction).

However, it is difficult using a pure vision-based approach to distinguish between dangerous observation and interaction and to differentiate between incidental and intended contact types not only for a machine but also for a human. Therefore, at this stage, considering both types of perception, vision and haptics, is of significance. As indicated above, this approach is able to increase the safety and can be like a supervisory unit to the vision part as the latter can fail due to occlusion effects.

To support our hypothesis, we first choose the approach of developing two separate networks for human action and contact recognition, which meet the requirements for human–robot collaboration and real-time capability. These networks will be examined and discussed with regard to their appropriateness and their results. As a first step, we want to determine in this paper whether a logical correlation of the outputs of the two networks is theoretically able to provide a reasonable expansion of the perception spectrum of a robot for human–robot collaboration. We want to find out what the additional information content is and how it can specifically be used to further increase the safety and with that possibly also additional performance parameters of HRC solutions such as short cycle time, low downtime, high efficiency, and high productivity. The concrete merging of the two networks in a common application represents an additional stage of our investigations, which is not a subject of this work. The results of the present investigation, however, shall provide evidence that the use of AI in robotics is able to open up significant new possibilities and enables robots to achieve their operational objectives in close cooperation with humans. Enhanced perceptional abilities of robots are future key features to shift the existing technological limits and open up new fields of application in industry and beyond.

2.2. Robotic Platform

The accessible platform used throughout this project is a Franka Emika robot (Panda), recognized as a suitable collaborative robot in terms of agility and contact sensitivity. The key features of this robot will be summarized hereafter; it consists of two main parts, arm and hand. The arm has 7 revolute joints and precise torque sensors (13 bits resolution) at every joint, is driven by high efficiency brushless dc motors, and has the possibility to be controlled by external or internal torque controllers at a 1 kHz frequency. The hand is equipped with a gripper, which can securely grasp objects due to a force controller. Generally, the robot has a total weight of approximately 18 kg and can handle payloads up to 3 kg.

2.3. Camera Systems

The vision system is based on a multi-sensor approach using two Kinect V2 cameras for monitoring the environment to tackle the risk of occlusion. The Kinect V2 has a depth camera with resolution of 512×424 pixels with a field of view (FoV) of $70.6° \times 60°$, and the color camera has a resolution of

1920×1080 px with a FoV of $84.1° \times 53.8°$. Therefore, this sensor as one of the RGB-D cameras can be used for human body and skeleton detection.

2.4. Standard Robot Collision Detection

A common collision detection approach is defined as Equation (1) [61].

$$cd(\mu(t)) = \begin{cases} TRUE, \ if \left|\mu(t)\right| > \epsilon_\mu \\ FALSE, \ if \left|\mu(t)\right| \leq \epsilon_\mu \end{cases} \tag{1}$$

where cd is the collision detection output function, which maps the selected monitoring signal $\mu(t)$ such as external torque into a collision state, true or false. ϵ_μ indicates a threshold parameter, which determines the sensitivity for detecting a collision.

2.5. Deep Learning Approach

A convolutional neural network (CNN) model performs classification in an end-to-end manner and learns data patterns automatically, which is different to the traditional approaches where the classification is done after feature extraction and selection [77]. In this paper, a combination of 3D-CNN for HAR and 1D-CNN for contact type detection has been utilized. The following subsections describe each network separately.

2.5.1. Human Action Recognition Network

Since human actions can be interpreted by analyzing the sequence of human body movements involving arms and legs and placing them in a situational context, the consecutive skeleton images are used as inputs for our 3D-CNN network, which was successfully applied for real-time action recognition. In this section, the 3D-CNN, which is shown in Figure 1, classifies HAR to five states, namely: Passing, Observation, Dangerous Observation, Interaction, and Fail. These categories are based on the most feasible situations which may occur during human–robot collaboration:

1. Passing: a human operator occasionally needs to enter the robot's workspace, which is specified due to the fix position of the robot but without any intention to actively intervene the task execution of the robot.
2. Interaction: a human operator wants to actively intervene the robot's task execution, which can be the case due to correct a Tool Center Point (TCP) path or to help the robot if it gets stuck.
3. Observation: the robot is working on its own and a human operator is about to observe and check the working process from within the robot's workspace.
4. Dangerous Observation: the robot is working on its own and a human operator is about to observe the working process. Due to the proximity of exposed body parts (head and upper extremities) to the robot in the shared workspace, there is a high potential of life-threatening injury in case of a collision.
5. Fail: one or all system cameras are not able to detect the human operator in the workspace due to occlusion by the robot itself or other artefacts in the working area.

The input layer has 4 dimensions, $N_{channel} \times N_{image\text{-}height} \times N_{image\text{-}width} \times N_{frame}$. The RGB image of Kinect V2 has a resolution of 1980×1080 pixels which is decreased to 320×180 for reducing the trainable parameters and network complexity. Therefore, $N_{channel}$, $N_{image\text{-}height}$, and $N_{image\text{-}width}$ are 3, 180, and 320, respectively. N_{frame} indicates the total number of frames in the image sequence, which is 3 in this research.

As shown in Figure 1, the proposed CNN is composed of fifteen layers, consisting of four convolutional layers, four pooling layers, three fully connected layers followed by three dropout layers and a SoftMax layer for predicting actions. Convolutional layers utilized for feature extraction by applying filters and pooling layers are specifically used to reduce the dimensionality of the input. This layer performs based on the specific function, such as max pooling or average pooling, which extracts the maximum or medium value in a particular region. Fully connected layers are like a neural network for learning non-linear features as represented by the output of convolutional layers. In addition, dropout layers as a regularization layer try to remove overfitting in the network. Over 10 million parameters have to be trained for establishing a map to action recognition.

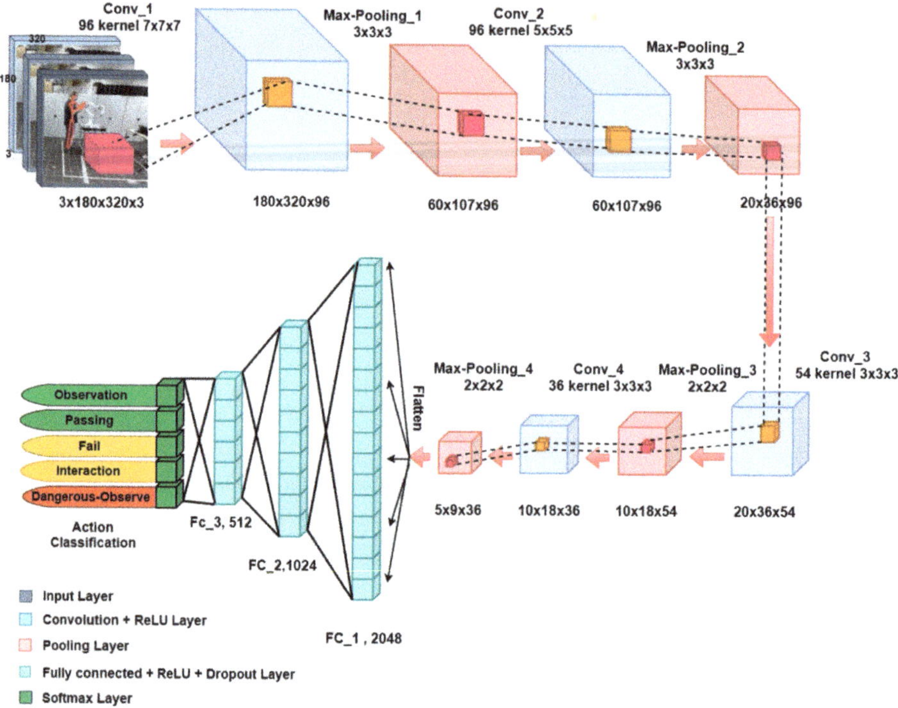

Figure 1. Three-dimensional convolutional neural networks (CNN) for human action recognition.

The input layer is followed by a convolution layer with 96 feature maps of size 7^3. Subsequently, the output is fed to the rectified linear unit (ReLU) activation function. ReLU is the most suitable activation function for this work, as it is specifically designed for image processing, and it can keep the most important features of the input. In addition, it is easier to train and usually achieves better performance, which is significant for real-time applications. The next layer is a max-pooling layer with size and stride of 3. The filter size of the next convolutional layers decreases to 5^3 and 3^3, respectively, with stride 1 and zero padding. Then, max-pooling windows decline to 2^3 with stride of 2. The output of the last pooling layer is flattened out for the fully connected layer input. The fully connected layers consist of 2024, 1024, 512 neurons, respectively. The last step is to use a SoftMax level for activity recognition.

2.5.2. Contact Detection Network

For contact detection, a deep network, which is shown in Figure 2, is proposed. In this scheme, a 1D-CNN, which is a multi-layered architecture with each layer consisting of few one-dimensional convolution filters, is used. In this research, just two links of the robot which are more likely to be used as contact points during physical human–robot collaboration, considered which indeed does not influence the general approach used in this paper. Therefore, it includes one network for classification of 5 states, which were defined as:

1. No-Contact: no contact is detected within the specified sensitivity.
2. Intentional_Link5: an intentional contact at robot link 5 is detected.
3. Incidental_Link5: a collision at robot link 5 is detected.
4. Intentional_Link6: an intentional contact at robot link 6 is detected.
5. Incidental_Link6: a collision at robot link 6 is detected.

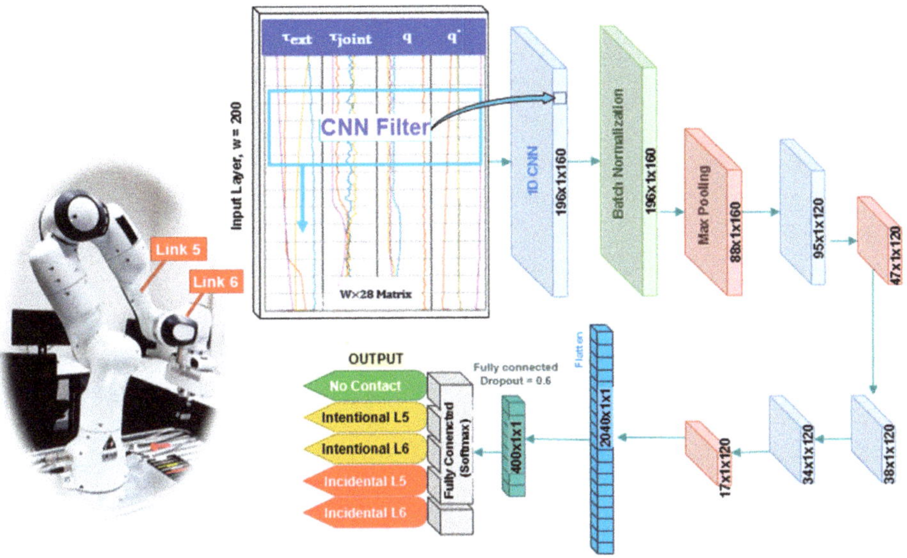

Figure 2. Contact detection network diagram.

In this paper, the input vector represents a time series of robot data as

$$
x = \begin{bmatrix}
\tau_J^0 & \tau_{ext}^0 & q^0 & \dot{q}^0 \\
\tau_J^1 & \tau_{ext}^1 & q^1 & \dot{q}^1 \\
\vdots & \vdots & \vdots & \vdots \\
\tau_J^W & \tau_{ext}^W & q^W & \dot{q}^W
\end{bmatrix}
\tag{2}
$$

and

$$
\tau_J^i = \begin{bmatrix} \tau_{j1}^i & \tau_{j2}^i & \tau_{j3}^i & \tau_{j4}^i & \tau_{j5}^i & \tau_{j6}^i & \tau_{j7}^i \end{bmatrix}
\tag{3}
$$

$$
\tau_{ext}^i = \begin{bmatrix} \tau_{ext1}^i & \tau_{ext2}^i & \tau_{ext3}^i & \tau_{ext4}^i & \tau_{ext5}^i & \tau_{ext6}^i & \tau_{ext7}^i \end{bmatrix}
\tag{4}
$$

$$
q^i = \begin{bmatrix} q_1^i & q_2^i & q_3^i & q_4^i & q_5^i & q_6^i & q_7^i \end{bmatrix}
\tag{5}
$$

$$
\dot{q}^i = \begin{bmatrix} \dot{q}_1^i & \dot{q}_2^i & \dot{q}_3^i & \dot{q}_4^i & \dot{q}_5^i & \dot{q}_6^i & \dot{q}_7^i \end{bmatrix}
\tag{6}
$$

where τ_J, τ_{ext}, q, and $\dot{q}$ indicate joint torque, external torque, joint position, and joint velocity, respectively. W is the size of a window over the collected signals, which stores time-domain samples as an independent instance for training the proposed models. Hence, the input vector is W × 28, and in this research, by selecting 100, 200, and 300 samples for W, three different networks were trained to compare the influence of this parameter.

As shown in Figure 2, the designed CNN is composed of eleven layers. In the first layer of this model, the convolution process maps the data with 160 filters. The kernel size of this layer is optimally considered 5 to obtain a faster and sensitive enough human contact status; a parameter higher than 5 led to an insufficient network's response, as it is more influenced by past data rather than near to present data. To normalize the data and avoid overfitting, especially due to the different maximum force patterns of every human, a batch normalization is used in the second layer. Furthermore, the size of all max pooling layers is chosen as 2, and ReLU function is considered as the activation function, due to reasons already mentioned before.

2.6. Data Collection

2.6.1. Human Action Recognition

The HAR data are collected simultaneously from different views by two Kinect V2 cameras recording the scene of an operator moving next to a robot performing repetitive motions. The human skeleton is detected using the Kinect library based on the random forest decision method [78]. As the Kinect V2 library in Linux is not precise and does not project human skeleton in RGB images, the 3D joint position in depth coordination was extracted and converted to RGB coordinates as follows:

$$x_{rgb} = x_d \times \frac{PD_{xrgb}}{PD_{xd}} + \frac{C_{xrgb} \times PD_{xd} - C_{xd} \times PD_{xrgb}}{PD_{xrgb} \times PD_{xd}} \tag{7}$$

$$y_{rgb} = y_d \times \frac{PD_{yrgb}}{PD_{yd}} + \frac{C_{yrgb} \times PD_{yd} - C_{yd} \times PD_{yrgb}}{PD_{yrgb} \times PD_{yd}} \tag{8}$$

where (C_{xrgb}, C_{yrgb}) and (C_{xd}, C_{yd}) are RGB and depth image centers, respectively. PD shows the number of pixels per degree for depth and RGB images, respectively equal to 7×7 and 22×20 [79,80]. Then, the RGB images, which are supplemented with the skeleton representation in each frame, are collected as a dataset. The sample rate by considering the required time for saving the images was 22 frames/second. Both cameras start collecting data once the human operator enters the environment, while it is assumed that the robot is stationary in a structured environment. The collected images are then sorted into 5 different categories and labeled accordingly based on the skeleton position and configuration and with respect to the fixed base position of the robot.

2.6.2. Contact Detection

The data acquired at the robot joints during a predefined motion with a speed of 0.5 m/s were collected in three states, contact-free, during interaction with the operator, and collision-like contacts, at a sampling rate of 200 Hz (one sample per 5 ms). In this part, collecting collision-like contact data is challenging, as the dedicated operator induces the collision intentionally [65]. However, the collision can be considered to happen in a normal situation where the human is standing with no motion and the robot is performing its task, while the impact happens. Indeed, a data analysis shows that it can be clearly distinguished from object and intentional contacts and therefore can be used at least as similar samples of real collision data. Then, a frame of W-window with 200 ms latency passed through the entire data gathered, preparing it to be used as training data for the input layer of the designed network. Thanks to the default cartesian contact detection ability of the Panda robot, those contact data are used as a trigger to stop recording data after contact occurrence. Consequently, the last W-samples of each collision trial data is considered as input for training the network. For assuring comprehensiveness of the gathered data, each trial is repeated 10 times with different scenes, including touched links, direction of motion, line of collision with the human operator, and contact type (intentional or incidental). Additionally, each sample is labeled according to the mentioned sequence.

2.7. Training Hardware and API Setup

In the training of a network by using Graphic Processor Units (GPU), memory plays an important role in reducing the training time. In this research, a powerful computer with NVIDIA Quadro P5000 GPU, Intel Xeon W-2155 CPUs, and 64 GB of RAM is employed for modeling and training the CNN networks using the Keras library of TensorFlow. To enable CUDA and GPU-acceleration computing, a GPU version of TensorFlow is used, and in consequence, the training process is speeded up. The total runtime of the vision network trained with 30,000 image sequences was about 12 h for 150 epochs, while it was less than 5 min for training contact networks.

2.8. Real Time Interface

The real-time interface for collecting the dataset and implementing the trained network on the system was provided by Robotics Operating System (ROS) on Ubuntu 18.04 LTS. Figure 3 shows the hardware and software structure used in this work. Two computers execute the vision networks for each camera separately and publish the action states at the rate of 200 Hz on ROS. Furthermore, CDN and CDM are executed on another PC at the same rate, connected to the robot controller for receiving the robot torque, velocity, and position data of joints 5 and 6.

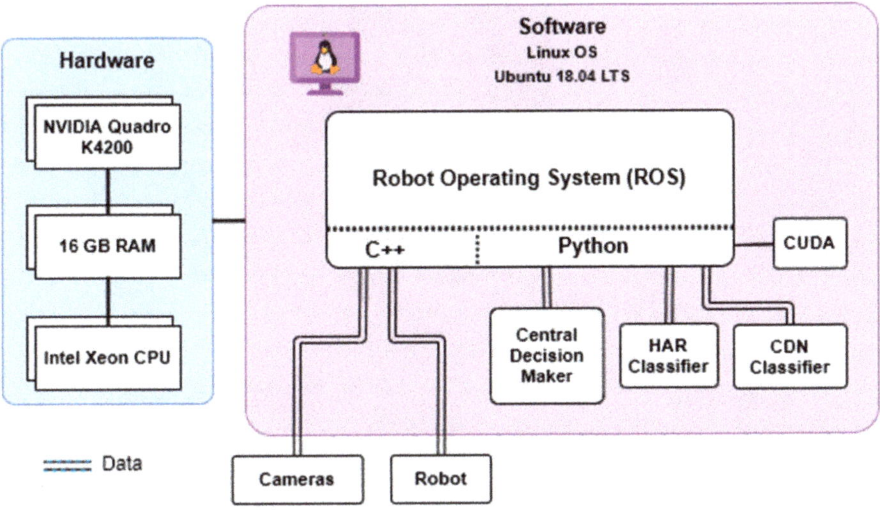

Figure 3. Real-time interface of complex system.

3. Results

In order to evaluate the performance of the proposed system, the following metrics are used. A first evaluation consists of an offline testing, for which the results are compared based on the key figures precision, recall, and accuracy, defined as follows:

$$Precision = \frac{tp}{tp + fp} \tag{9}$$

$$Recall = \frac{tp}{tp + fn} \tag{10}$$

$$Accuracy = \frac{t_p + t_n}{t_p + f_n + t_n + f_p} \tag{11}$$

where t_p is the amount of the predicted true positive samples, t_n is the number of data points labeled as negative correctly, f_p represents the amount of the predicted false positive samples, and f_n is the count of predicted false positive classes.

The second evaluation is based on real-time testing; the tests have shown promising results in early trials. The YouTube video (https://www.youtube.com/watch?v=ED_wH9BFJck) gives an impression of the performance (due to safety reasons, the velocity of the robot is reduced to an amount, which is considered to be safe according to ISO 10218).

3.1. Dataset

Regarding the vision category, the dataset consisting of 33,050 images is divided into five classes, including Interaction, Observation, Passing, Fail, and Dangerous Observation, with Figure 4 representing the different possible actions of a human operator during robot operation. The contact detection dataset [81] with 1114 samples is subdivided into five classes, namely No-Contact, Intentional_Link5, Intentional_Link6, Incidental_Link5, Incidental_Link6, which determine the contact state on the last two links including their respective type, incidental or intentional.

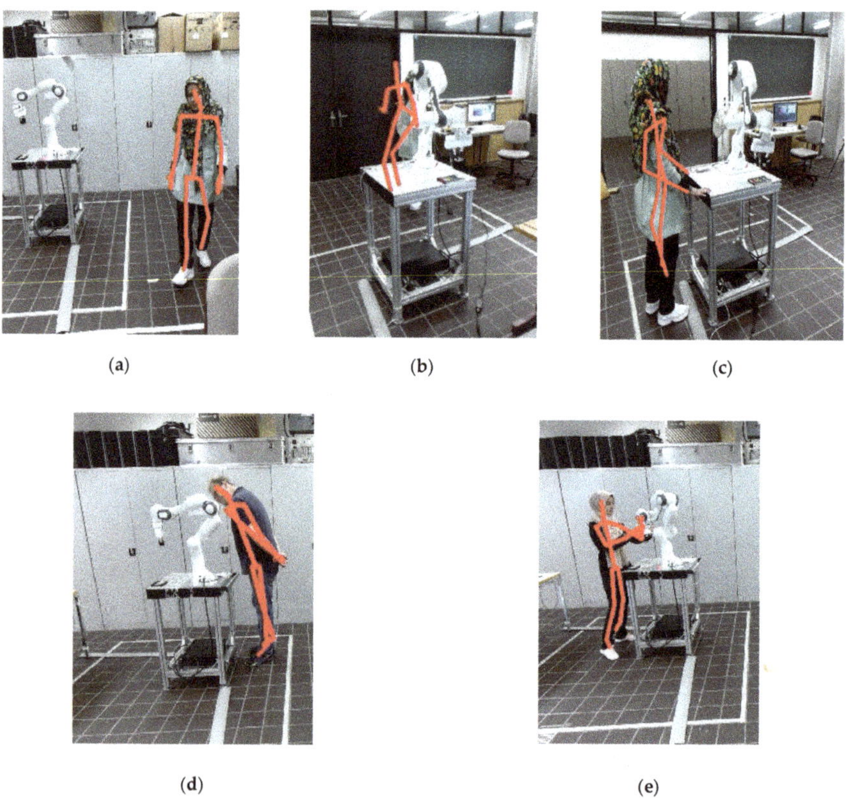

Figure 4. Type of human actions: (**a**) Passing: operator is just passing by, without paying attention to the robot. (**b**) Fail: blind spots or occlusion of the visual field may happen for a camera, in this situation the second camera takes over detection. (**c**) Observation: operator enters the working zone, without any interaction intention and stands next to the robot. (**d**) Dangerous Observation: operator proximity is too close, especially his head is at danger of collision with the robot. (**e**) Interaction: operator enters the working zone and prepares to work with the robot.

3.2. Comparison between Networks

3.2.1. Human Action Recognition

For optimizing efficiency in HAR, two different networks, 2D and 3D, were tested, the latter indicating a significant outcome in both real-time and off-line testing cases. These two networks are compared with respect to the results of 150 training epochs, in Table 1. The confusion matrix can be considered as a good measurement to deliver the overall performance in the multi-category classification systems. As it is shown in Table 2, each row of the table represents the actual label, and each column indicates the predicted labels, which can also show the number of failed predictions in every class. As shown in Table 1, both networks have promising result in classifying "Interaction", "Passing", and "Fail" states. However, these networks have lower, but sufficient, accuracy in classifying the "Dangerous Observation" category due to the lack of third dimensional (depth) information in the network input. By considering the confusion matrix shown in Table 2, it is obvious that the networks did not precisely distinguish between "Interaction", "Observation", and "Dangerous Observation" caused by the similarity of these three classes. With regard to the condition of the experimental setup where the location of cameras and robot base are fixed, the current approach has enough accuracy, but for a real industry case, we need to add a true 3D representation of the human skeleton and the robot arm in our network input.

Table 1. Precision and recall of two trained networks for human action recognition.

Network	2D		3D	
	Precision	Recall	Precision	Recall
Observation	0.99	0.99	1.00	1.00
Interaction	1.00	1.00	1.00	1.00
Passing	1.00	1.00	1.00	1.00
Fail	1.00	1.00	1.00	1.00
Dangerous Observation	0.98	0.96	0.98	0.99
Accuracy	0.9956		0.9972	

Table 2. Confusion Matrix for different classes in HRC.

Network	2D					3D				
	Observation	Interaction	Passing	Fail	Dangerous Observation	Observation	Interaction	Passing	Fail	Dangerous Observation
Observation	3696	7	2	0	5	3751	6	2	1	7
Interaction	13	4130	0	0	1	8	4030	0	0	0
Passing	2	0	1145	0	0	1	0	1160	0	0
Fail	0	0	0	593	0	0	0	0	588	0
Dangerous Observation	12	1	0	0	313	2	0	0	0	359

(Row labels grouped under "True Labels")

3.2.2. Contact Detection

To evaluate the influence of the size of the sampling window (w) on the precision of the trained networks, three different size dimensions of 100, 200, and 300 unity are selected, corresponding to 0.5, 1, 1.5 s of sampling period duration. Seventy percent of the dataset is selected for training and 30% for testing. Each network is trained with 300 epochs, and the results are shown in Tables 3 and 4. Window size of 200 and 300 unities provide a good precision for identifying the contact status, in contrast to w = 100, which is not satisfactory. Furthermore, by comparing the result of the 200-window and 300-window networks, the 200-window network provides a better precision and recall.

Table 3. Precision and recall of trained networks for contact detection with different window size.

w	100	200	300	100	200	300
	Precision			Recall		
No-Contact	0.94	0.99	0.98	0.94	1.00	1.00
Intentional_Link5	0.74	0.91	0.89	0.84	0.91	0.84
Intentional_Link6	0.68	0.97	0.91	0.64	0.90	0.91
Incidental_Link5	0.61	0.89	0.83	0.61	0.93	0.89
Incidental_Link6	0.69	0.96	0.96	0.57	0.96	0.93
Accuracy	0.78	0.96	0.93			

Table 4. Confusion matrix of trained networks for contact detection with different window size.

Window Size	100					200					300				
	No-Contact	Intentional_Link5	Intentional_Link6	Incidental_Link5	Incidental_Link6on	No-Contact	Intentional_Link5	Intentional_Link6	Incidental_Link5	Incidental_Link6	No-Contact	Intentional_Link5	Intentional_Link6	Incidental_Link5	Incidental_Link6
No-Contact	166	0	9	0	1	242	0	3	0	1	167	0	3	0	0
Intentional_Link5	0	86	12	19	0	0	93	4	4	1	0	86	5	5	1
Intentional_Link6	8	1	59	2	17	0	3	83	0	0	0	5	84	0	3
Incidental_Link5	0	15	1	33	5	0	6	0	50	0	0	10	0	48	0
Incidental_Link6	3	0	11	0	31	0	0	2	0	52	0	1	0	1	50

(True Labels indicated along the rows.)

3.2.3. Mixed Perception Safety Monitoring

Every perception system designed separately to detect human intention according to Figure 5a,b is regarded as a preliminary condition for the mixed perception system shown in Figure 5c. As shown in Figure 5, for proper safety monitoring, the robot is programmed to categorize human safety into three levels—Safe, Caution, and Danger—with its respective color codes green/yellow, orange, and red. Safe level consist of two states, indicating whether the cobot has physical contact with human (yellow) or not (green). Considering only visual perception or only tactile perception in determining the safety level does not provide sufficient information compared to the mixed perception system. For instance, in green Safe state of mixed perception, the robot can have a higher speed and in consequence, increased productivity, while in the other perception systems, green Safe does not give this confidence to the robot to be faster; consequently, it should be more conservative about possible collisions. Thus, this higher information content can increase human safety and the robot's productivity of pHRC systems. Already a simple logical composition of the results (Figure 5c) shows a significantly higher information content and thus a possible increase in safety and productivity in human–robot collaboration. However, it might be that the mixed perception approach will have multiple effects on the safety of HRC. Therefore, we will examine in detail the influence of the two subsystems on the overall performance and quality of the entire system at a later stage.

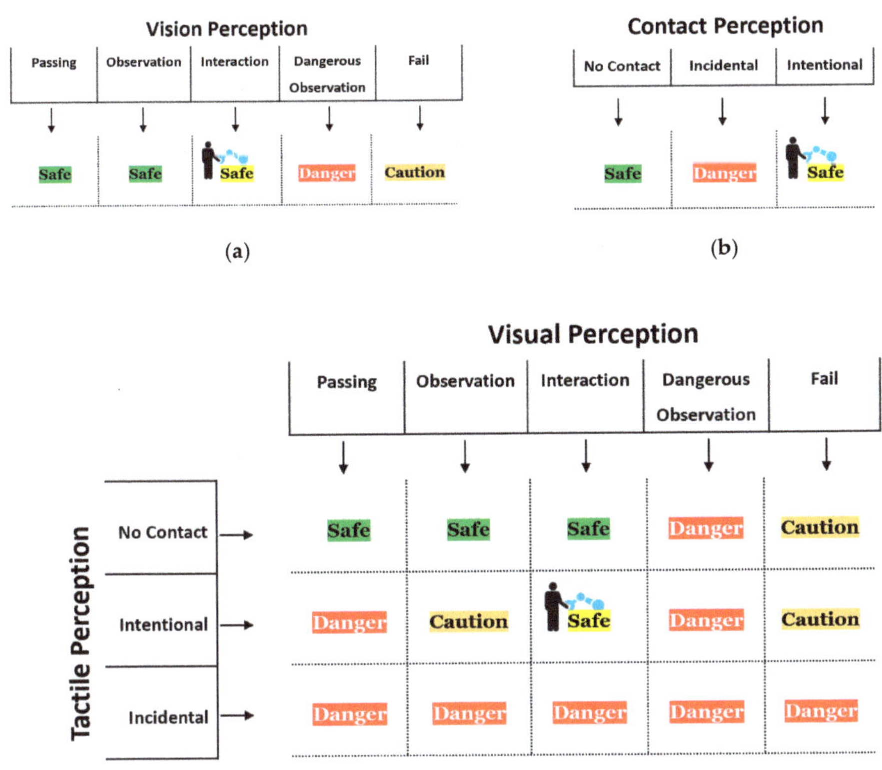

Figure 5. Safety perception spectrum in (**a**) visual perception, (**b**) contact perception, (**c**) mixed perception safety systems.

4. Discussion

Human–robot collaboration has recently gained a lot of interest and received many contributions on both theoretical and practical aspects, including sensor development [82], design of robust and adaptive controllers [83,84], learning robots force-sensitive manipulation skills [85], human interfaces [86,87], and similar. Besides, some companies attempt to introduce collaborative robots in order for HRC to become more suited to enter manufacturing applications and production lines. However, cobots available on the market have limited payload/speed capacities because of safety concerns, which limits HRC application to some light tasks with very limited productivity. On the other hand, according to the norms for HRC operations [88], it is not essential to observe a strict design or to limit the power of operations if human safety can be ensured in all its aspects. In this regard, an intelligent safety system as the mixed perception approach has been proposed in this research to detect hazardous situations to take care of the human safety from entering the shared workspace to physical interaction in order to jointly accomplish a task by taking advantage of visual and tactile perceptions. Visual perception detects human actions in the shared workspace. Meanwhile, tactile perception identifies human–robot contacts. One relevant piece of research in human action recognition focusing on industrial assembly application is mentioned in [88]. By taking advantage of RGB image and 3D-CNN network, the authors of the mentioned paper classified human action during assembly and achieved 82% accuracy [89], while our visual perception system shows higher accuracy of 99.7% by adding a human skeleton to the RGB series as the network input. Although our skeleton detection using Kinect library can be slightly affected by lighting conditions, it detects the human skeleton in near 30 FPS, which is essential for fast human detection in real-time HRC applications [90]. Indeed, using deep learning approaches such as OpenPose [91] and AlphaPose [92] can omit lightening problems [93]. However, their respective detection rates are 22 [91] and 23 FPS [92], which needs more researches to be faster and applicable in safety monitoring systems. Besides, among contact detection approaches in the state of the art, there are two similar works investigating collision detection using 1D-CNN. The authors of [94] compared both approaches, CollisionNet [65] and FMA [94], where the accuracy was 88% and 90%, respectively, featuring a detection delay of 200ms [94]. While our procedure in tactile perception (what is called collision detection in the state-of-the-art literature [61,65,75,76]) reached 99% with 80ms detection delay. For detecting contact type and robot joint, the accuracy was higher than 89% up to 96%, which in turn, needs more research to achieve a higher accuracy.

In this study, combining the result of both abovementioned intelligent systems is presented using a safety perception spectrum to examine the potential of the mixed perception approach in safety monitoring of collaborative workspaces. The result shows that even with a simple combination of both systems, the performance of safety monitoring can be improved as each system separately does not have enough perception of the collaborative workspace. Furthermore, this research suggests that the different forms of collaboration, such as coexistence, cooperation, etc., with their different safety requirements can be reduced to a single scenario using mixed perception as the robot would be able to "see" humans and "percept" external contacts.

As a result of this safety scenario, the robot reacts by being able to detect human intention, determining human safety level, and thus ensuring safety in all work situations. Another advantage of the proposed system is that the robot would be smart enough to take care about safety norms depending on the conditions and, consequently, could operate at an optimum speed during HRC applications. In other words, current safety requirements in most cases stop or drastically slow down the robot when a human enters a shared workspace. However, with the proposed safety system, based on the robots' awareness using the presented mixed perception approach, it is possible to implement a reasonable trade-off between safety and productivity, which will be discussed in more detail in our future research.

In this research, there are two limitations concerning data collection: the collision occurred intentionally, and we did not gather data when the human and/or the robot move at high speed, which can be extremely dangerous for the human operator. As can be proved, the speed of the robot has an insignificant influence on the result, since the model has learned the dynamics of the robot in the presence or absence of human contact with normalized input data. On the other hand, if the human operator wants to grab the robot at high speed with the intention of working with it, this could be classified as a collision by the model due to its clear difference between contactless and intentional data patterns. However, this only increases the false positive error of the collision class (i.e., this would then be mistakenly perceived as a collision by the robot), which does not represent a safety problem in this case.

In addition, the current work focuses on a structured environment with fixed cameras and a stationary robot base position, which has yet to be generalized for an unstructured environment. In principle; however, this does not restrict the generality of this approach, since for cobots, only the corresponding position of the robot base has to be determined for the proximity detection to a human operator. In our ongoing work, we are trying to use some methods to tackle these problems. Moreover, with the current software and hardware, a sampling rate of HAR and contact detection networks are 30 Hz and 200Hz, respectively, while for the mixed perception system, there is a need for synchronization of the result of both systems.

5. Conclusions

The efficiency of safety and productivity of cobots in HRC can be improved if cobots are able to easily recognize complex human actions and can differentiate between multitude contact types. In this paper, a safety system using a mixed perception is proposed to improve the productivity and safety in HRC applications by making the cobot aware of human actions (visual perception), with the ability to distinguish between intentional and incidental contact (tactile perception). The vision perception system is based on a 3D-CNN algorithm for human action recognition, which unlike the latest HAR methods, was able to achieve 99.7% accuracy in an HRC scenario. The HAR system is intended to detect human action once the latter enters the workspace and only in case of hazardous situations, the robot would adapt its speed or stop accordingly, which can lead to higher productivity. On the other hand, the tactile perception, by focusing on the contact between robot and human, can decide about the final situation during pHRC. The contact detection system, by taking advantage of the contact signal patterns and 1D-CNN network, was able to distinguish between the incidental and intentional contact and recognize the impacted cobot's link. According to the experimental result, with respect to traditional and new methods, our proposed model is obtained the highest accuracy of 96% in tactile perception.

Yet, based on our experimental results, visual and tactile perceptions are not sufficient enough separately for intrinsically safe robotic applications, since each system exhibits some lack of information, which may cause less productivity and safety. By considering this fact, the mixed perception, by taking advantage of both visual and tactile perception, can enhance productivity and safety. Although a simple safety perception spectrum of the mixed perception is proposed, which needs more research to enhance its intelligence, it shows higher resolution in compared to each single perception system.

As future work for our system, we will extend our research regarding to multi-contact and multi-person detection, which is highly beneficial for the latest Industry 4.0 safety considerations.

Author Contributions: Conceptualization, F.M.A., M.R., H.W.v.d.V. and H.K.; methodology, F.M.A., M.R. and H.W.v.d.V.; software, F.M.A. and M.R.; validation, F.M.A., M.R., H.W.v.d.V. and H.K.; formal analysis, H.W.v.d.V. and H.K.; investigation, F.M.A. and M.R.; resources, F.M.A., M.R., H.W.v.d.V. and H.K.; data curation, F.M.A. and M.R.; writing—original draft preparation, F.M.A. and M.R.; writing—review and editing, F.M.A., M.R., H.W.v.d.V. and H.K.; visualization, F.M.A. and M.R.; supervision, H.W.v.d.V. and H.K.; project administration, H.W.v.d.V. and H.K.; funding acquisition, H.W.v.d.V. and H.K. All authors have read and agreed to the published version of the manuscript.

Funding: This research was partly funded by a Seed Money Grant of the Leading House South Asia and Iran of Zurich University of Applied Sciences (ZHAW, Leading House South Asia and Iran, ZHAW School of Management and Law, Theaterstrasse 17, CH-8401 Winterthur/Switzerland, leadinghouse@zhaw.ch). The APC was funded by ZHAW University Library due to its MDPI Open Access membership (ZHAW University Library, Open Access, openaccess@zhaw.ch).

Acknowledgments: This research was a part of an international project between Zurich University of Applied Science ZHAW and University of Isfahan. Authors acknowledge the Leading House of South Asia and Iran at Zurich University of Applied Science for funding this research. The persons recognizable in the pictures have consented to private and/or commercial use—publication, distribution, use, processing, and transfer—in digital and print form by the photographer or by third parties.

Conflicts of Interest: Authors declare no conflict of interest.

References

1. Robla-Gomez, S.; Becerra, V.M.; Llata, J.R.; Gonzalez-Sarabia, E.; Torre-Ferrero, C.; Perez-Oria, J. Working Together: A Review on Safe Human–Robot Collaboration in Industrial Environments. *IEEE Access* **2017**, *5*, 26754–26773. [CrossRef]
2. Nikolakis, N.; Maratos, V.; Makris, S. A cyber physical system (CPS) approach for safe human–robot collaboration in a shared workplace. *Robot. Comput. Integr. Manuf.* **2019**, *56*, 233–243. [CrossRef]
3. Villani, V.; Pini, F.; Leali, F.; Secchi, C. Survey on human–robot collaboration in industrial settings: Safety, intuitive interfaces and applications. *Mechatronics* **2018**, *55*, 248–266. [CrossRef]
4. Safety Fence Systems—F.EE Partner Für Automation. Available online: https://www.fee.de/en/automation-robotics/safety-fence-systems.html (accessed on 7 October 2020).
5. PILZ Safety Sensors PSEN. Available online: https://www.pilz.com/en-INT/products/sensor-technology (accessed on 7 October 2020).

6.   Safe Camera System SafetyEYE—Pilz INT. Available online: https://www.pilz.com/en-INT/eshop/00106002207042/SafetyEYE-Safe-camera-system (accessed on 7 October 2020).

7.   Virtual Fence. Available online: https://www.densorobotics-europe.com/product-overview/products/robotics-functions/virtual-fence (accessed on 7 October 2020).

8.   Losey, D.P.; McDonald, C.G.; Battaglia, E.; O'Malley, M.K. A Review of Intent Detection, Arbitration, and Communication Aspects of Shared Control for Physical Human–Robot Interaction. *Appl. Mech. Rev.* **2018**, *70*. [CrossRef]

9.   Zhang, H.B.; Zhang, Y.X.; Zhong, B.; Lei, Q.; Yang, L.; Du, J.X.; Chen, D.S. A comprehensive survey of vision-based human action recognition methods. *Sensors* **2019**, *19*, 1005. [CrossRef] [PubMed]

10.   Otim, T.; Díez, L.E.; Bahillo, A.; Lopez-Iturri, P.; Falcone, F. Effects of the Body Wearable Sensor Position on the UWB Localization Accuracy. *Electronics* **2019**, *8*, 1351. [CrossRef]

11.   Moschetti, A.; Cavallo, F.; Esposito, D.; Penders, J.; Di Nuovo, A. Wearable sensors for human–robot walking together. *Robotics* **2019**, *8*, 38. [CrossRef]

12.   Otim, T.; Díez, L.E.; Bahillo, A.; Lopez-Iturri, P.; Falcone, F. A Comparison of Human Body Wearable Sensor Positions for UWB-based Indoor Localization. In Proceedings of the 10th International Conference Indoor Positioning Indoor Navigat, Piza, Italy, 30 September 2019; pp. 165–171.

13.   Rosati, S.; Balestra, G.; Knaflitz, M. Comparison of different sets of features for human activity recognition by wearable sensors. *Sensors* **2018**, *18*, 4189. [CrossRef]

14.   Xu, Z.; Zhao, J.; Yu, Y.; Zeng, H. Improved 1D-CNNs for behavior recognition using wearable sensor network. *Comput. Commun.* **2020**, *151*, 165–171. [CrossRef]

15.   Xia, C.; Sugiura, Y. Wearable Accelerometer Optimal Positions for Human Motion Recognition. In Proceedings of the 2020 IEEE 2nd Global Conference on Life Sciences and Technologies (LifeTech), Kyoto, Japan, 10–12 March 2020; pp. 19–20.

16.   Qian, H.; Mao, Y.; Xiang, W.; Wang, Z. Recognition of human activities using SVM multi-class classifier. *Pattern Recognit. Lett.* **2010**, *31*, 100–111. [CrossRef]

17.   Reddy, K.K.; Shah, M. Recognizing 50 human action categories of web videos. *Mach. Vis. Appl.* **2013**, *24*, 971–981. [CrossRef]

18.   Manosha Chathuramali, K.G.; Rodrigo, R. Faster human activity recognition with SVM. In Proceedings of the International Conference on Advances in ICT for Emerging Regions (ICTer2012), Colombo, Sri Lanka, 12–15 December 2012; pp. 197–203.

19.   Sharma, S.; Modi, S.; Rana, P.S.; Bhattacharya, J. Hand gesture recognition using Gaussian threshold and different SVM kernels. In Proceedings of the Communications in Computer and Information Science, Dehradun, India, 20–21 April 2018; pp. 138–147.

20.   Berg, J.; Reckordt, T.; Richter, C.; Reinhart, G. Action recognition in assembly for human–robot-cooperation using hidden Markov Models. *Procedia CIRP* **2018**, *76*, 205–210. [CrossRef]

21.   Le, H.; Thuc, U.; Ke, S.-R.; Hwang, J.-N.; Tuan, P.V.; Chau, T.N. Quasi-periodic action recognition from monocular videos via 3D human models and cyclic HMMs. In Proceedings of the 2012 International Conference on Advanced Technologies for Communications, Hanoi, Vietnam, 10–12 October 2012; pp. 110–113.

22.   Hasan, H.; Abdul-Kareem, S. Static hand gesture recognition using neural networks. *Artif. Intell. Rev.* **2014**, *41*, 147–181. [CrossRef]

23.   Cho, W.H.; Kim, S.K.; Park, S.Y. Human action recognition using hybrid method of hidden Markov model and Dirichlet process Gaussian mixture model. *Adv. Sci. Lett.* **2017**, *23*, 1652–1655. [CrossRef]

24.   Piyathilaka, L.; Kodagoda, S. Gaussian mixture based HMM for human daily activity recognition using 3D skeleton features. In Proceedings of the 2013 IEEE 8th Conference on Industrial Electronics and Applications (ICIEA), Melbourne, Australia, 19–21 June 2013; pp. 567–572.

25.   Wang, P.; Liu, H.; Wang, L.; Gao, R.X. Deep learning-based human motion recognition for predictive context-aware human–robot collaboration. *CIRP Ann. Manuf. Technol.* **2018**, *67*, 17–20. [CrossRef]

26.   Ullah, A.; Ahmad, J.; Muhammad, K.; Sajjad, M.; Baik, S.W. Action Recognition in Video Sequences using Deep Bi-Directional LSTM with CNN Features. *IEEE Access* **2017**, *6*, 1155–1166. [CrossRef]

27.   Zhao, Y.; Man, K.L.; Smith, J.; Siddique, K.; Guan, S.U. Improved two-stream model for human action recognition. *Eurasip J. Image Video Process.* **2020**, *2020*, 1–9. [CrossRef]

28. Gao, J.; Yang, Z.; Sun, C.; Chen, K.; Nevatia, R. TURN TAP: Temporal Unit Regression Network for Temporal Action Proposals. In Proceedings of the 2017 IEEE International Conference on Computer Vision (ICCV), Venice, Italy, 22–29 October 2017; pp. 3648–3656.

29. Wang, P.; Li, W.; Ogunbona, P.; Wan, J.; Escalera, S. RGB-D-based human motion recognition with deep learning: A survey. *Comput. Vis. Image Underst.* **2018**, *171*, 118–139. [CrossRef]

30. Gu, Y.; Ye, X.; Sheng, W.; Ou, Y.; Li, Y. Multiple stream deep learning model for human action recognition. *Image Vis. Comput.* **2020**, *93*, 103818. [CrossRef]

31. Srihari, D.; Kishore, P.V.V.; Kiran Kumar, E.; Anil Kumar, D.; Teja, M.; Kumar, K.; Prasad, M.V.D.; Raghava Prasad, C. A four-stream ConvNet based on spatial and depth flow for human action classification using RGB-D data. *Multimed. Tools Appl.* **2020**, *79*, 11723–11746. [CrossRef]

32. Yan, S.; Xiong, Y.; Lin, D. Spatial temporal graph convolutional networks for skeleton-based action recognition. In Proceedings of the 32nd AAAI Conference on Artificial Intelligence, New Orleans, LA, USA, 2–7 February 2018; pp. 7444–7452.

33. Si, C.; Jing, Y.; Wang, W.; Wang, L.; Tan, T. Skeleton-Based Action Recognition with Spatial Reasoning and Temporal Stack Learning. *Lect. Notes Comput. Sci. (Incl. Subser. Lect. Notes Artif. Intell. Lect. Notes Bioinform.)* **2018**, *11205 LNCS*, 106–121. [CrossRef]

34. Cheng, J. Skeleton-Based Action Recognition with Directed Graph Neural Networks. In Proceedings of the Conference on Computer Vision and Pattern Recognition (CVPR), Long Beach, CA, USA, 15–20 June 2019; pp. 7912–7921.

35. Trăscău, M.; Nan, M.; Florea, A.M. Spatio-temporal features in action recognition using 3D skeletal joints. *Sensors* **2019**, *19*, 423. [CrossRef]

36. Baccouche, M.; Mamalet, F.; Wolf, C.; Garcia, C.; Baskurt, A. Sequential deep learning for human action recognition. In Proceedings of the 2nd International Workshop on Human Behavior Understanding (HBU), Amsterdam, The Netherlands, 16 November 2011; pp. 29–39.

37. Ji, S.; Xu, W.; Yang, M.; Yu, K. 3D Convolutional neural networks for human action recognition. *IEEE Trans. Pattern Anal. Mach. Intell.* **2013**, *35*, 221–231. [CrossRef]

38. Latah, M. Human action recognition using support vector machines and 3D convolutional neural networks. *Int. J. Adv. Intell. Inform.* **2017**, *3*, 47–55. [CrossRef]

39. Almaadeed, N.; Elharrouss, O.; Al-Maadeed, S.; Bouridane, A.; Beghdadi, A. A Novel Approach for Robust Multi Human Action Detection and Recognition based on 3-Dimentional Convolutional Neural Networks. *arXiv* **2019**, arXiv:1907.11272.

40. Arunnehru, J.; Chamundeeswari, G.; Bharathi, S.P. Human Action Recognition using 3D Convolutional Neural Networks with 3D Motion Cuboids in Surveillance Videos. *Procedia Comput. Sci.* **2018**, *133*, 471–477. [CrossRef]

41. Asghari-Esfeden, S.; Io, S.; Sznaier, M.; Camps, O. Dynamic Motion Representation for Human Action Recognition. In Proceedings of the IEEE/CVF Winter Conference on Applications of Computer Vision (WACV), Snowmass Village, CO, USA, 2–5 March 2020; pp. 557–566.

42. Fan, H.; Luo, C.; Zeng, C.; Ferianc, M.; Que, Z.; Liu, S.; Niu, X.; Luk, W. F-E3D: FPGA-based acceleration of an efficient 3D convolutional neural network for human action recognition. In Proceedings of the 2019 IEEE 30th International Conference on Application-specific Systems, Architectures and Processors (ASAP), New York, NY, USA, 15–17 July 2019; pp. 1–8.

43. Recognition of Human Actions. Available online: https://www.csc.kth.se/cvap/actions/ (accessed on 8 October 2020).

44. Hoang, V.D.; Hoang, D.H.; Hieu, C.L. Action recognition based on sequential 2D-CNN for surveillance systems. In Proceedings of the IECON 2018—44th Annual Conference of the IEEE Industrial Electronics Society, Washington, DC, USA, 21–23 October 2018; pp. 3225–3230.

45. Kim, J.H.; Won, C.S. Action Recognition in Videos Using Pre-Trained 2D Convolutional Neural Networks. *IEEE Access* **2020**, *8*, 60179–60188. [CrossRef]

46. Singh, I.; Zhu, X.; Greenspan, M. Multi-Modal Fusion With Observation Points For Skeleton Action Recognition. In Proceedings of the 2020 IEEE International Conference on Image Processing (ICIP), IEEE, Abu Dhabi, UAE, 25–28 October 2020; pp. 1781–1785.

47. Weng, J.; Luo, D.; Wang, Y.; Tai, Y.; Wang, C.; Li, J.; Huang, F.; Jiang, X.; Yuan, J. Temporal Distinct Representation Learning for Action Recognition. *arXiv* **2020**, arXiv:2007.07626.

48. Akkaladevi, S.C.; Heindl, C. Action recognition for human robot interaction in industrial applications. In Proceedings of the 2015 IEEE International Conference on Computer Graphics, Vision and Information Security (CGVIS), Bhubaneswar, India, 2–3 November 2015; pp. 94–99.

49. Roitberg, A.; Perzylo, A.; Somani, N.; Giuliani, M.; Rickert, M.; Knoll, A. Human activity recognition in the context of industrial human–robot interaction. In Proceedings of the 2014 Asia-Pacific Signal and Information Processing Association Annual Summit and Conference (APSIPA), Chiang Mai, Thailand, 9–12 December 2014. [CrossRef]

50. Olatunji, I.E. Human Activity Recognition for Mobile Robot. *J. Phys. Conf. Ser.* **2018**, *1069*, 4–7. [CrossRef]

51. Duckworth, P.; Hogg, D.C.; Cohn, A.G. Unsupervised human activity analysis for intelligent mobile robots. *Artif. Intell.* **2019**, *270*, 67–92. [CrossRef]

52. Cao, P.; Gan, Y.; Dai, X. Model-based sensorless robot collision detection under model uncertainties with a fast dynamics identification. *Int. J. Adv. Robot. Syst.* **2019**, *16*, 172988141985371. [CrossRef]

53. Haddadin, S.; Albu-Schaffer, A.; De Luca, A.; Hirzinger, G. Collision Detection and Reaction: A Contribution to Safe Physical Human–Robot Interaction. In Proceedings of the 2008 IEEE/RSJ International Conference on Intelligent Robots and Systems, IEEE, Nice, France, 22–26 September 2008; pp. 3356–3363.

54. de Luca, A.; Mattone, R. Sensorless Robot Collision Detection and Hybrid Force/Motion Control. In Proceedings of the 2005 IEEE International Conference on Robotics and Automation, IEEE, Barcelona, Spain, 18–22 April 2005; pp. 999–1004.

55. Xiao, J.; Zhang, Q.; Hong, Y.; Wang, G.; Zeng, F. Collision detection algorithm for collaborative robots considering joint friction. *Int. J. Adv. Robot. Syst.* **2018**, *15*. [CrossRef]

56. Cao, P.; Gan, Y.; Dai, X. Finite-Time Disturbance Observer for Robotic Manipulators. *Sensors* **2019**, *19*, 1943. [CrossRef]

57. Luca, A.; Albu-Schaffer, A.; Haddadin, S.; Hirzinger, G. Collision Detection and Safe Reaction with the DLR-III Lightweight Manipulator Arm. In Proceedings of the 2006 IEEE/RSJ International Conference on Intelligent Robots and Systems, IEEE, Beijing, China, 9–15 October 2006; pp. 1623–1630.

58. Ren, T.; Dong, Y.; Wu, D.; Chen, K. Collision detection and identification for robot manipulators based on extended state observer. *Control Eng. Pract.* **2018**, *79*, 144–153. [CrossRef]

59. Min, F.; Wang, G.; Liu, N. Collision Detection and Identification on Robot Manipulators Based on Vibration Analysis. *Sensors* **2019**, *19*, 1080. [CrossRef]

60. Haddadin, S. *Towards Safe Robots: Approaching Asimov's 1st Law*; Springer: Berlin/Heidelberg, Germany, 2014; ISBN 9783642403088. [CrossRef]

61. Haddadin, S.; De Luca, A.; Albu-Schaffer, A. Robot Collisions: A Survey on Detection, Isolation, and Identification. *IEEE Trans. Robot.* **2017**, *33*, 1292–1312. [CrossRef]

62. Sharkawy, A.-N.; Koustoumpardis, P.N.; Aspragathos, N. Neural Network Design for Manipulator Collision Detection Based Only on the Joint Position Sensors. *Robotica* **2019**, 1–19. [CrossRef]

63. Sharkawy, A.-N.; Koustoumpardis, P.N.; Aspragathos, N. Human–robot collisions detection for safe human–robot interaction using one multi-input–output neural network. *Soft Comput.* **2019**, 1–33. [CrossRef]

64. Liu, Z.; Hao, J. Intention Recognition in Physical Human–Robot Interaction Based on Radial Basis Function Neural Network. Available online: https://www.hindawi.com/journals/jr/2019/4141269/ (accessed on 8 June 2020).

65. Heo, Y.J.; Kim, D.; Lee, W.; Kim, H.; Park, J.; Chung, W.K. Collision Detection for Industrial Collaborative Robots: A Deep Learning Approach. *IEEE Robot. Autom. Lett.* **2019**, *4*, 740–746. [CrossRef]

66. El Dine, K.M.; Sanchez, J.; Ramón, J.A.C.; Mezouar, Y.; Fauroux, J.-C. *Force-Torque Sensor Disturbance Observer Using Deep Learning*; Springer: Cham, Switzerland, 2018. [CrossRef]

67. Pham, C.; Nguyen-Thai, S.; Tran-Quang, H.; Tran, S.; Vu, H.; Tran, T.H.; Le, T.L. SensCapsNet: Deep Neural Network for Non-Obtrusive Sensing Based Human Activity Recognition. *IEEE Access* **2020**, *8*, 86934–86946. [CrossRef]

68. Gao, X.; Shi, M.; Song, X.; Zhang, C.; Zhang, H. Recurrent neural networks for real-time prediction of TBM operating parameters. *Autom. Constr.* **2019**, *98*, 225–235. [CrossRef]

69. Yang, K.; Wang, X.; Quddus, M.; Yu, R. Predicting Real-Time Crash Risk on Urban Expressways Using Recurrent Neural Network. In Proceedings of the Transportation Research Board 98th Annual Meeting, Washington, DC, USA, 13–17 January 2019.

70. Masood, S.; Srivastava, A.; Thuwal, H.C.; Ahmad, M. *Real-Time Sign Language Gesture (Word) Recognition from Video Sequences Using CNN and RNN*; Springer: Singapore, 2018; pp. 623–632.

71. Li, S.; Li, W.; Cook, C.; Zhu, C.; Gao, Y. Independently Recurrent Neural Network (IndRNN): Building a Longer and Deeper RNN. In Proceedings of the IEEE Conference on Computer Vision and Pattern Recognition, Salt Lake City, UT, USA, 18–22 June 2018; pp. 5457–5466.

72. Zhang, B.; Wang, L.; Wang, Z.; Qiao, Y.; Wang, H. Real-Time Action Recognition with Enhanced Motion Vector CNNs. In Proceedings of the IEEE Conference on Computer Vision and Pattern Recognition, Las Vegas, NV, USA, 27–30 June 2016; pp. 2718–2726.

73. Jin, C.-B.; Li, S.; Do, T.D.; Kim, H. *Real-Time Human Action Recognition Using CNN Over Temporal Images for Static Video Surveillance Cameras*; Springer: Cham, Switzerland, 2015; pp. 330–339.

74. Pathak, D.; El-Sharkawy, M. Architecturally Compressed CNN: An Embedded Realtime Classifier (NXP Bluebox2.0 with RTMaps). In Proceedings of the 2019 IEEE 9th Annual Computing and Communication Workshop and Conference (CCWC), Las Vegas, NV, USA, 7–9 January 2019; pp. 0331–0336.

75. Birjandi, S.A.B.; Kuhn, J.; Haddadin, S. Observer-Extended Direct Method for Collision Monitoring in Robot Manipulators Using Proprioception and IMU Sensing. *IEEE Robot. Autom. Lett.* **2020**, *5*, 954–961. [CrossRef]

76. BaradaranBirjandi, S.A.; Haddadin, S. Model-Adaptive High-Speed Collision Detection for Serial-Chain Robot Manipulators. *IEEE Robot. Autom. Lett.* **2020**. [CrossRef]

77. Ullah, I.; Hussain, M.; Qazi, E.-H.; Aboalsamh, H. An automated system for epilepsy detection using EEG brain signals based on deep learning approach. *Expert Syst. Appl.* **2018**, *107*, 61–71. [CrossRef]

78. Ho, T.K. Random Decision Forests Tin Kam Ho Perceptron training. In Proceedings of the 3rd International Conference on Document Analysis and Recognition, Quebec, QC, Canada, 14–16 August 1995; pp. 278–282.

79. Kinect for Windows v2. Available online: https://docs.depthkit.tv/docs/kinect-for-windows-v2 (accessed on 10 March 2020).

80. Kim, C.; Yun, S.; Jung, S.W.; Won, C.S. Color and depth image correspondence for Kinect v2. *Adv. Multimed. Ubiquitous Eng.* **2015**, *352*, 111–116. [CrossRef]

81. Rezayati, M.; van de Venn, H.W. Collision Detection in Physical Human Robot Interaction. *Mendeley Data* **2020**, *V1*. [CrossRef]

82. Cirillo, A.; Cirillo, P.; De Maria, G.; Natale, C.; Pirozzi, S. A Distributed Tactile Sensor for Intuitive Human–Robot Interfacing. *J. Sens.* **2017**. [CrossRef]

83. Khoramshahi, M.; Billard, A. A dynamical system approach to task-adaptation in physical human–robot interaction. *Auton. Robots* **2019**, *43*, 927–946. [CrossRef]

84. Xiong, G.L.; Chen, H.C.; Xiong, P.W.; Liang, F.Y. Cartesian Impedance Control for Physical Human–Robot Interaction Using Virtual Decomposition Control Approach. *Iran. J. Sci. Technol. Trans. Mech. Eng.* **2019**, *43*, 983–994. [CrossRef]

85. Johannsmeier, L.; Gerchow, M.; Haddadin, S. A Framework for Robot Manipulation: Skill Formalism, Meta Learning and Adaptive Control. In Proceedings of the 2019 International Conference on Robotics and Automation (ICRA), Montreal, QC, Canada, 20–24 May 2019; pp. 5844–5850.

86. Yang, C.; Zeng, C.; Liang, P.; Li, Z.; Li, R.; Su, C.Y. Interface Design of a Physical Human–Robot Interaction System for Human Impedance Adaptive Skill Transfer. *IEEE Trans. Autom. Sci. Eng.* **2018**, *15*, 329–340. [CrossRef]

87. Weistroffer, V.; Paljic, A.; Callebert, L.; Fuchs, P. A Methodology to Assess the Acceptability of Human–Robot Collaboration Using Virtual Reality. In Proceedings of the 19th ACM Symposium on Virtual Reality Software and Technology, Singapore, 6 October 2013; pp. 39–48.

88. ISO/TS 15066:2016. *Robots and Robotic Devices—Collaborative Robots*; International Organization for Standardization: Geneva, Switzerland, 2016.

89. Wen, X.; Chen, H.; Hong, Q. Human Assembly Task Recognition in Human–Robot Collaboration based on 3D CNN. In Proceedings of the 2019 IEEE 9th Annual International Conference on CYBER Technology in Automation, Control, and Intelligent Systems (CYBER), Suzhou, China, 29 July–2 August 2019; pp. 1230–1234.

90. Liu, H.; Wang, L. Gesture recognition for human–robot collaboration: A review. *Int. J. Ind. Ergon.* **2018**, *68*, 355–367. [CrossRef]

91. Cao, Z.; Simon, T.; Wei, S.-E.; Sheikh, Y. OpenPose: Realtime Multi-Person 2D Pose Estimation using Part Affinity Fields. *arXiv* **2018**, arXiv:1812.08008. [CrossRef]

92. Fang, H.-S.; Xie, S.; Tai, Y.-W.; Lu, C.; Jiao Tong University, S.; YouTu, T. RMPE: Regional Multi-Person Pose Estimation. In Proceedings of the 2017 IEEE International Conference on Computer Vision (ICCV), Venice, Italy, 22–29 October 2017; pp. 2380–7504.
93. Robertini, N.; Bernard, F.; Xu, W.; Theobalt, C. Illumination-invariant robust multiview 3d human motion capture. In Proceedings of the 2018 IEEE Winter Conference on Applications of Computer Vision (WACV), Lake Tahoe, NV, USA, 12–15 March 2018; pp. 1661–1670.
94. Anvaripour, M.; Saif, M. Collision detection for human–robot interaction in an industrial setting using force myography and a deep learning approach. In Proceedings of the 2019 IEEE International Conference on Systems, Man and Cybernetics (SMC), Bari, Italy, 6–9 October 2019; pp. 2149–2154.

Publisher's Note: MDPI stays neutral with regard to jurisdictional claims in published maps and institutional affiliations.

Article

Development and Application of a Tandem Force Sensor

Zhijian Zhang, Youping Chen and Dailin Zhang *

State Key Laboratory of Digital Manufacturing Equipment & Technology, School of Mechanical Science and Engineering, Huazhong University of Science and Technology, Wuhan 430074, China; zhijian516@hust.edu.cn (Z.Z.); ypchen@hust.edu.cn (Y.C.)
* Correspondence: mnizhang@hust.edu.cn; Tel.: +86-2787543555

Received: 2 September 2020; Accepted: 20 October 2020; Published: 23 October 2020

Abstract: In robot teaching for contact tasks, it is necessary to not only accurately perceive the traction force exerted by hands, but also to perceive the contact force at the robot end. This paper develops a tandem force sensor to detect traction and contact forces. As a component of the tandem force sensor, a cylindrical traction force sensor is developed to detect the traction force applied by hands. Its structure is designed to be suitable for humans to operate, and the mechanical model of its cylinder-shaped elastic structural body has been analyzed. After calibration, the cylindrical traction force sensor is proven to be able to detect forces/moments with small errors. Then, a tandem force sensor is developed based on the developed cylindrical traction force sensor and a wrist force sensor. The robot teaching experiment of drawer switches were made and the results confirm that the developed traction force sensor is simple to operate and the tandem force sensor can achieve the perception of the traction and contact forces.

Keywords: tandem force sensor; traction force sensor; human–robot interaction; contact task; imitation learning

---

1. Introduction

Imitation learning or learning by demonstration is one of the promising approaches for non-experts to develop a task control method or a policy in a straightforward and feasible manner [1,2]. Within imitation learning, a task control model or policy is learned from the task demonstrations, one of which is a sequence of state-action pairs recorded during the teacher's demonstration. After the teacher demonstrates how to complete the task several times, learning algorithms utilize the state-action pairs in these demonstrations to derive a mapping model of the state and action, namely the policy.

To obtain the state-action pairs in demonstrations, the robot needs to sense the environment information and the actions taken by the teacher simultaneously during the task demonstration. The environment information depends on the task to be learned. In non-contact tasks of industrial robots, such as spraying and welding, the state only contains the robot motion parameters, target position, and posture, etc. [3,4]. In the contact tasks of industrial robots, the contact force needs to be included [5–9]. The actions taken by a teacher can be perceived by sensors, such as visual sensors to capture a teacher's body movements [10,11] or recognize a teacher's gestures [12], wearable sensors, and force sensors to perceive a teacher's behavioral intentions [13–15]. Compared with visual sensors, wearable sensors, etc., force sensor-based kinesthetic teaching is suitable for non-professionals to tell the robot the action needed to be taken in current state in a simple and intuitive way [5–7,13–16].

In the robot teaching for contact tasks, force sensors need to detect not only traction force, but also the contact force. However, there is only one perceptual unit in a wrist force sensor, which makes it impossible to detect the traction and contact forces synchronously. In the imitation learning of peg-in-hole tasks, references [17–19] adopted kinesthetic teaching to guide the robot to carry out assembly tasks, in which a wrist force sensors was used to measure the traction force exerted by human hands and the contact status between peg and holes. However, this force sensor installed at the end flange of robot cannot distinguish between contact force and traction force, which makes the force data used for the policy learning inaccurate. To avoid this problem, Abu-Dakka [18] repeated the demonstration trajectory to collect the net contact force, which is complicated. Different from reference [18], in reference [13], Zeng grasped the end-point of Baxter robot to guide the robot motion, and the force sensor installed at the end flange of robot just detected the contact status. However, this method is only suitable for collaborative robots equipped with joint torque sensors rather than common ones. One method to obtain traction and contact forces is to adopt two wrist force sensors mounted in parallel, which can complicate the robot's end structure [20,21]. For example, the last two joints of the robot in reference [20] cannot move freely within their motion range, which limits the adjustable range of the robot's attitude. Therefore, for the kinesthetic teaching of robot contact tasks, simultaneous detection of traction and contact forces is still an important issue to be solved.

The main contribution of the paper is that a tandem force sensor is developed, which helps robots to learn human skills of opening and closing a drawer. A cylindrical traction force sensor that can be connected with a contact force sensor in series is developed, which is different from the common wrist force sensors [22–26]. Compared with these common wrist force sensors, the main novelty of the cylindrical traction force sensor is that the sensor's side surface is sensitive to external forces rather than the lateral end surface. Besides, in the cylindrical traction force sensor, there is a central column coaxial to and within the elastic structural body (ESB), which allows other device to be connected with this sensor without influencing the measurement of the traction force. Compared with the force sensor in reference [27], the developed traction force sensor is easier to be operated by human hands and suitable for drawer switch teaching.

2. Introduction to the Tandem Force Sensor

2.1. The Ideal Tandem Force Sensor

To realize the perception of traction and contact forces, a tandem force sensor consisting of two perceptual units connected in series is designed, as shown in Figure 1a. Figure 1a shows an ideal tandem force sensor, which helps to understand the basic perception principle of the tandem force sensor. In the ideal tandem force sensor, one perceptual unit is connected with its side surface, and the other is connected with its end surface. In the kinesthetic teaching of the robot contact tasks, the end effector is connected to the end surface of the tandem force sensor, and the human hand guides the robot's motion by grasping the side surface of the tandem force sensor. The traction force applied to the side surface is detected by the perception unit (i.e., traction force sensor) connected with it, and another perception unit (i.e., contact force sensor) connected with the end surface is used to measure the contact force between the end-effector and external environment. Therefore, the side surface and end surface are sensitive to the traction and contact forces, respectively.

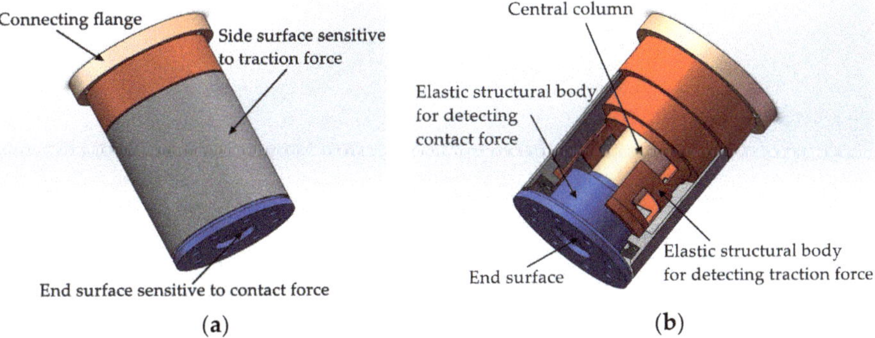

Figure 1. Schematic diagram of the ideal tandem force sensor: (**a**) structure of the ideal tandem force sensor; (**b**) the inner structure of the ideal tandem force sensor.

Each perceptual unit in the tandem force sensor is composed of an elastic structural body, strain type sensors pasted on the ESB, etc., and the two ESBs in it are shown in Figure 1b. The two ESBs in the tandem force sensor are connected in series, and the serial connection mode can be explained by Figure 2. The free end of the ESB for detecting traction force is connected to the side surface, and the end surface is fixed to the free end of the ESB for detecting contact force. The fixed end of the former is directly connected to the connecting flange, while the fixed end of the latter is indirectly fixed to the connecting flange through the central column. In addition, all the connections are made by screw fastening. In application, the traction force applied to the side surface will transmitted to the ESB for detecting traction force and ultimately to the connecting flange, as shown in Figure 2. The contact force exerted on the end surface will flow to the ESB for detecting contact force and to the connecting flange through central column. By adopting the connection mode shows in Figure 2, the traction and contact forces can be detected by the corresponding ESBs, and do not interfere with each other. Finally, the tandem force sensor can achieve the perception of traction and contact forces in decoupled manner.

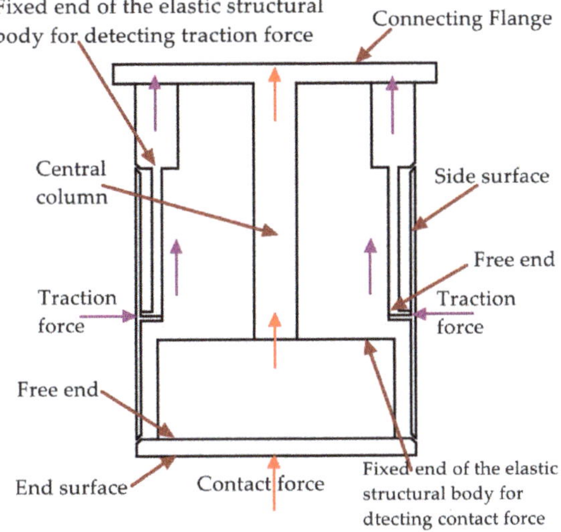

Figure 2. Simplified schematic diagram of series connection mode of the tandem force sensor.

The two perceptual units shown in Figure 2 are connected in serial structure. In principle, the two perceptual units are independent of each other, which is similar to the measurement principle of two wrist force sensors in Figure 3. The two wrist force sensors shown in Figure 3 are connected in parallel structure, which is a currently adopted method to realize the measurement of traction and contact forces. Different form this method, the two perceptual units in the tandem force sensor are connected in series so we have named the sensor shown in Figure 1 as tandem force sensor. Compared with the perception method of the traction and contact forces shown in Figure 3, the tandem force sensor is compact in structure and does not require the handle to be fixed to the sensor. Therefore, the effect of the handle gravity on the measurement accuracy of the traction force is eliminated. Moreover, the tandem force sensor does not increase the transverse structural complexity of the robot end and will not limit the motion range of the last two joints of a six degree of freedom (6-DOF) industrial robot.

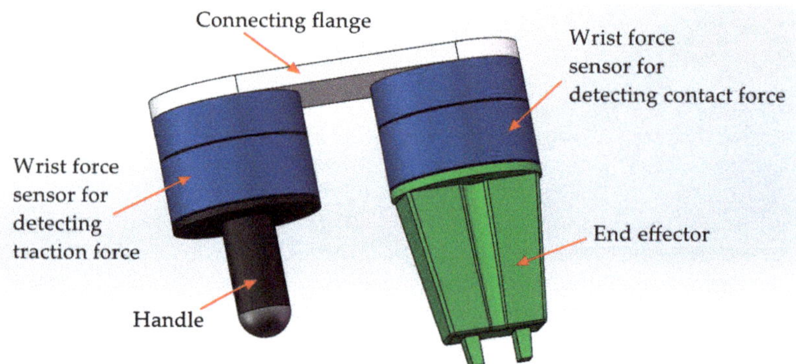

Figure 3. Two wrist force sensors installed in parallel for detecting the traction and contact forces.

2.2. The Developed Tandem Force Sensor

In order to simplify the realization difficulty of the tandem force sensor, this paper proposes and designs a tandem force sensor, as shown in Figure 4a. Both the wrist force sensor and the contact force sensor in Figure 1a use the end surface to sense external forces, so the wrist force sensor is used as the contact force sensor. Based on this idea, the developed tandem force sensor is different from the ideal tandem force sensor in appearance. However, the perception principle of the developed tandem force sensor is the same as the ideal tandem force sensor, that is, the perception of the traction and contact forces are achieved by the perception units connected to the side surface and end surface of the develop tandem force sensor. Moreover, the series connection mode of the two perception units in the developed tandem force sensor is consistent with the ideal tandem force sensor, as shown in Figure 4b. The difference of the traction force sensor in the developed tandem force sensor from that of in the ideal tandem force sensor lies in that the its central column is longer. Unlike the ideal tandem force sensor, limited by the size of the contact force sensor, the contact force sensor is not surrounded by the side surface of traction force sensor. Similarly, the connections of different components of the developed tandem force sensor are made by screw fastening. Besides, to achieve the series connection of the contact force sensor and traction force sensor, an intermediate connection flange is added.

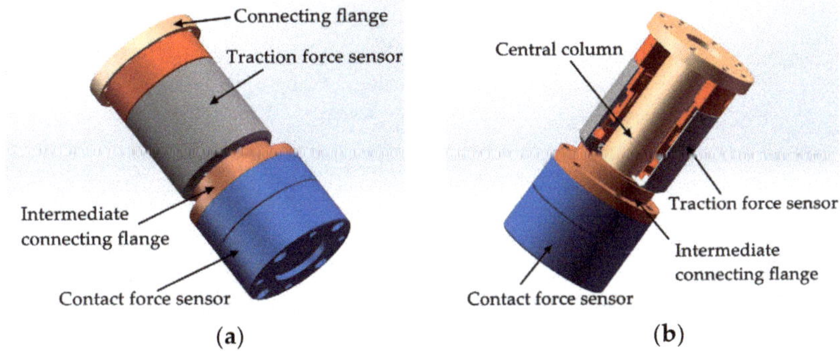

Figure 4. Schematic diagram of the developed tandem force sensor: (**a**) structure of the developed tandem force sensor; (**b**) the inner structure of the developed tandem force sensor.

To realize the tandem force sensor, we design and develop a cylindrical traction force sensor firstly. Compared with common wrist force sensors, the unique features of the cylindrical traction force sensor are that it senses external force applied to the side surface and its internal space provides adequate space for the central column. The basic structure of the ESB of the cylindrical traction force sensor is a thin-walled cylinder. The free end of the thin-walled cylinder-shaped ESB is connected with the side surface of the traction force sensor, and its fixed end is fixed to the connecting flange, as shown in Figures 2 and 4b. The internal space of the ESB is not valuable for the detection of traction force. However, it is significant for the realization of the tandem force sensor. In the tandem force sensor, the central column is not only used to connect the contact force sensor, but also provides rigid support for the contact force sensor and the end effector mounted on it. Hence, the diameter of the central column should not be small, which is 32 mm in this paper. By selecting reasonable structural parameters of the cylinder-shaped ESB, enough space can be provided for the central column, which is one of the main advantages of the cylinder-shaped ESB. In addition, the internal space is also important for the arrangement of the contact force sensor and for the signal lines of the contact force sensor.

3. Development of the Cylindrical Traction Force Sensor

3.1. Architecture of the Cylindrical Traction Force Sensor

Referring the force sensor in literature [28], the cylindrical traction force sensor is designed, as shown in Figure 5a. The cylindrical traction force sensor is consisting of a cylinder-shaped elastic structural body, a connecting fitting and a shell. The cylinder-shaped ESB shown in Figure 5b is the core of the traction force sensor, and it has layer A (black area), layer B (red area) and layer C (blue area). Compared with the diaphragm type ESB [29], cross beam type ESB [30–32], parallel type ESB [22,33], etc., the cylinder-shaped ESB is hollow, and the free space inside can be used as the connection channel between the contact force sensor and the traction force sensor. The layer A consists of *A1*, *A2*, *A3*, and *A4*, and layer C is composed by *C1*, *C2*, *C3*, and *C4* (Figure 6). *A1*, *A2*, *A3*, and *A4* are uniformly distributed along the circumference, the *C1*, *C2*, *C3*, and *C4* are also uniformly distributed along the circumference. In addition, the angle between *A1* and *C1* is 45 degrees, and the angle between the slots in layer A and the slots in layer C are 45 degrees or times of 45 degrees. The fixed end of cylinder-shaped ESB is fixed to the connecting fitting shown in Figure 5c by screw fastening, and the contact force sensor is fixed to the central column of it by screw fastening. Then, the connecting fitting can be fixed to the end flange of a robot and provide rigid support for the ESB and the contact force sensor. The shell shown in Figure 5d is secured to the free end of the ESB by screw fastening, and it can transfer the traction force exerted by human hands to the free end of ESB, as shown in Figure 2.

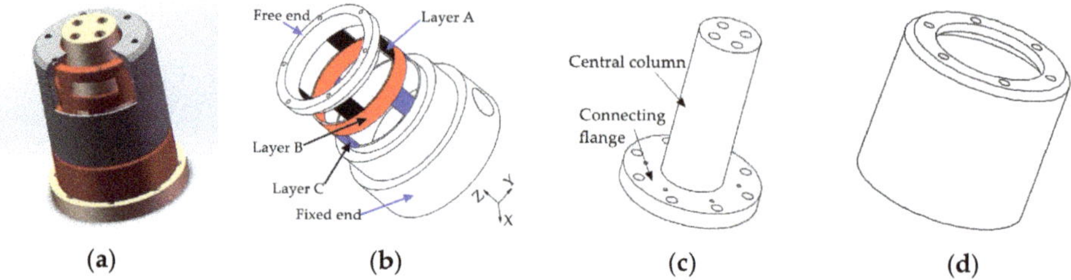

Figure 5. Schematic diagram of the composition of the cylindrical traction force sensor: (**a**) the basic architecture of the cylindrical traction force sensor; (**b**) cylinder-shaped elastic structural body; (**c**) connecting fitting; (**d**) shell.

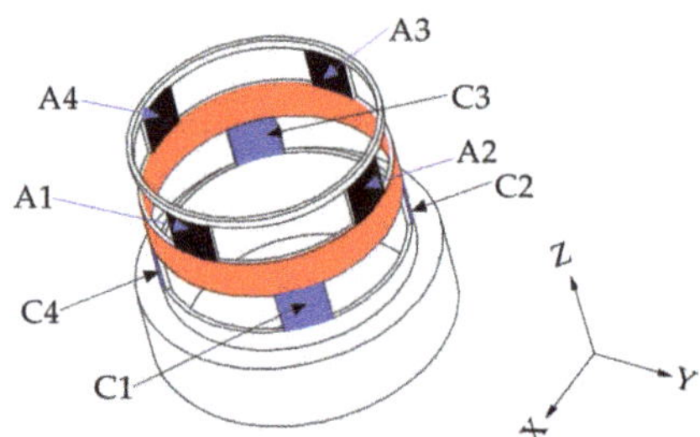

Figure 6. Basic structure of the cylinder-shaped elastic structural body.

3.2. Basic Force Measurement Principle of the Cylindrical Traction Force Sensor

The basic structure of the cylinder-shaped ESB can be illustrated by Figure 6. Under the traction force, the ESB will produce bending deformation and shear deformation, which will lead to the occurrence of normal stress and shear stress in the ESB. The normal stress mainly exists in layer A and layer C, which is relatively small. Therefore, the traction force sensor uses shear stress to measure traction force.

The layer A of ESB, which is used to measure the force F_X along the X-axis and the force F_Y along the Y-axis, consists of *A1*, *A2*, *A3*, and *A4*. When the force F_X is applied on the ESB, the *A2* and *A4* will produce shear stress. The strain values of the two points on the same diameter in the outside surface of *A2* and *A4* have the same signs, as shown in Figure 7a. Besides, under the moment M_Z, the *A1*, *A2*, *A3*, and *A4* will produce shear deformation. The strain values of the two points on the same diameter in the outside surface of *A2* and *A4*, respectively, have apposite signs, and the strain values of the two points on the same diameter in the outside surface of *A1* and *A3* respectively have apposite signs, as shown in Figure 7b. When the moment M_Z act on the cylinder-shaped ESB, the sum of strain values of the two points on the same diameter in the outside surface of *A2* and *A4* respectively is zero. Then, by measuring the sum of strain values of the points in the outside surface of *A2* and *A4*, respectively and using this characteristic, the force F_X can obtain. Similar to F_X, the force F_Y can be measured by measuring the sum of strain values of the points in the outside surface of *A1* and *A3*, respectively.

Figure 7. The shear stress' direction of the points in the outside surface of layer A: (**a**) under the force F_X; (**b**) under the torque M_Z.

The layer C of ESB used for the measurement of moment M_Z is composed by *C1*, *C2*, *C3*, and *C4*. Under the moment M_Z, the strain values of the two points on the same diameter in the outside surface of *C1* and *C3*, respectively, own different signs, and the sign of strain values of the points in the outside surface of *C2* is opposite to that of the points on the same diameter in *C4*, as shown in Figure 8c. When the force F_X or the force F_Y or the combination of both is acting on the ESB, the sign of strain values of the points located at the outside surface of *C1* is the same as that of the point on the same diameter in *C3*, the same as the *C2* and *C4* (Figure 8a,b). Then, under the force F_X or the force F_Y or the combination of both, the difference of strain values of the two points on the same diameter in the outside surface of *C1* and *C3* or *C2* and *C4*, respectively, is zero. However, when the F_X, F_Y, and M_Z act on the cylinder-shaped ESB, the difference of strain values of the two points on the same diameter in the outside surface of *C1* and *C3*, respectively, is not zero, same thing with *C2* and *C4*. By using this property, the moment M_Z can be detected by measuring the difference of strain values of the points in the outside surface of *C1*, and *C3* and the difference of strain values of the points in the outside surface of *C2* and *C4*.

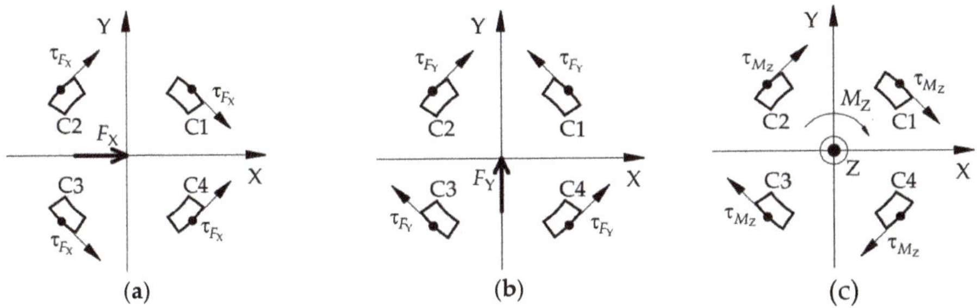

Figure 8. The shear stress' direction of the points in the outside surface of layer C: (**a**) under the force F_X; (**b**) under the force F_Y; (**c**) under the torque M_Z.

The layer B of the ESB is a ring connected to layer A and layer C respectively, and it can measure the force F_Z, the moment M_X and the moment M_Y. Under the force F_Z, layer A bears axial pressure (Figure 9a). When this axial pressure is transmitted to the layer B, there is the axial shear stress in layer B, as shown in Figure 10. Figure 10 illustrates the basic constitutional unit of the ESB, and the expansion diagram of ESB is shown in Figure 11. The axial pressure induced by force F_Z causes shear deformation of *B1*, *B2*, *B3*, *B4*, *B5*, *B6*, *B7*, and *B8* (*B1−B8*), and then generate axial shear stress in the axial cross section of *B1−B8*. Moreover, the sign of strain values of the points in the outside surface of *B1−B8* are the same. Besides, under the moment M_X, *A2* and *A4* are subjected to the axial pressure in opposite direction, respectively (Figure 9b), which causes shear deformation in *B3*, *B4*, *B7*, and *B8*.

The sign of strain values of the points in the outside surface of *B3* and *B4*, respectively, is opposite to that of in *B7* and *B8*. Then, by using this property, the force F_Z can be measured by measuring the sum of strain values of the points in the outside surface of *B1−B8*, and the moment M_X can be measured by the difference between the strain values of the points in the outside surface of *B3* and *B7* and that of in *B4* and *B8*. Similar to moment M_X, moment M_Y can also be measured.

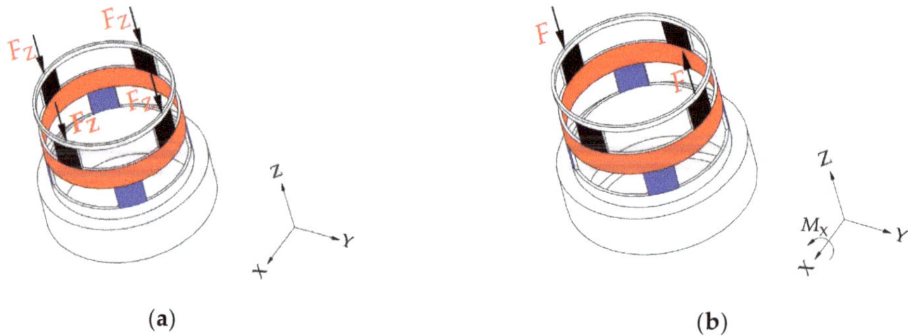

Figure 9. Force diagram of the cylindrical traction force sensor: (**a**) under the force F_Z; (**b**) under the torque M_X.

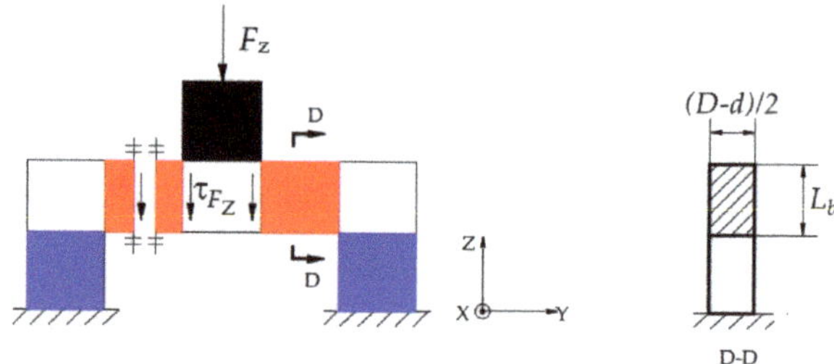

Figure 10. Basic constitutional unit of cylinder-shaped elastic structural body.

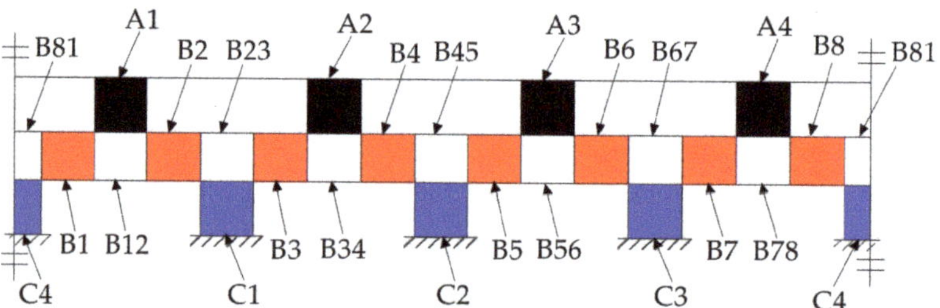

Figure 11. The unfold of the cylinder-shaped elastic structural body.

3.3. Mechanical Model of the Cylindrical Traction Force Sensor

To meet the design requirement of traction force sensor, the selection of ESB structural sizes should be carried out on the basis of theoretical analysis. Therefore, based on theory of mechanics, we analyze the mechanical properties of the ESB and establish the mechanical model of the ESB, which is of great significance for the determination of structural sizes of ESB and for the understanding of the mechanism of force perception and the mechanical properties of the ESB.

3.3.1. The Mechanics Analysis under the F_X

When the traction force F_X exerts on ESB, the circular ring between layer A and the free end of the ESB will produce shear deformation along the force direction. According to the mechanics of materials, the direction of the shear stress of a point on the excircle of the circular ring coincides with the tangential direction of the excircle of it, and the angle between its direction vector and the direction of force F_X is an acute angle, as shown in Figure 12. According to the calculation method of shear stress, the shear stress of point e can be calculated by using the following equation.

$$\tau_{F_X} = \frac{F_X \cdot S_z}{I_z \cdot (D-d)/2} = \frac{4F_X \cdot \sin \alpha}{\pi D (D-d)} \tag{1}$$

where $S_z = D^2(D-d)\sin \alpha/8$ is the static moment of $\hat{c}e$ segment ring with regard to Z-axis, $I_z = \pi D^3(D-d)/16$ is the moment of inertia with respect to Z-axis, D is the diameter of the excircle of the ESB, d is the diameter of the inner circle of the ESB, α is the acute angle between point a and point e about the Z-axis.

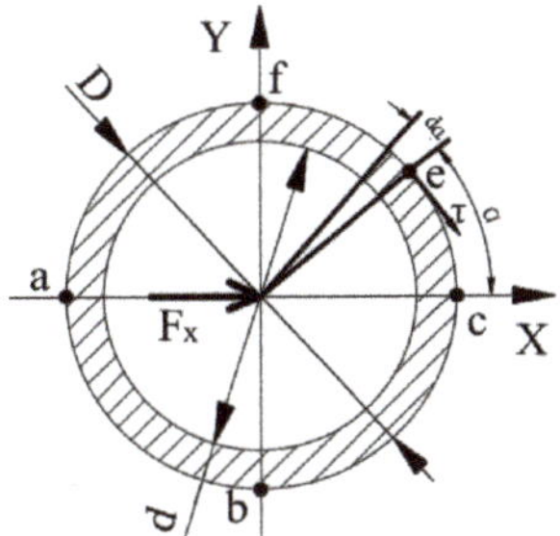

Figure 12. The shear stress analysis of the circular ring.

According to Equation (1), the shear stresses of point a and point c are zero, and the shear stresses of point b and point f are the largest. Then, the distribution of shear stress values of points in the outer surface of the circular ring is shown in Figure 13.

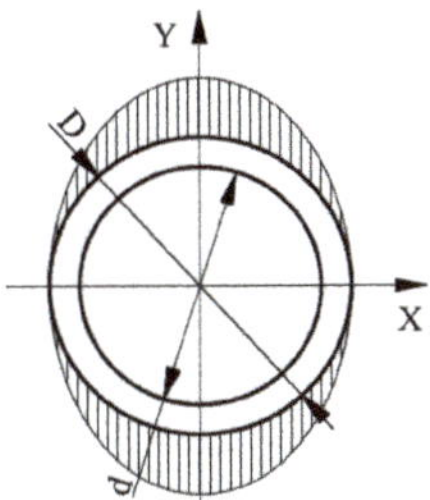

Figure 13. The distribution of shear stress values of the points in the outside surface of the circular ring.

Based on Equation (1) and Figure 13, without considering stress concentration, the distribution of shear stress values of the points in the outside surface of layer A is shown in Figure 14a. According to Figure 14a, the shear stress in *A1* and *A3* is small, while that of *A2* and *A4* is large. Therefore, the shear stress of the points in *A2* and *A4* can be utilized to measure F_X. In addition, the largest shear stress value in *A2* and *A4* caused by force F_X is as follows.

$$\tau_{F_X} = \frac{4F_X \cdot \sin(\pi/2)}{\pi D (D-d)} = \frac{4F_X}{\pi D (D-d)} \tag{2}$$

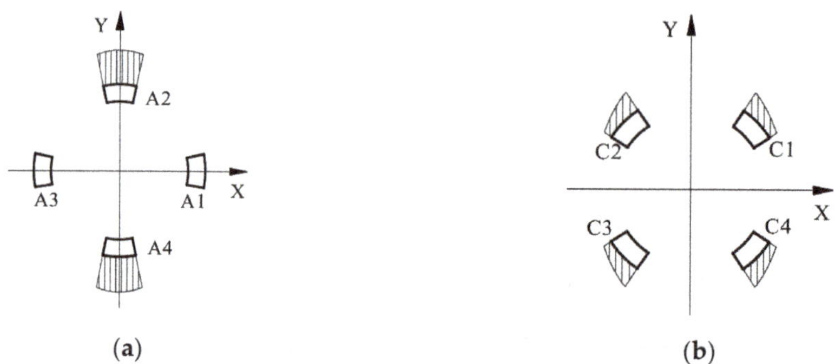

Figure 14. The distribution of shear stress values of the points in the outside surface of layer A and layer C: (**a**) layer A; (**b**) layer C.

Similar to layer A, the distribution of shear stress values of the points in the outside surface of layer C can be obtained, as shown in Figure 14b. According to Figure 14b, the values of shear stress of the points in *C1*, *C2*, *C3*, and *C4* are not too large nor too small. Besides, the direction of shear stress at points in *C2* is the same as that of the points in *C4* and the direction of shear stress at points in *C1* is the same as that of the points in *C3* (Figure 8a).

Unlike layer A and layer C, under the F_X, the layer B bears no shear stress. For layer B, the shear stress in *A2* and *A4* transmits to *B34* and *B78* that connects with layer A, which induces the occurrence of the normal stress in layer B, as shown in Figure 15. This paper utilizes the shear stress in the ESB to measure the traction force. Therefore, the normal stress in layer B will not affect the measurement of M_X, M_Y and F_Z.

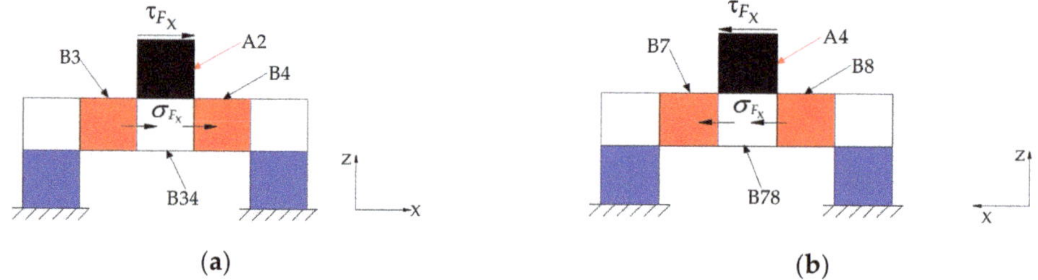

Figure 15. The stress in layer B induced by F_X: (**a**) the stress transmitted to B34 by A2; (**b**) the stress transmitted to B78 by A4.

3.3.2. The Mechanics Analysis under the F_Y

According to the basic structure of the ESB, under the F_Y, the deformation of the ESB is similar to that of under the F_X. Similar to Equation (2), the following formula is important for the measurement of force F_Y.

$$\tau_{F_Y} = \frac{4F_Y}{\pi D(D-d)} \tag{3}$$

However, unlike under force F_X, the points with the largest shear stress are in *A1* and *A3* and the points that owns zero shear stress value exist in *A2* and *A4*. Hence, the indirect measurement of force F_Y can be achieved by using the shear stress values of the points in *A1* and *A3*.

3.3.3. The Mechanics Analysis under the Force F_Z

Under the force F_Z, the $A1$, $A2$, $A3$, and $A4$ bear axial pressure, the $C1$, $C2$, $C3$, and $C4$ also under axial pressure. Therefore, the shear stress of the points in the outside surface of layer A and layer C is zero. According to Figures 10 and 11, under the F_Z, the cross sections along Z-axis of $B1$–$B8$ will bear shear force, and the shear stress of the points in $B1$–$B8$ can be calculated using the following equation.

$$\tau_{F_Z} = \frac{F_Z}{A} = \frac{2F_Z}{L_b(D-d)} \tag{4}$$

where $A = L_b(D-d)/2$ is the area of the cross section along Z-axis of layer B (Figure 10), $(D-d)/2$ is the wall thickness of layer B, L_b is the height of layer B.

Then, based on Equation (4), the force F_Z can be measured by detecting the shear stress values of the points in $B1$–$B8$.

3.3.4. The Mechanics Analysis under the Moment M_X

When the moment M_X acts on the cylinder-shaped ESB, the force/moment applied on $A1$, $A2$, $A3$, and $A4$ can be simplified as shown in Figure 9b, which leads to the occurrence of normal stress in the $A1$, $A2$, $A3$, and $A4$. In addition, under the M_X, $C1$, $C2$, $C3$, and $C4$ will also produce normal stress, but no shear stress. When the M_X is positive, $A2$ bears the largest tension and $A4$ understands the largest pressure. However, the normal stress in $A1$ and $A3$ is close to zero, because the central axis of twist goes through $A1$ and $A3$. For the layer B, the tension applied on $A2$ will transmit to $B34$, and the tension in $B34$ will cause shear stress in the outside surface of $B3$ and $B4$. Similarly, the outside surface of $B7$ and $B8$ will also produce shear stress. Because the normal stress in $A1$ and $A3$ is approximately zero, the tensions/pressures applied on $B3$ and $B4$ or $B7$ and $B8$ induced by moment M_X approximate to $F_{M_X} = M_X/D$. Then, the largest shear stress value of the points in the outside face of $B3$, $B4$, $B7$ and $B8$ can be figured out.

$$\tau_{M_X} = \frac{F_{M_X}}{A} = \frac{M_X/D}{L_b(D-d)} = \frac{M_X}{L_b(D-d)D} \tag{5}$$

Although both F_Z and M_X cause shear stress in $B3$, $B4$, $B7$, and $B8$, the sign of shear stress incurred by M_X in $B3$ and $B4$ is apposite to that of $B7$ and $B8$, the sign of shear stress caused by F_Z in $B3$ and $B4$ is the same as that of $B7$ and $B8$. Then, the shear stress of the points in the outside surface of $B3$ and $B4$ minus the shear stress of the points in the outside surface of $B7$ and $B8$ is the shear stress caused by M_X. On the contrary, the shear stress of the points in the outside surface of $B3$ and $B4$ add the shear stress of the points in the outside surface of $B7$ and $B8$ is the shear stress caused by F_Z. Therefore, by using this property, F_Z and M_X can be measured, respectively.

In addition, the F_Y applied on the ESB generates the moment around X-axis at layer B, as shown in Figure 16. Therefore, the moment measured by using the shear stress in $B3$, $B4$, $B7$, and $B8$ is the superposition of the true moment M_X and the moment caused by F_Y. However, the true moment M_X applied on the ESB is the moment value we need to measure. The force F_Y is measurable by using the shear stress in $A1$ and $A3$, and the moment arm of the moment caused by F_Y is available. Then, the moment caused by F_Y can be calculated out, after which the true moment M_X is obtainable.

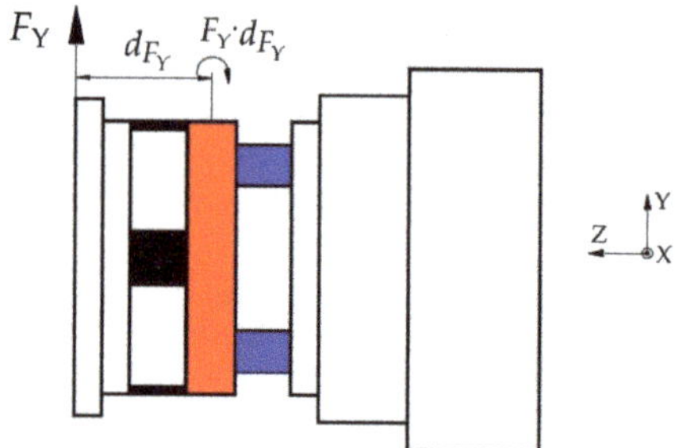

Figure 16. The moment applied on layer B caused by F_Y.

3.3.5. The Mechanics Analysis under the M_Y

Under the M_Y, the deformation of ESB is similar to that of under the M_X. Therefore, similar to Equation (5), the following equation can be obtained.

$$\tau_{M_Y} = \frac{M_Y}{L_b(D-d)D} \tag{6}$$

Unlike under moment M_X, under the M_Y, $A1$, and $A3$ bear the largest tension or pressure, and the normal stress in $A2$ and $A4$ is close to zero. Then, the points in the outside surface of $B1$, $B2$, $B5$, and $B6$ produce relatively large shear stress. Under the combined action of M_Y and F_Z, both will cause shear stress in $B1$, $B2$, $B5$, and $B6$. The sign of shear stress incurred by M_Y in $B1$ and $B2$ is apposite to that of $B5$ and $B6$, the sign of shear stress caused by F_Z in $B1$ and $B2$ is the same as that of $B5$ and $B6$. By using this property, the F_Z and M_X can be measured respectively. In addition, the F_X applied on the shell will also produce moment around Y-axis. Therefore, the measurement of true moment M_Y applied on the ESB also needs to wipe off the moment around Y-axis caused by F_X.

3.3.6. The Mechanics Analysis under the Moment M_Z

When the moment M_Z act on the cylinder-shaped ESB, the $A1$, $A2$, $A3$, and $A4$ all produce shear stress and the value of shear stress can be calculated using the following equation.

$$\tau_{M_Z} = \frac{M_Z}{R \cdot A'} = \frac{M_Z}{D/2 \cdot \pi D(D-d)/2(r+1)} = \frac{4M_Z(r+1)}{\pi D^2(D-d)} \tag{7}$$

where $A' = \pi D(D-d)/2(r+1)$ is the area of the cross section of layer A perpendicular to Z-axis, $R = D/2$ is the radius of the excircle of the ESB, r is the ratio between the arc length of four grooves and the arc length of $A1$, $A2$, $A3$ and $A4$.

Under the M_Z, the direction of shear stress of the points in the outer surface of $A1$ and $A2$ is opposite to that of $A3$ and $A4$ respectively, as shown in Figure 7b. However, under the F_X, the direction of shear stress of the points in $A2$ and $A4$ is the same (Figure 7a). Similarly, under the F_Y, the direction of shear stress of the points in $A1$ and $A3$ is the same. Therefore, the measurement of F_X or F_Y that using the shear stress of the outside surface of $A2$ and $A4$ or $A1$ and $A3$ respectively will not be affected by M_Z.

For layer B, the shear stress in $A1$, $A2$, $A3$, and $A4$ transmits to $B12$, $B34$, $B56$, and $B78$ that connects with layer A. Then, $B1$–$B8$ under the normal stress, which does not affect the measurement of M_X, M_Y and F_Z. The normal stress in $B1$–$B8$ causes stress in $B23$, $B45$, $B67$, and $B81$, which induces the shear stress in $C1$, $C2$, $C3$, and $C4$. The values of the shear stress in the outside surface of $C1$, $C2$, $C3$, and $C4$ are the same as that of layer A, which can be figured out by using Equation (7). Similarly, the direction of the shear stress of the points in the outside surface of $C1$ is opposite to that of $C3$, the direction of the shear stress of the points in the outside surface of $C2$ is opposite to that of $C4$ (Figure 8c). Besides,

F_X and F_Y also affect the shear stress of the points in the outside surface of $C1$, $C2$, $C3$, and $C4$. However, according to Figure 8a,b, under F_X and F_Y, the direction of the shear stress of the points in the outside surface of $C1$ is the same as that of $C3$ and the shear stress of the points in the outside surface of $C2$ is the same as that of $C4$. Hence, the shear stress in the outside surface of $C1$, $C2$, $C3$, and $C4$ can be used to detect M_Z without being affected by F_X and F_Y.

3.4. Parameter Selection and Strength Check of Cylinder-Shaped Elastic Structural Body

3.4.1. Sensitivity and Parameter Selection of the Elastic Structural Body

Strain values under unit forces and torques can reflect the sensitivities of a force sensor. The microstrain measured by the strain gauge is $\varepsilon = \tau/E$, where E is the elasticity modulus and τ is the shear stress caused by unit force/moment. Aluminum alloy 7075 is selected to machine the cylinder-shaped ESB, and the elasticity modulus of aluminum alloy 7075 is $E = 71.7\ Gpa$. Under the unit traction force, the micro strains measured by the strain gauges pasted in the outside surface of the ESB are as follows.

$$\begin{cases} S_{Fx} = S_{Fy} = 4/\pi D(D-d)E \\ S_{Fz} = 2/l_b(D-d)E \\ S_{Mx} = S_{My} = 1/l_b(D-d)DE \\ S_{Mz} = 4(r+1)/\pi D^2(D-d)E \end{cases} \tag{8}$$

where $S_{Fx}, S_{Fy}, S_{Fz}, S_{Mx}, S_{My}, S_{Mz}$ are the sensitivities of ESB with respect to traction forces/torques $F_X, F_Y, F_Z, M_X, M_Y, M_Z$ respectively.

According to Equation (8), the smaller D, d and l_b are, the larger r is, and the larger sensitivities will be. However, based on the design criteria, the parameters of the ESB must not be too small. Considering the convenience of mechanical processing of the ESB and the pasting of strain gauges, the selected parameters are $D = 50\ mm$, $d = 48\ mm$, $l_b = 9\ mm$, $r = 3$ and the heights of layer A and layer C are $l_a = 10\ mm$, $l_c = 9\ mm$, respectively. Substituting the parameters into Equation (8), the theoretical sensitivities of the ESB can be obtained, as shown in Table 1.

Table 1. Theoretical sensitivity of the elastic structural body.

Forces/Moments	Sensitivity	Unit
F_X	0.177	$\mu\varepsilon/N$
F_Y	0.177	$\mu\varepsilon/N$
F_Z	1.549	$\mu\varepsilon/N$
M_X	0.155	$\mu\varepsilon/N{\cdot}cm$
M_Y	0.155	$\mu\varepsilon/N{\cdot}cm$
M_Z	0.142	$\mu\varepsilon/N{\cdot}cm$

3.4.2. Strength Check of the Cylinder-Shaped ESB

In order to prevent the overload damage of the traction force sensor, it is necessary to obtain the maximum force that the cylinder-shaped ESB can withstand, which can be calculated by the following equation.

$$\begin{cases} F_{X-max} = F_{Y-max} = (\pi D(D-d)/4)[\tau] \\ F_{Z-max} = (L_b(D-d)/2)[\tau] \\ M_{X-max} = M_{Y-max} = L_b(D-d)D[\tau] \\ M_{Z-max} = (\pi D^2(D-d)/4(r+1))[\tau] \end{cases} \tag{9}$$

where $[\tau]$ is the permissible shear stress of the 7075 aluminum alloy used to machine the cylinder-shape ESB, $[\tau] = 0.5[\sigma]$, $[\sigma] = \sigma_s/2.5 = 182\ N/mm^2$, $\sigma_s = 455\ N/mm^2$ is the yield stress of 7075 aluminum alloy, $F_{X-max}, F_{Y-max}, F_{Z-max}, M_{X-max}, M_{Y-max}$, and M_{Z-max} are the largest F_X, F_Y, F_Z, M_X, M_Y, and M_Z that the ESB can bear, respectively.

Substituting the parameters into Equation (9), the maximum forces/moments that the ESB can withstand can be figured out, and they are $F_{X-max} = 2857.4$ N, $F_{Y-max} = 2857.4$ N, $F_{Z-max} = 819$ N, $M_{X-max} = 8190$ N·cm, $M_{Y-max} = 8190$ N·cm and $M_{Z-max} = 8929.38$ N·cm, respectively. In the kinesthetic teaching of robot, humans will not to use large forces to guide the movement of robots. Therefore, for human, the theoretical maximum forces/moments that the ESB can bear are very large, which are enough to prevent the traction force sensor from damaging.

3.5. Measurement of the Traction Force

Given the true sensitivities of the traction force sensor, the traction force can be calculated according to the measured strain values. By combining Equations (2)–(8), the traction force can be figured out, as follows.

$$
\begin{bmatrix} F_X \\ F_Y \\ F_Z \\ M'_X \\ M'_Y \\ M_Z \end{bmatrix} = \begin{bmatrix} 1/S_{F_X} & 0 & 0 & 0 & 0 & 0 \\ 0 & 1/S_{F_Y} & 0 & 0 & 0 & 0 \\ 0 & 0 & 1/S_{F_Z} & 0 & 0 & 0 \\ 0 & 0 & 0 & 1/S_{M_X} & 0 & 0 \\ 0 & 0 & 0 & 0 & 1/S_{M_Y} & 0 \\ 0 & 0 & 0 & 0 & 0 & 1/S_{M_Z} \end{bmatrix} \begin{bmatrix} \varepsilon_{F_X} \\ \varepsilon_{F_Y} \\ \varepsilon_{F_Z} \\ \varepsilon_{M'_X} \\ \varepsilon_{M'_Y} \\ \varepsilon_{M_Z} \end{bmatrix} \tag{10}
$$

where ε_{F_X}, ε_{F_Y}, ε_{F_Z}, $\varepsilon_{M'_X}$, $\varepsilon_{M'_Y}$ and ε_{M_Z} are the shear strain values caused by F_X, F_Y, F_Z, M'_X, M'_Y, and M_Z, respectively.

In order to measure the shear strains caused by external forces/moments, the strain gauges need to be pasted to the ESB in a manner of $\pm 45°$ with the axis of the ESB and the strain gauges pasted in different regions are formed into six electric bridges. The output of an electric bridge is voltage, not strain value. Then, Equation (10) can be rewrote to exhibit the mapping relation between the voltage changes of electric bridges and the external forces.

$$
[F]_{6\times1} = [S]_{6\times6} \cdot [K]_{6\times6} \cdot [\Delta v]_{6\times1} = [P]_{6\times6}[\Delta v]_{6\times1} \tag{11}
$$

where $[F]_{6\times1}$ is $[F_X, F_Y, F_Z, M'_X, M'_Y, M_Z]^T$; $[S]_{6\times6}$ is the diagonal matrix in Equation (10); $[K]_{6\times6}$ is the coefficient matrix of strain transfer of electric bridge, the elements in $[K]_{6\times6}$ is the strain values corresponding to unit voltage; the elements in $[\Delta v]_{6\times1}$ are the change values in the output voltage of the electric bridges; $[P]_{6\times6}$ is equal to $[S]_{6\times6}[K]_{6\times6}$.

Owing to the moment M'_X in $[F]_{6\times1}$ includes the moment caused by F_Y and the moment M'_Y contains the moment induced by F_X, the amendment is necessary to get the real moment M_X and M_Y applied on the traction force sensor. The following equation can eliminate the errors in M'_X and M'_Y.

$$
\begin{cases} M_X = M'_X - F_Y \cdot d_{F_Y} \\ M_Y = M'_Y - F_X \cdot d_{F_X} \end{cases} \tag{12}
$$

where d_{F_X} is the moment arm from the application point of the force F_X to the moment measuring point, d_{F_Y} is the moment arm from the application point of the force F_Y to the moment measuring point, and in ideal circumstances, d_{F_X} is equal to d_{F_Y}.

3.6. The Realization of Traction Force Sensor

After the traction force sensor is machined, it is necessary to paste strain gauges for the shear stress measurement on the surface of the cylinder-shaped ESB. To measure the traction force, we pasted 48 miniature strain gauges on the outer surface of the cylinder-shaped ESB, and the distribution diagram of these strain gauges is shown in Figure 17. The blue rectangles in Figure 17 represent strain gauges, and the red squares in Figure 17 represent the connecting terminals of strain gauges. In order to measure the shear stress of one point, two strain gauges are pasted on the same area at an angle of 90°, and the angle between the two strain gauges and the direction of shear stress is 45° and −45° respectively. Therefore. One strain gauge is used to detect tensile stress caused by shear stress, and the other is used to measure compression stress induced by shear stress. Moreover, the strain gauges pasted in *A1, A2, A3, A4, C1, C2, C3,* and *C4* should be pasted in the area that bears the largest shear stress under the F_X, F_Y, and M_Z, that is, the middle position of these areas. However, based on the analysis in Sections 3.3.3–3.3.5, the strain gauges pasted in *B1–B8* can be arranged as Figure 17 shows. After the strain gauges were pasted, the cylinder-shaped ESB is shown in Figure 19a.

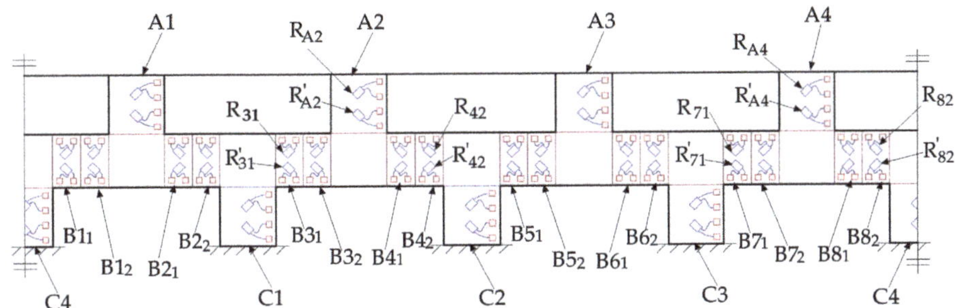

Figure 17. Distribution of strain gauges pasted on the cylinder-shaped elastic structural body.

After the pasting of the strain gauges, the strain gauges pasted in different areas are connected to form six electric bridges. The four strain gauges pasted in *A2* and *A4* are connected to form the 1st electric bridge to measure the strain caused by the force F_X. The strain gauges pasted in *A1* and *A3* are connected to form the 2nd electric bridge to measure the strain caused by the force F_Y. The strain gauges pasted in $B1_2$, $B2_1$, $B3_2$, $B4_1$, $B5_2$, $B6_1$, $B7_2$, and $B8_1$ are connected to form the third electric bridge to measure the strain induced by the force F_Z. Similarly, the strain caused by the moment M'_X can be measured by the fourth electric bridge made up of strain gauges stuck in $B3_1$, $B4_2$, $B7_1$, and $B8_2$, the indirect measurement of the moment M'_Y is obtainable by the fifth electric bridge made up of strain gauges pasted in $B1_1$, $B2_2$, $B5_1$, and $B6_2$, the strain caused by the moment M_Z can be measured by the sixth electric bridge made up of strain gauges pasted in *C1, C2, C3,* and *C4*.

As presented in Sections 3.2 and 3.3, this paper utilizes the sum or the difference of strain values of the points in the outside surface of the ESB to measure the traction force. The sum of strain values of the points in the outside surface of *A2* and *A4* respectively is used to represent the force F_X. Therefore, the connection mode of the four strain gauges pasted in *A2* and *A4* is shown in Figure 18a. ΔR_{F_X} and ΔR_{M_Z} represent the changes in the resistance values of the strain gauges caused by the force F_X and the moment M_Z respectively. In addition, the minus sign and plus sign of ΔR_{F_X} and ΔR_{M_Z} indicate that the strain gauge is compressed and stretched respectively. According to the measurement principle of electric bridges, when ΔR_{F_X} is zero, the output voltage is zero even if ΔR_{M_X} is not equal to zero. However, the output voltage is not zero when ΔR_{M_X} is equal to zero and ΔR_{F_X} is not zero. Therefore, the 1st electric bridge can measure the force F_X. In order to measure the forces F_Y and F_Z, the connection mode of the second and the third electric bridges is basically the same as that of the 1st electric bridge.

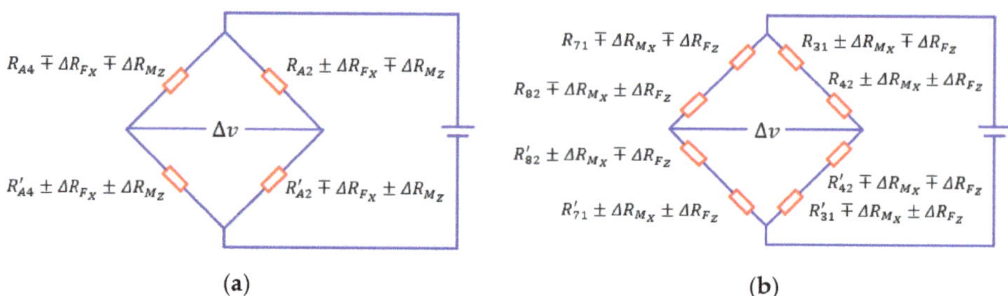

Figure 18. The connection mode of strain gauges in electric bridge: (**a**) the first electric bridge; (**b**) the fourth electric bridge.

According to Sections 3.2 and 3.3.4, the difference between the strain values of the points in the outside surface of *B3* and *B4* and that in *B7* and *B8* is used to represent the moment M_X. Therefore, the connection mode of the eight strain gauges pasted in $B3_1$, $B4_2$, $B7_1$, and $B8_2$ is shown in Figure 18b. ΔR_{F_Z} and ΔR_{M_X} represent the changes in the resistance values of the strain gauges caused by the force F_Z and the moment M_X respectively. According to the measurement principle of electric bridges, when ΔR_{M_X} is zero, the output voltage is zero even if ΔR_{F_Z} is not equal to zero. However, the output voltage is not zero when ΔR_{F_Z} is equal to zero and ΔR_{M_X} is not zero. Therefore, the fourth electric bridge can measure the moment M_X. In order to measure the moments M_Y and M_Z, the connection mode of the fifth and the sixth electric bridges is basically the same as that of the fourth electric bridge.

According to the structure of the traction force sensor shown in Figure 5, the cylinder-shaped ESB, connecting fitting and shell were assembled into a traction force sensor by screw fastening, as shown in Figure 19b. The central column in the connecting fitting attaches the contact force sensor to the end of the traction force sensor and form a tandem force sensor. The output signal of the traction force sensor is voltage, and we developed a 12-channel signal acquisition instrument to realize the signal acquisition (Figure 19c). The 6-channel in the signal acquisition instrument is used for the signal acquisition of the traction force sensor, and the other 6-channel is used for the information acquisition of the contact force sensor.

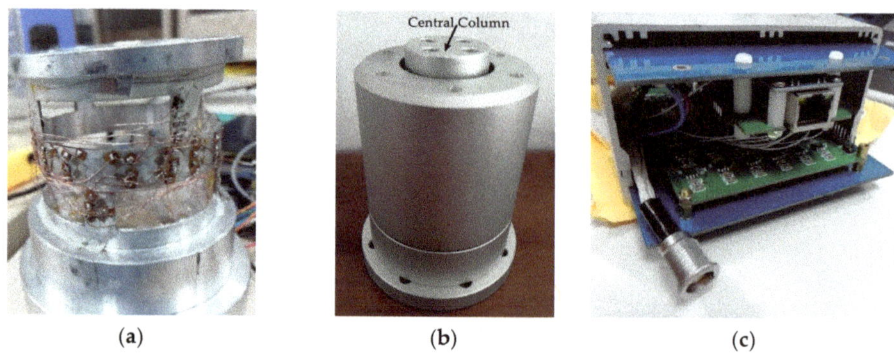

Figure 19. Cylinder-shaped elastic structural body and traction force sensor: (**a**) cylinder-shaped elastic structural body; (**b**) traction force sensor; (**c**) signal acquisition instrument.

3.7. Calibration Experiment of Cylindrical Traction Force Sensor

Equations (11) and (12) exhibit that the traction force can be detected by measuring variation values of voltages of six electric bridges. To obtain the real matrix $[P]_{6\times6}$ in Equation (11) and the moment arm in Equation (12), calibration experiment is necessary. In the calibration experiment of traction force sensor, we use a 6-DOF industrial robot to finish the calibration experiment, as shown in

Figure 20. In the calibration experiment, the robot remains stationary during the calibration process to provide a rigid support for the sensor, and forces and torques are applied to the sensor by mounting weights on the loading structure. Moreover, the attitude of the traction force sensor can be changed by adjusting the posture of the robot so that the forces/moments in different directions can be applied to the sensor. After applying forces/torques to the sensor, the self-developed signal acquisition instrument collects the output voltages of the sensor.

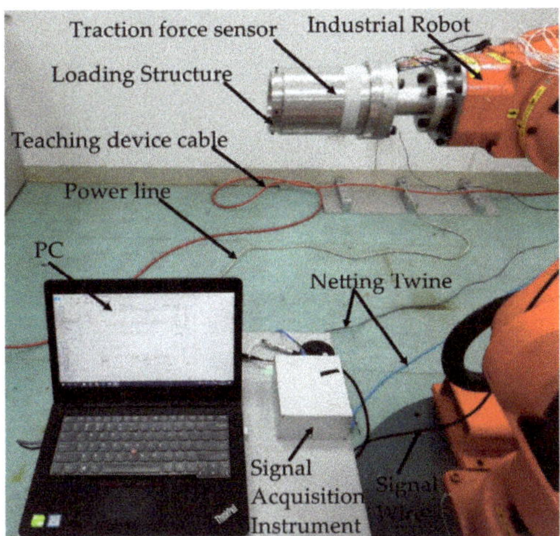

Figure 20. Industrial robot used in calibration experiments.

In the calibration experiment, small force/moment ranges are adopted because humans like to guide robot with small forces/torques. During the calibration process, F_X and F_Y adopt interval load of $\pm60\,N \times 10\,N$, F_Z adopts interval load of $60\,N \times 10\,N$, M_X and M_Y adopt interval load of $\pm201\,N{\cdot}cm \times (10\,N \times 3.35\,cm)$ and M_Z adopts interval load of $\pm192\,N{\cdot}cm \times (10\,N \times 3.2\,cm)$. The moments applied to the sensor were achieved by mounting weights on the loading structure. Therefore, when moments were applied to the sensor, the weights will also exert forces on the sensor. After each loading, the output values of electric bridge were recorded. Each calibration experiment was repeated three times to ensure the availability and repeatability of the experimental data. Under the external force, the changes of output voltage value are shown in Figure 21, and CH1, CH2, CH3, CH4, CH5, and CH6 represent the output voltage values of the first, second, third, fourth, fifth, and sixth electric bridge, respectively. Under the F_X and M_Z, CH1, CH5, and CH6 have significant output, and this certifies that F_X will induce the occur of moment around Y-axis; under the force F_Y and moment M_Z, CH2, CH4 and CH6 have significant output, and this certifies that F_Y will induce the occur of moment around X-axis. In addition, Figure 21 shows CH1 is mainly sensitive to F_X, CH2 is mainly sensitive to F_Y, CH3 is mainly sensitive to F_Z, CH4 is mainly sensitive to M_X, CH5 is mainly sensitive to M_Y and CH6 is mainly sensitive to M_Z. All of this certifies the theoretical analysis in Section 3.2.

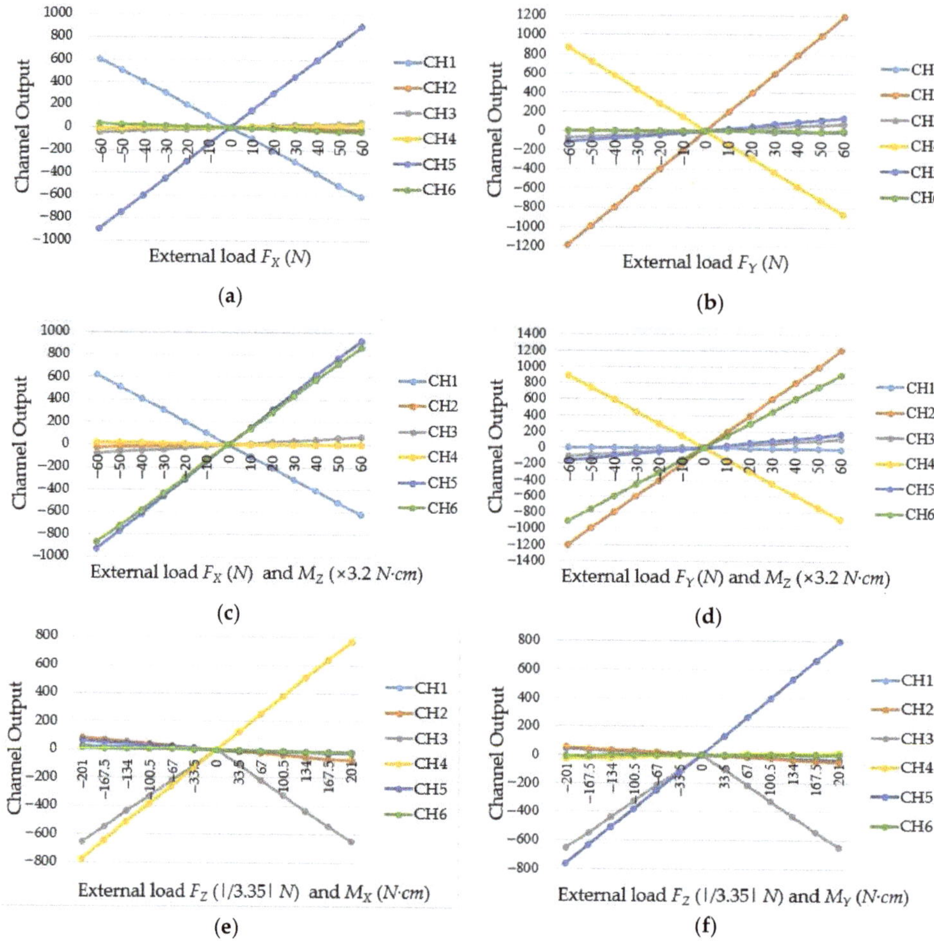

Figure 21. The output voltage changes of electric bridges: (**a**) under the force F_X; (**b**) under the force F_Y; (**c**) under the force F_X and moment M_Z; (**d**) under the force F_Y and moment M_Z; (**e**) under the force F_Z and moment M_X; (**f**) under the force F_Z and moment M_Y.

After the calibration experiment, the least square method was used to calculate the calibration matrix $[P]_{6\times6}$, as follows.

$$[P]_{6\times6} = \begin{pmatrix} -1.07 \times 10^{-1} & 5.21 \times 10^{-3} & -6.22 \times 10^{-3} & -2.26 \times 10^{-2} & 5.23 \times 10^{-1} & 1.37 \times 10^{-1} \\ -2.45 \times 10^{-3} & 5.44 \times 10^{-2} & -1.80 \times 10^{-3} & 2.20 \times 10^{-1} & -1.74 \times 10^{-1} & -4.14 \times 10^{-2} \\ -6.06 \times 10^{-4} & 3.13 \times 10^{-4} & -9.07 \times 10^{-2} & -3.50 \times 10^{-2} & 5.61 \times 10^{-2} & 1.61 \times 10^{-2} \\ -3.67 \times 10^{-3} & 5.91 \times 10^{-3} & -1.03 \times 10^{-2} & 2.62 & 1.43 \times 10^{-1} & -4.05 \times 10^{-2} \\ -5.64 \times 10^{-3} & 3.65 \times 10^{-3} & 2.88 \times 10^{-4} & -5.48 \times 10^{-2} & 2.59 & 8.91 \times 10^{-3} \\ -6.65 \times 10^{-4} & -7.34 \times 10^{-4} & 2.89 \times 10^{-3} & 6.99 \times 10^{-2} & -9.18 \times 10^{-2} & -2.12 \end{pmatrix} \quad (13)$$

Plug the calibration matrix into Equation (11) and using Equation (12), the calculated forces/torques can be obtained, which are presented in Figure 22. Then, the interference errors of the cylindrical traction force sensor are shown in Table 2, which shows that most of the errors are not larger than 1.0%, and the measurement ranges are $-60 \leq F_X \leq 60$ N, $-60 \leq F_Y \leq 60$ N, $0 \leq F_Z \leq 60$ N, $-201 \leq M_X \leq 201$ N·cm, $-201 \leq M_Y \leq 201$ N·cm, $-192 \leq M_Z \leq 192$ N·cm, respectively.

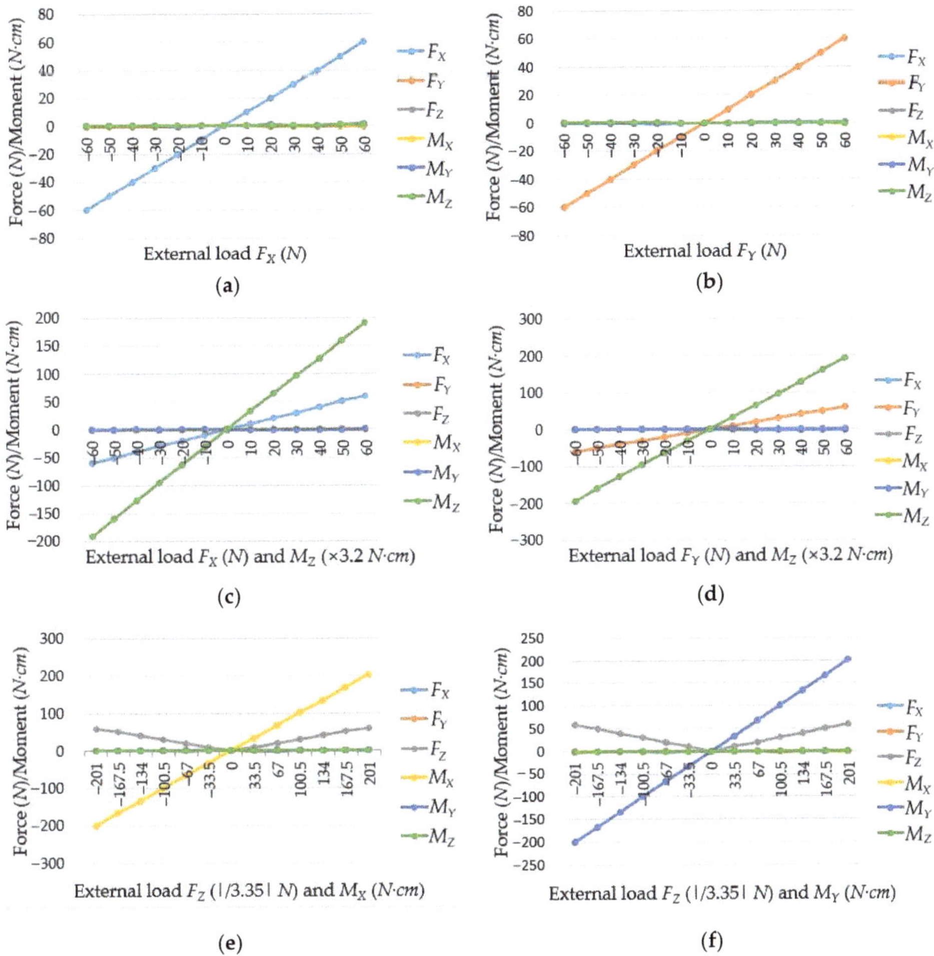

Figure 22. Force/Torque obtained by the cylindrical traction force sensor: (**a**) the force F_X acts on the sensor; (**b**) the force F_Y acts on the sensor; (**c**) under the force F_X and torque M_Z; (**d**) under the force F_Y and torque M_Z; (**e**) under the force F_Z and torque M_X; (**f**) under the force F_Z and torque M_Y.

Table 2. Interference error of the cylindrical traction force sensor.

Interference Error (%F.S.)	F_X	F_Y	F_Z	M_X	M_Y	M_Z
F_X	-	1.13	0.14	0.03	2.17	1.20
F_Y	1.09	-	0.22	0.94	0.14	1.52
F_Z	0.08	0.11	-	1.10	0.86	0.10
M_X	0.05	0.05	-	-	2.94×10^{-4}	1.54×10^{-5}
M_Y	0.02	0.03	-	0.05	-	8.53×10^{-5}
M_Z	-	0.35	0.14	3.43×10^{-4}	2.58×10^{-4}	-

Non-linear errors (NLES), hysteresis errors (HES) and repeatability errors (RES) are important indexes to show the static performance of a sensor. Five of the six NLES of the cylindrical traction force sensor are not larger than 0.70%, five of the six HES are not larger than 0.85% and four of the six RES are not larger than 0.80%, as Table 3 shows. To demonstrate the measurement error visually of the sensor, several load and measurement experiments of forces/moments were conducted, and Table 4 compares the calculated values with the actual values. The measurement errors in Table 4 verified that the cylindrical traction force sensor can detect the external forces/torques applied to it, and the measurement errors are small.

Table 3. Static performance indices of the cylindrical traction force sensor.

Forces/Moments	F_X	F_Y	F_Z	M_X	M_Y	M_Z
Non-Linear Error (%F.S.)	0.02	0.11	1.54	0.46	0.29	0.69
Hysteresis Error (%F.S.)	0.19	0.24	2.19	0.37	0.83	0.34
Repeatability Error (%F.S.)	0.30	0.51	0.75	1.24	1.98	0.31

Table 4. Calculated and real values when forces/torques are applied on traction force sensor.

	F_X	F_Y	F_Z	M_X	M_Y	M_Z
Applied	60.00	0.00	0.00	0.00	0.00	192.00
Measured	60.06	−0.56	−0.11	−0.86	0.26	191.50
Error (%F.S.)	0.10	0.93	0.18	0.43	0.13	0.26
Applied	0.00	60.00	0.00	0.00	0.00	192.00
Measured	0.55	60.07	−0.07	−0.36	0.00	193.76
Error (%F.S.)	0.92	0.12	0.12	0.22	0.17	0.92
Applied	0.00	0.00	60.00	201.00	0.00	0.00
Measured	−0.10	−0.10	59.79	201.52	1.19	−0.37
Error (%F.S.)	0.17	0.17	0.35	0.25	0.57	0.19
Applied	0.00	0.00	60.00	0.00	201.00	0.00
Measured	−0.03	−0.03	59.40	0.26	201.03	−0.10
Error (%F.S.)	0.05	0.05	1.00	0.12	0.01	0.05

4. The Realization and Application of the Tandem Force Sensor

4.1. The Tandem Force Sensor Based on the Developed Cylindrical Traction Force Sensor

According to the schematic diagram of the structure of the tandem force sensor shown in Figure 4a and the series connection mode shown in Figure 4b, a tandem force sensor is developed, as shown in Figure 23. The tandem force sensor is composed of a developed cylindrical traction force sensor and a contact force sensor connected in series, and the contact force sensor is connected with the cylindrical traction force sensor by an intermediate connecting flange. In addition, all connections are made by screw fastening. In the application, the tandem force sensor is connected to the robot end through the connection flange, and the end-effector can be fixed to the end of the tandem force sensor. In the kinesthetic teaching of robot contact tasks, the human hand exerts the traction force by grasping the shell of the traction sensor to guide the robot's motion, while the contact force sensor can accurately perceive the contact force between the robot's end-effector and the environment. Then, the traction and contact forces can be simultaneously perceived by the developed tandem force sensor in a decoupled manner.

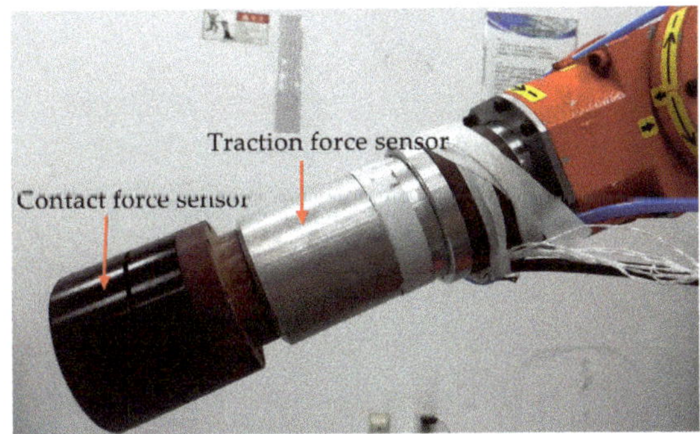

Figure 23. The developed tandem force sensor.

4.2. Application of the Developed Tandem Force Sensor

To further test the feasibility of the developed tandem force sensor, this paper designs drawer switch experiment based on human–robot interaction. In daily work and life, people can easily open a variety of drawers. However, it is not an easy task for robot to open and close diversified drawers like what human does. Human–robot interaction helps to transmit experience to robot and inform the robot of the method of opening and closing drawers, and then robot can learn the method to open and close drawers.

With the developed tandem force sensor, the drawer switch experiment can complete with human–robot interaction without damaging the drawer, and the robot can obtain several effective demonstrations. In the human–robot interaction to finish the drawer switch experiment, the tandem force sensor is mounted at the end of the robot and vacuum chuck, which allows the robot to control the opening and closing of drawers, is attached to the contact force sensor. In human–robot interaction, the teacher chooses a drawer in the locker and selects the adsorption area of the drawer. People guide the robot move from the initial point to the selected drawer, and control the suction cup to hold the drawer. Then, the human guides the robot to open the drawer to the maximum and finally, the human guides the robot to close the drawer, as shown in Figure 24. In this process, the tandem force sensor detects the traction force and the contact force between vacuum chuck and drawer, which allows the robot to act according to human intentions and its contact state with the object being operated on, not just human intentions. During the experiment, the data sampled by the tandem force sensor and the action taken by the teacher are saved as state-action pairs. Then, the robot can learn the policy of drawer switch task and perform drawer opening and closing by itself (Figure 25), which confirms the feasibility and effectiveness of the tandem force sensor.

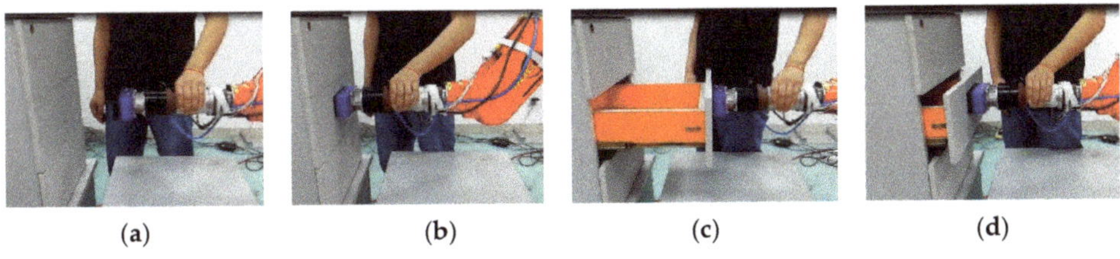

(a)   (b)   (c)   (d)

Figure 24. Human-robot cooperate to finish drawer switch experiment: (**a**) approach to the target; (**b**) grab the target; (**c**) open switch (**d**) close switch.

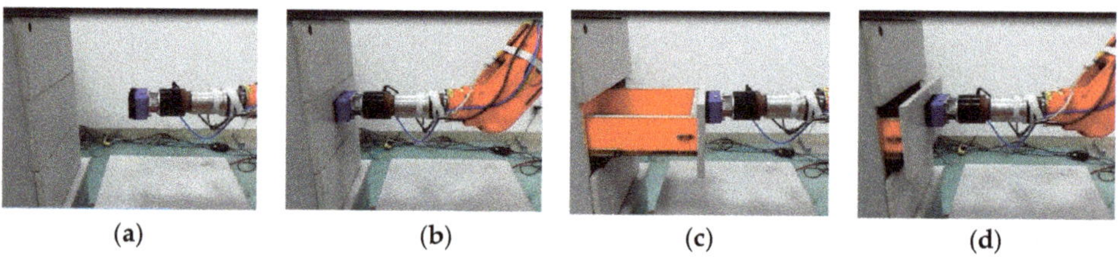

Figure 25. Robot finish drawer switch with the method human teaches: (**a**) approach to the target; (**b**) grab the target; (**c**) open switch; (**d**) close switch.

In the human–robot interaction, to complete the drawer switch experiment, the change curves of F_z^T and F_z^C (superscript T and C represent the traction force and contact force sensors, respectively) are shown in Figure 26. Owing to the inaccuracy of manual operation, the numerical fluctuation of the traction force F_z^T is high, while the numerical fluctuation of the contact force F_z^C is lower than F_z^T. In order to simulate the force curve in the kinesthetic teaching of drawer switch experiment based on single wrist force sensor, the resultant force of the traction force F_z^T and the contact force F_z^C has been calculated, as shown in Figure 27. By comparing Figures 26b and 27, it can be seen that the resultant force cannot accurately represent the contact state between the robot and the drawer. If the task policy is learned based on the resultant force, it cannot make the right decision. For example, when the drawer switch task policy is learned based on the data shown in Figure 27, the learned policy only outputs effective action instructions when the absolute value of contact force is about 20 N, instead of 0 N. Therefore, the net contact force obtained by the tandem force sensor is necessary for learning effective contact task policy.

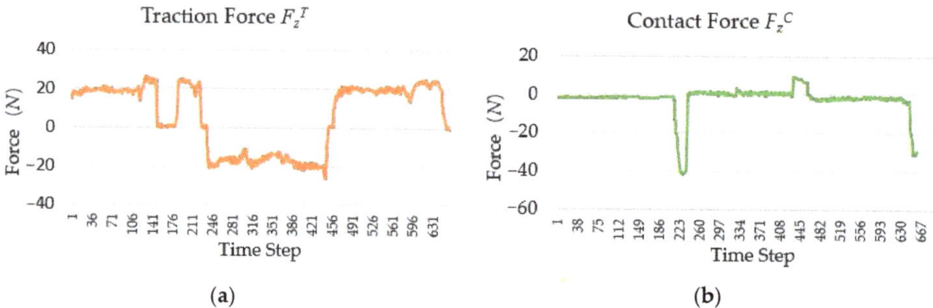

Figure 26. The change curve of the traction and contact forces in the drawer switch experiment: (**a**) Traction force F_z^T; (**b**) contact force F_z^C.

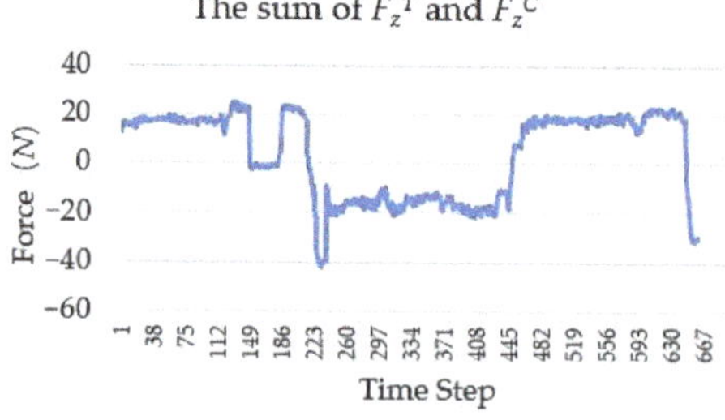

Figure 27. The change curve of the sum of the traction force F_z^T and the contact force F_z^C.

5. Conclusions

A tandem force sensor for measuring the traction and contact forces is introduced in this paper. In cases that a wrist force sensor is used as the contact force sensor, a cylindrical traction force sensor that is easy to handle by hand, has been designed. As the core of the cylindrical traction force sensor, the cylinder-shaped elastic structural body is designed, and the force measurement theory of it is analyzed in detail. Calibration experiments verify the theoretical analysis of the cylinder-shaped elastic structural body and the good static characteristics of the traction force sensor. Then, a wrist force sensor is mounted to the developed cylindrical traction force sensor to realize the tandem force sensor.

To verify whether the tandem force sensor can meet the original intention, the drawer switch experiment based on the tandem force sensor has been carried out. The traction force sensor in drawer switch experiment transmits the human intention to the robot, and the contact force sensor detects the contact status between the robot and the drawer. Human–robot interaction experiment shows that the tandem force sensor can sense the way and skill of a teacher and the contact force between robot and environment, so that the human and robot can cooperate to complete the task, which is the basis of how robots learn to accomplish contact tasks.

The traction force sensor can be combined with the contact force sensor as a tandem force sensor, or can be used alone. Note that, although we have only applied the tandem force sensor to the drawer switch experiment, this sensor can also be applied to a wide range of contact tasks that needs human–robot collaboration, such as assembly, grinding, polishing, and deburring. Moreover, the traction force sensor can be used alone for the non-contact tasks that need human–robot collaboration, such as paint spraying and track teaching. In these tasks, the most important advantage of the traction force sensor over the common wrist force sensor is that the gravity of the end-effector does not affect its measurement, which simplifies the sensor's gravity compensation.

Because the contact force sensor adopted in the tandem force sensor is a commercial wrist force sensor, and its structure is not optimized for the tandem force sensor, which makes the appearance of the developed tandem force sensor less graceful and complex. In the future, the structure of the tandem force sensor will be optimized, which will make it very graceful and close to the ideal tandem force sensor. By that time, the tandem force sensor can be utilized to robotic contact tasks in a very elegant way.

Author Contributions: Conceptualization, Z.Z., Y.C., and D.Z.; methodology, Z.Z.; software, Z.Z.; validation, Z.Z., and D.Z.; formal analysis, Z.Z.; investigation, Z.Z.; resources, Y.C.; data curation, Z.Z.; writing—original draft preparation, Z.Z.; writing—review and editing, Z.Z., Y.C., and D.Z.; visualization, Z.Z.; supervision, D.Z.; project administration, Y.C.; funding acquisition, Y.C. All authors have read and agreed to the published version of the manuscript.

Funding: This work was funded by the Science and Technology Support Project of the National Science Foundation of China, grant number 51775215.

Acknowledgments: We thank Jiming Sa (Wuhan University of Technology) and Jiangyu Hu (Wuhan University of Technology) for they assistance in the data collection. We also acknowledge the anonymous reviewers for taking time out of their busy schedules to review this paper.

Conflicts of Interest: The authors declare no conflict of interest.

References

1. Argall, B.D.; Chernova, S.; Veloso, M.; Browning, B. A Survey of Robot Learning from Demonstration. *Robot. Auton. Syst.* **2009**, *57*, 469–483. [CrossRef]
2. Oudeyer, P.Y. Socially Guided Intrinsic Motivation for Robot Learning of Motor Skills. *Auton. Robot.* **2013**, *36*, 273–294.
3. Erden, M.S.; Billard, A. Robotic Assistance by Impedance Compensation for Hand Movements While Manual Welding. *IEEE Trans Cybern* **2016**, *46*, 2459–2472. [CrossRef] [PubMed]
4. Song, C.; Liu, G.; Zhang, X.; Zang, X.; Xu, C.; Zhao, J. Robot Complex Motion Learning Based on Unsupervised Trajectory Segmentation and Movement Primitives. *ISA Trans* **2020**, *97*, 325–335. [CrossRef] [PubMed]

5. Xu, J.; Hou, Z.; Wang, W.; Xu, B.; Zhang, K.; Chen, K. Feedback Deep Deterministic Policy Gradient with Fuzzy Reward for Robotic Multiple Peg-in-Hole Assembly Tasks. *IEEE Trans. Ind. Inform.* **2019**, *15*, 1658–1667. [CrossRef]

6. Li, F.; Jiang, Q.; Zhang, S.; Wei, M.; Song, R. Robot Skill Acquisition in Assembly Process Using Deep Reinforcement Learning. *Neurocomputing* **2019**, *345*, 92–102. [CrossRef]

7. Racca, M.; Pajarinen, J.; Montebelli, A.; Kyrki, V. Learning in-Contact Control Strategies from Demonstration. In Proceedings of the 2016 IEEE/RSJ International Conference on Intelligent Robots and Systems (IROS), Daejeon, Korea, 9–14 October 2016; pp. 688–695.

8. Savarimuthu, T.R.; Buch, A.G.; Schlette, C.; Wantia, N.; Roßmann, J.; Martínez, D.; Torras, C.; Ude, A.; Nemec, B.; Kramberger, A.; et al. Teaching a Robot the Semantics of Assembly Tasks. *IEEE Trans. Syst. Man Cybern. Syst.* **2018**, *48*, 670–692. [CrossRef]

9. Fazeli, N.; Oller, M.; Wu, J.; Wu, Z.; Tenenbaum, J.B.; Rodriguez, A. See, Feel, Act: Hierarchical Learning for Complex Manipulation Skills with Multisensory Fusion. *Sci. Robot.* **2019**, *4*, 1–10. [CrossRef]

10. Duque, D.A.; Prieto, F.A.; Hoyos, J.G. Trajectory Generation for Robotic Assembly Operations Using Learning by Demonstration. *Robot. Comput. Integr. Manuf.* **2019**, *57*, 292–302. [CrossRef]

11. Zhang, T.; Louie, W.Y.; Nejat, G.; Benhabib, B. Robot Imitation Learning of Social Gestures with Self-Collision Avoidance Using a 3d Sensor. *Sensors* **2018**, *18*, 2355. [CrossRef]

12. Jha, A.; Chiddarwar, S.S.; Bhute, R.Y.; Alakshendra, V.; Nikhade, G.; Khandekar, P.M. Imitation Learning in Industrial Robots: A Kinematics Based Trajectory Generation Framework. *Artif. Intell. Rev.* **2017**, *15*, 1–6.

13. Zeng, C.; Yang, C.; Zhong, J.; Zhang, J. Encoding Multiple Sensor Data for Robotic Learning Skills from Multimodal Demonstration. *IEEE Access* **2019**, *7*, 145604–145613. [CrossRef]

14. Kang, G.; Oh, H.S.; Seo, J.K.; Kim, U.; Choi, H.R. Variable Admittance Control of Robot Manipulators Based on Human Intention. *IEEE/ASME Trans. Mechatron.* **2019**, *24*, 1023–1032. [CrossRef]

15. Dong, J.; Xu, J.; Zhou, Q.; Hu, S. Physical Human–Robot Interaction Force Control Method Based on Adaptive Variable Impedance. *J. Frankl. Inst.* **2020**, *357*, 7864–7878. [CrossRef]

16. Massa, D.; Callegari, M.; Cristalli, C. Manual Guidance for Industrial Robot Programming. *Ind. Robot Int. J.* **2015**, *42*, 457–465. [CrossRef]

17. Vergara, C.; De Schutter, J.; Aertbeliën, E. Combining Imitation Learning with Constraint-Based Task Specification and Control. *IEEE Robot. Autom. Lett.* **2019**, *4*, 1892–1899.

18. Abu-Dakka, F.J.; Nemec, B.; Kramberger, A.; Buch, A.G.; Krüger, N.; Ude, A. Solving Peg-in-Hole Tasks by Human Demonstration and Exception Strategies. *Ind. Robot: Int. J.* **2014**, *41*, 575–584. [CrossRef]

19. Song, H.C.; Kim, Y.L.; Song, J.B. Guidance Algorithm for Complex-Shape Peg-in-Hole Strategy Based on Geometrical Information and Force Control. *Adv. Robot.* **2016**, *30*, 552–563. [CrossRef]

20. Kim, W.Y.; Ko, S.Y.; Park, J.O.; Park, S. 6-Dof Force Feedback Control of Robot-Assisted Bone Fracture Reduction System Using Double F/T Sensors and Adjustable Admittances to Protect Bones against Damage. *Mechatronics* **2016**, *35*, 136–147. [CrossRef]

21. Lee, S.; Song, C.; Kim, K. Design of Robot Direct-Teaching Tools in Contact with Hard Surface. In Proceedings of the 2009 IEEE International Symposium on Assembly and Manufacturing, Suwon, Korea, 17–20 November 2009; pp. 231–234.

22. Yao, J.; Zhu, J.; Wang, Z.; Xu, Y.; Zhao, Y. Measurement Theory and Experimental Study of Fault-Tolerant Fully Pre-Stressed Parallel Six-Component Force Sensor. *IEEE Sens. J.* **2013**, *13*, 3472–3482. [CrossRef]

23. Lee, D.H.; Kim, U.; Jung, H.; Choi, H.R. A Capacitive-Type Novel Six-Axis Force/Torque Sensor for Robotic Applications. *IEEE Sens. J.* **2016**, *16*, 2290–2299. [CrossRef]

24. Xiong, L.; Jiang, G.; Guo, Y.; Liu, H. A Three-Dimensional Fiber Bragg Grating Force Sensor for Robot. *IEEE Sens. J.* **2018**, *18*, 3632–3639. [CrossRef]

25. Liu, J.; Li, M.; Qin, L.; Liu, J. Active Design Method for the Static Characteristics of a Piezoelectric Six-Axis Force/Torque Sensor. *Sensors* **2014**, *14*, 659–671. [CrossRef]

26. Lee, Y.R.; Neubauer, J.; Kim, K.J.; Cha, Y. Multidirectional Cylindrical Piezoelectric Force Sensor: Design and Experimental Validation. *Sensors* **2020**, *20*, 4840. [CrossRef]

27. Zhang, Z.; Chen, Y.; Zhang, D.; Xie, J.; Liu, M. A Six-Dimensional Traction Force Sensor Used for Human-Robot Collaboration. *Mechatronics* **2019**, *57*, 164–172. [CrossRef]

28. Song, W. Optimization Design and Static-Dynamic Characteristic Study of Six-axis Force Sensor. Master's Thesis, Anhui University of Science and Technology, Huainan, China, 2010.

29. Hu, G.; Gao, Q.; Cao, H.; Pan, H.; Shuang, F. Decoupling Analysis of a Six-Dimensional Force Sensor Bridge Fault. *IEEE Access* **2018**, *6*, 7029–7036. [CrossRef]
30. Hu, S.; Wang, H.; Wang, Y.; Liu, Z. Design of a Novel Six-Axis Wrist Force Sensor. *Sensors* **2018**, *18*, 3120. [CrossRef] [PubMed]
31. Kebede, G.A.; Ahmad, A.R.; Lee, S.C.; Lin, C.Y. Decoupled Six-Axis Force-Moment Sensor with a Novel Strain Gauge Arrangement and Error Reduction Techniques. *Sensors* **2019**, *19*, 3012. [CrossRef]
32. Lin, C.Y.; Ahmad, A.R.; Kebede, G.A. Novel Mechanically Fully Decoupled Six-Axis Force-Moment Sensor. *Sensors* **2020**, *20*, 395. [CrossRef]
33. Noh, Y.; Bimbo, J.; Sareh, S.; Wurdemann, H.; Fras, J.; Chathuranga, D.S.; Liu, H.; Housden, J.; Althoefer, K.; Rhode, K. Multi-Axis Force/Torque Sensor Based on Simply-Supported Beam and Optoelectronics. *Sensors* **2016**, *16*, 1936. [CrossRef]

Publisher's Note: MDPI stays neutral with regard to jurisdictional claims in published maps and institutional affiliations.

approach for social human-robot interaction. Worth noting is that humans perform sequential decision-making in daily life where sequential decision making describes problems that require successive observations, i.e., cannot be solved with a single action [6]. Consequently, much of social human-robot interactions can be formulated as sequential decision-making tasks, i.e., RL problems. The goal of the robot in these types of interactions would be to learn an action-selection strategy in order to optimize some performance metric, such as user satisfaction.

Before outlining the research related to reinforcement learning in social robots, first it is important to establish the definition of a social robot in the context of this article. A variety of definitions for a social robot have been proposed in the literature [7–12]. Within each of these definitions, there is a wide spectrum of characteristics. However, two important aspects become prominent in these definitions that are considered in this paper, namely, embodiment and interaction/communication capability. One example can be found in Bartneck and Forlizzi [10] where they define a social robot as an "... autonomous or semi-autonomous robot that interacts and communicates with humans by following the behavioral norms expected by the people with whom the robot is intended to interact." Following this definition, the authors stress that a social robot must have a physical embodiment. Based on the presented definitions in [7–12], we consider social robots as embodied agents that can interact and communicate with humans. Figure 1 shows some of the social robots that are used in the reviewed papers.

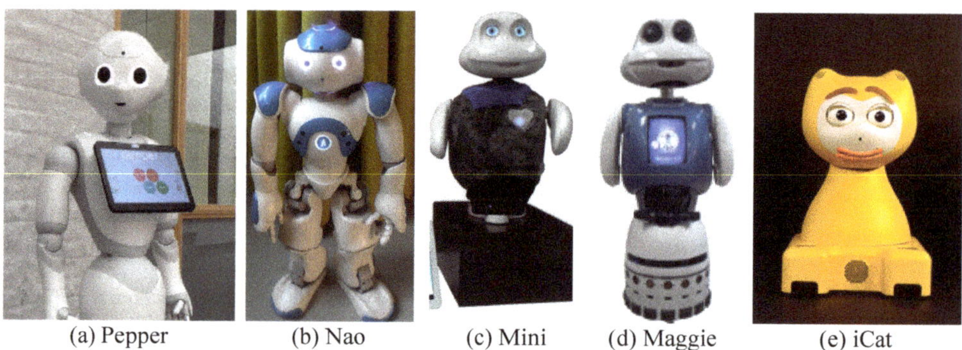

Figure 1. Some of the social robots platforms referenced within the reviewed papers. (The pictures of (**a**) Pepper robot, and (**b**) Nao robot were taken by the authors. (**c**) Mini robot, the figure is adapted from [13]—licensed under the Creative Commons Attribution, (**d**) Maggie robot, the figure is from https://robots.ros.org/maggie/, accessed on 20 March 2020—licensed under the Creative Commons Attribution, (**e**) iCat robot, the figure is from https://www.bartneck.de/wp-content/uploads/2009/08/iCat02.jpg, accessed on 22 March 2020—used with permission, photo credit to Christoph Bartneck.)

This article presents a survey on RL approaches in social robotics. As such, it is important to emphasize that the scope of this survey is focused on studies that include physically embodied robots and real-world interactions. Considering the definition of [10] given above, this paper excludes studies with simulations and virtual agents where no physical embodiment is present. The presented review also excludes studies with industrial robots and studies that do not include any interaction with humans. Rather, this review exclusively focuses on papers that comprise both a social robot(s) and human input/user studies. It is worth noting that studies which use simulations for training and test on physical robot deployment with user studies fall within the selection criteria. Likewise, studies that use explicit or implicit human input in the learning process are also included.

Due to the complexity of the social interactions and the real-world, most of the studies applying RL are trained and tested in simulation environments. However, real-world interactions are extremely important not only for social robots but also for understanding the full potential of reinforcement learning. It is mentioned in [14] (p. 391), that "the full potential of reinforcement learning requires reinforcement learning agents to be embedded into the flow of real-world experience, where they act, explore, and learn in our world, and not just in their worlds." Generally speaking, the overall goal of an RL agent is to maximize the expected cumulative reward over time, as stated in the "reward hypothesis" [14] (p. 42). The reward in RL is used as a basis for discovering an optimal behavior. Hence, reward design is extremely important to elicit desired behaviors in RL-based systems. The choice of reward function is crucial in robotics, where the problem is also referred to as the "curse of goal specification" [15]. Therefore, in this paper, we provide a categorization based on reward design which is crucial for RL to be successful. Moreover, since communication capability is a distinctive feature of social robots, we discuss communication mediums utilized for reward design together with RL algorithms.

Finally, it is also worth noting that in the general field of robotics there is a plethora of research in RL. There also exist review papers on the topic of RL in robotics such as applications of RL in robotics in general [15,16], policy search in robot learning [17], safe RL [18], and Deep Reinforcement Learning (DRL) in soft robotics [19]. Indeed, RL has been applied to a variety of scenarios and domains within social robotics, with growing popularity. While the field of social robotics deserves a survey on its own, to the best of our knowledge, there exists no such survey on this particular research field. Thus, the main purpose of this work is to serve as a reference guide that provides a quick overview of the literature for social robotics researchers who aim to use RL in their research. Depending on the target user group, the application domain or the experimental scenario, different types of rewards, problem formulations or algorithms can be more suitable. In that sense, we believe that this survey paper will be beneficial for social robotics researchers.

Overview of the Survey

After surveying research on RL and social robotics, we analyze and categorize the studies based on four different criteria: (1) RL type, (2) the utilized communication mediums for reward function formulation, (3) the nature of the reward function, (4) the evaluation methodologies of the algorithms. These categorizations aim to facilitate and guide the choice of a suitable algorithm by social robotics researchers in their application domain. For that purpose, we elaborate on the different methods that are tested in real-world scenarios with a physical robot.

Categorization based on RL type includes bandit-based methods, value-based methods, policy-based methods, and deep RL (see Section 4). The utilized communication mediums are verbal communication, nonverbal communication, affective communication, tactile communication, and additional communication medium between the robot and the human. Moreover, there are studies in which higher interaction dynamics are used for reward formulation such as engagement, comfort, and attention. There are also other studies that do not use any communication medium at all for reward formulation. In the categorization based on the design of the reward mechanisms, three major themes emerged:

1. Interactive reinforcement learning: In these methods, humans are involved in the learning process either by providing reward or guidance to the agent (Section 5.1). This approach, in which the human delivers explicit or implicit feedback to the agent, is known as Interactive Reinforcement Learning (IRL).
2. Intrinsically motivated methods: There are different intrinsic motivations in the literature on RL [20], however, the most frequently used approaches in social robotics depend on the robot maintaining an optimal internal state by considering both internal and external circumstances (Section 5.2).
3. Task performance driven methods: In these methods, the reward the robot receives depends on either the robot's task performance or the human interactant's task performance, or a combination of both (Section 5.3).

The evaluation methodologies include (1) the algorithm point of view, (2) the user experience point of view, and (3) evaluation of both learning algorithm-related factors and user experience-related factors.

To formulate the social interactions as a reinforcement learning problem, researchers need to consider some key concepts such as input data, state representation, robot actions, and reward function. Moreover, after the implementation of RL, it should be decided how the evaluation will be performed. Therefore, we extract from each of the cited works the following key points (1) the input data, state space and action space (2) the reward function (3) the communication medium in the HRI scenario (4) the main experimental results (5) the experimental scenario and its validation. Therefore, the contributions of this paper include: (i) analysing and categorising the relevant literature in terms of type of RL used; (ii) analysing and categorising the relevant literature based on the reward function; (iii) analysing the relevant literature in terms of evaluation methodologies.

The paper is organized as follows: In Section 2, we discuss the benefits and challenges of applying RL in the social robotics domain. In Section 3, we present a background on reinforcement learning. Following the formal presentation of the methods, in Section 4, we present the applications of these methods in social robotics. Later, we present the categorization based on reward functions in Section 5. Evaluation methods are discussed in Section 6. In Section 7, we discuss the current approaches in the view of real-world RL challenges and proposed solutions. The section further includes the points that remain to be explored, and the approaches that have thus far received less attention. Finally, in Section 8, we conclude the paper.

2. RL in Social Robotics—Benefits and Challenges

Applications of social robots are numerous and range from entertainment to eldercare. The robot tasks in such cases involve interactive elements such as human-robot cooperation, collaboration, and assistance. To achieve longitudinal interaction with social robots, it is important for such robots to learn incrementally from interactions, often with non-expert end-users. In consideration of continuously evolving interactions where user needs and preferences change over time, hand-coded rules are labor-intensive. Even though rule-based systems are deterministic, it can be difficult to create rules for complex interaction patterns. Machine learning is bound to play an important role in a wide range of domains and applications including robotics. However, the social robot learning problem differs from the traditional ML setting in which there is a need for collected datasets or assumptions about the distribution of input data [21]. Often, social robots should be able to learn new tasks and task refinements in domestic (unstructured) environments. Furthermore, social robotics researchers need to deal with a particular challenge of learning in real-time from human-robot interactions. ML paradigms such as supervised learning and unsupervised learning are not designed for learning from real-time social interactions. On the contrary, RL represents an active process. Unlike other ML methods, it does not need to be provided desired outputs instead, it trains interactively based on reward signals and refines its behavior throughout the interaction. Moreover, interaction is a key component for social robots which makes RL a suitable approach. RL also provides a possibility to learn from natural interaction patterns by utilizing the various social elements in the learning process. Consideration of all these points suggests that socially guided machine learning [22] could be a more suitable approach than traditional ML approaches for social HRI.

In general, combining human and machine intelligence may be effective for solving computationally hard problems [23]. The term "socially guided machine learning" was first used by Thomaz et al. [22] and refers to approaches that include social interaction between a user and a machine in the learning process. Studies using IRL in social robotics can be considered as socially guided machine learning since they make use of human feedback in different forms in the learning process. The feedback provided by the human can be used for shaping the action policy (the human is involved in the action selection mechanism), or shaping the reward function [24]. It can be treated either as reward, in that the feedback is given based on the agent's past actions indicating "how good the taken action was", or policy feedback in which human feedback affects action selection or modification thereby indicating "what to do".

The majority of studies included in this review paper use IRL which may suggest that IRL could be the best suited approach in social robotics. However, IRL has its own challenges. Human teachers tend to give less frequent feedback (due to boredom and/or fatigue) as learning progresses, resulting in diminished cumulative reward [25]. Likewise, human teachers tend to provide more positive reward than punishment [26,27]. Yet another problem in IRL is the transparency issues that might arise during the training of a physical robot via human reward [28,29]. Reference [29] used an audible alarm to alert the trainer about the robot's loss of sense. Suay et al. [30] observed that experts could teach the defined task in a predefined time frame, whereas the same amount of time was not enough for inexperienced users. One solution suggested for this was algorithmic transparency during training, which shows the internal policy to the human teacher. However, the presentation of the model of the agent's internal policy might be obscure for naive human teachers. Therefore, this information should be presented in a straight-forward way that is easy to understand to avoid causing confusion. To exemplify, in [28] human trainers waited for the Leonardo robot to establish eye contact with them before they continued teaching. The eye contact was considered as the robot being ready for the next action. These kinds of transparent behaviors in which the robot communicates the internal state of the learning process should be taken into account for guiding human trainers in IRL. As noted in several studies, in IRL, the human teacher's positive and negative reward can be much more deliberate than a simple 'good' or 'bad' feedback [28,31]. The learning agent should be aware of the subtle meanings of these feedback signals. As an example, human trainers tend to have a positive bias [28,31].

In addition, there are a variety of technical challenges to address when implementing RL in social robotics and social HRI. One of the drawbacks of online learning through interaction with a human is the requirement of long interaction time, which can be tedious and impractical for the users, resulting in fatigue and a loss of interest. A considerable amount of interaction time can wear out the robot's hardware. An alternative is using a simulated world to train the algorithm and subsequently deploying it on the real robot. Using a simulated setting has several advantages. It allows the agent to carry out learning repeatedly, which would otherwise be very expensive in the real-world. Simulated environments can also run much faster than the real-world, thus permitting the learning agent to make proportionately more learning experiences. Bridging the gap between the simulated and the real-world is not a simple task. It may be achieved by randomizing the simulator and learning a policy that shows success across many simulators and can ultimately be robust enough to work in the real world. However, simulating the real-world can be very difficult, especially with regards to modeling relevant human behaviors. Simulating the human requires a predictive model of human interactive behaviors and social norms as well as modeling the uncertainty of the real-world. Furthermore, the use of RL in social robotics poses other challenges such as devising proper reward functions and policies, as well as dealing with the sparseness of the reward signals.

The exploration-exploitation dilemma is a well-known problem in RL and refers to the choice of actions to discover the environment or taking actions that have already proven to be effective in producing reward [14]. RL practitioners use different approaches to deal with the trade-off between exploration and exploitation, such as epsilon-greedy policy [32], epsilon-decreasing policy [33] and Boltzmann distribution [34]. The epsilon-greedy strategy exploits knowledge for maximizing rewards (greedily choosing the current best option), otherwise to select a random action with probability $\varepsilon \in [0,1]$ [14]. The epsilon-decreasing strategy decreases ε over time, thereby progressing towards exploitative behavior [14]. Boltzmann exploration uses Boltzmann distribution to select the action to execute. A temperature parameter balances between exploration and exploitation (high-temperature value for selecting actions randomly and low-temperature value for selecting actions greedily) [14].

Despite the mentioned challenges, there are also advantages of using RL in social robotics. One of the main advantages is that the robot can learn a personalized adaptation for different interactants, i.e., a different policy for each user. Social robots can learn social skills from their own actions without demonstrations through uncontrolled interaction experiences. This is especially true given that interaction dynamics are difficult to model and sometimes even humans cannot explain why they behave in a certain way. Therefore, RL may enable social robots to adapt their behaviors according to their human partners for natural human-robot interaction. In IRL, the immediate reward provided by the human teacher has the potential to improve the training by reducing the number of required interactions. Human teachers' guidance significantly reduces the number of states explored, and the impact of teacher guidance is proportional to the size of the state space, i.e., it increases as the size of the state space grows [26]. In RL, how to achieve a goal is not specified, instead the goal is encoded and the agent can devise its own strategy for achieving that goal. Intrinsically motivated reward signals might be useful in many real-world scenarios, where sparse rewards make the goal-directed behavior challenging. Approaches using human social signals have the advantage of utilizing signals that the user exhibits naturally during the interaction. It does not require an extra effort to collect the reward. However, the change in social signals would not be so sudden, which would very much affect the time for convergence. The role of human social factors deserves extra attention in online learning methods. Combination of RL with deep neural networks has shown success in many application areas. DRL is also a trending technique in social robotics as we see increasing work in recent years. It has the advantage of not needing manual feature engineering [35] and resulting in human-like behavior for social robots [36].

3. Reinforcement Learning

Reinforcement learning [5] is a framework for decision-making problems. Markov Decision Processes (MDPs) are mathematical models for describing the interaction between an agent and its environment. Formally, an MDP is denoted as a tuple of five elements $\langle \mathcal{S}, \mathcal{A}, \mathcal{P}, \mathcal{R}, \gamma \rangle$ where $\mathcal{S}$ represents the state space (i.e., the set of possible states), $\mathcal{A}$ represents the action space (i.e., the set of possible actions), $\mathcal{P} : \mathcal{S} \times \mathcal{A} \times \mathcal{S} \to [0, 1]$ represents the probability of transitioning from one state to another state given a particular action, $\mathcal{R} : \mathcal{S} \times \mathcal{A} \times \mathcal{S} \to \mathbb{R}$ represents the reward function, and γ is the discount factor that determines the importance of future rewards, $\gamma \in [0, 1]$. The agent interacts with its environment in discrete time steps, $t = 0, 1, 2, ...$; at each time step t, the agent gets a representation of the environmental state $S_t \in \mathcal{S}$, takes an action $A_t \in \mathcal{A}$, moves to next state S_{t+1}, and receives a scalar reward $R_{t+1} \in \mathcal{R}$. Figure 2 depicts the standard RL framework.

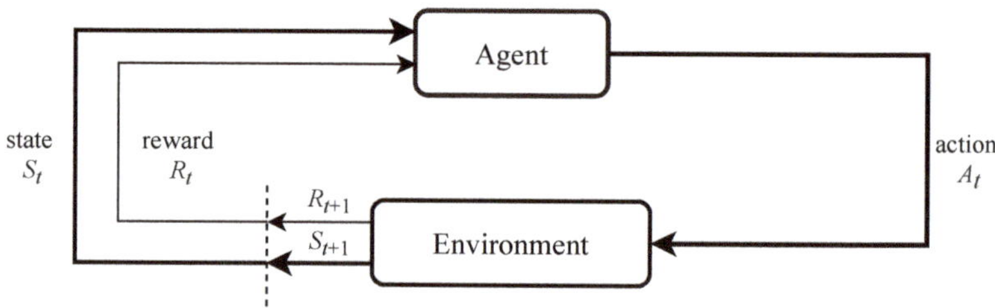

Figure 2. A standard reinforcement learning framework (reproduced from [14] (p. 38)).

The agent's behavior that maps states to actions is described as a policy, $\pi : \mathcal{S} \times \mathcal{A}$ where $\pi(s|a) = Pr(A_t = a|S_t = s)$ is the probability of taking action $a \in \mathcal{A}$ given state s. The agent's goal is to maximize the expected cumulative discounted reward, in other words *return* which is denoted as G_t:

$$G_t = \sum_{k=0}^{\infty} \gamma^k R_{t+k+1} \tag{1}$$

where γ is the discount factor and usually $\gamma \in [0, 1]$. The optimal behavior that is taking the best action at each state to maximize the reward over time is called optimal policy, π^*.

There exists a large variety of approaches in RL. They can be most broadly distinguished as model-based and model-free. Model-free approaches can be further subdivided into value-based and policy-based approaches. A shortened version of a RL taxonomy can be seen in Figure 3.

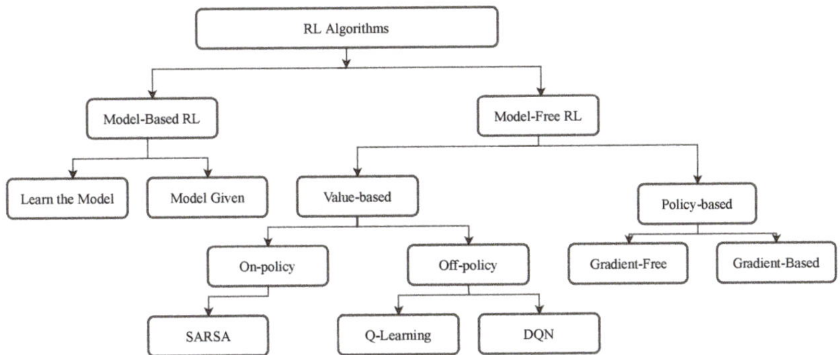

Figure 3. Taxonomy of Reinforcement Learning algorithms (reproduced and shortened from [37]).

3.1. Model-Based and Model-Free Reinforcement Learning

RL algorithms can be divided into two main categories, model-free RL and model-based RL, depending on whether the agent does or does not use a model of the environment dynamics, which can be either provided or learned. The model describes the transition function, $\mathcal{P}$, and the reward function, $\mathcal{R}$. The model-based methods can be divided into two categories: those that use a given model, i.e., the models of the transition and the reward function can be accessed by the agent, and the methods in which the agent learns the model of the environment [37]. In the latter approach, the agent learns a model, which it subsequently uses during policy improvement. The agent can collect samples from the environment by taking actions. From those samples state transitions and reward can be predicted through supervised learning. Planning methods can be used directly on the environment model. In the model-free approach, there is no effort to build a model of the environment, instead the agent searches for the optimal policy through trial and error interactions with the environment. Model-free methods are easier to implement in comparison with model-based methods. These methods can be advantageous over more complex methods when building a sufficiently accurate model is difficult [14] (p. 10).

3.2. Value-Based Methods

The value of policy π, namely the value function, is used to evaluate the states based on the total reward the agent receives over time. RL methods that approximate the value function through temporal difference (TD) learning instead of directly learning the policy π are called value-based methods. For each learned policy π, there are two related value functions: the state-value function, $v_\pi(s)$, and state-action value function (quality function), $q_\pi(s, a)$. The equations for $q_\pi(s, a)$ and $v_\pi(s)$ are given in Equations (2) and (3) respectively. E_π in Equations (2) and (3) means the agent follows policy π in each step.

$$q_\pi(s, a) = E_\pi[R_{t+1} + \gamma R_{t+2} + \gamma^2 R_{t+3} + ...|S_t = s, A_t = a] = E_\pi\left[\sum_{k=0}^{\infty} \gamma^k R_{t+k+1}\bigg|S_t = s, A_t = a\right] \quad (2)$$

$$v_\pi(s) = E_\pi[R_{t+1} + \gamma R_{t+2} + \gamma^2 R_{t+3} + ...|S_t = s] = E_\pi\left[\sum_{k=0}^{\infty} \gamma^k R_{t+k+1}\bigg|S_t = s\right]. \quad (3)$$

The value functions are expressed via the Bellman equation [38]. The Bellman equation for v_π and q_π is given in Equations (4) and (5) where s' indicates the next states from the set $\mathcal{S}$.

$$v_\pi(s) = \sum_a \pi(a|s) \sum_{s',r} p(s',r|s,a)[r + \gamma v_\pi(s')] \tag{4}$$

$$q_\pi(s,a) = \sum_{s'} p(s'|s,a)\left[r(s,a,s') + \gamma \sum_{a'} \pi(a'|s')q_\pi(s',a') \right]. \tag{5}$$

Comparing policies, a policy π is better than or equal to a policy π' if:

$$\pi \geq \pi' \text{ if } \forall s \in \mathcal{S} : v_\pi(s) \geq v_{\pi'}(s). \tag{6}$$

There exists always at least one optimal policy π^* whose expected return is greater than or equal to the other policy/policies for all states. Optimal policies share the same state-value function, defined as $v^*(s) = \max_\pi v_\pi(s)$ for all $s \in \mathcal{S}$, and action-value function, defined as $q^*(s,a) = \max_\pi q_\pi(s,a)$ for all $s \in \mathcal{S}$ and $a \in \mathcal{A}(s)$. The Bellman optimality equation for $q^*(s,a)$ is given in Equation (7).

$$q^*(s,a) = \sum_{s',r} p(s',r|s,a)\left[r + \gamma \max_{a'} q^*(s',a') \right]. \tag{7}$$

Another distinction in RL methods comes from the perspective of policy: on-policy vs. off-policy learning. On-policy methods learn the value of the policy that is used to make decisions. In the on-policy setting, the target policy and the behavior policy are the same. The target policy is the policy that is learned about, and the behavior policy is the policy that is used to generate behavior. The state-action-reward-state-action (SARSA) algorithm [39] is one of the on-policy methods in which the agent interacts with the environment, selects an action based on the current policy, then updates the current policy. The Q function update in SARSA is done using Equation (8). A transition from one state-action pair to the next is expressed as $(S_t, A_t, R_{t+1}, S_{t+1}, A_{t+1})$ which gives rise to the name SARSA. The update given in Equation (8) is done after every transition from a non-terminal state S_t.

$$Q(S_t, A_t) \leftarrow Q(S_t, A_t) + \alpha\left[R_{t+1} + \gamma\, Q(S_{t+1}, A_{t+1}) - Q(S_t, A_t)\right]. \tag{8}$$

In the off-policy methods, the target policy is different from the behavior policy. In these methods, the policy that is evaluated and improved does not match the policy that is used to generate data. Off-policy methods can re-use the experience from old policies or other agents' interaction experience to improve the policy. One example of an off-policy algorithm is Q-learning [40]. It is one of the most popular RL algorithms using discounted reward [41]. The Q-learning rule is defined by:

$$Q(S_t, A_t) \leftarrow Q(S_t, A_t) + \alpha\left[R_{t+1} + \gamma \max_a Q(S_{t+1}, a) - Q(S_t, A_t)\right]. \tag{9}$$

The Q-learning algorithm iteratively applies the Bellman optimality equation (given in Equation (7)). As shown in Equation (9), the main difference between Q-learning and SARSA (see Equation (8)) is that in the former the target value is not dependent on the policy being used and only depends on the state-action function.

3.3. Policy-Based Methods

Policy-based methods, also known as direct policy search methods, do not use value function models. In these methods, the policy is parameterized with θ and written as π_θ. They operate in the space of policy parameters Θ and $\theta \in \Theta$ [17]. The goal is still to maximize the accumulative return. The agent updates its policy by exploring various behaviors and exploiting the ones that perform well in regard to some predefined utility function $J(\theta)$. In many robot control tasks the state space, which includes both internal states and external states, is high-dimensional. The policy of the robot π_θ can be defined as a controller. For any state of the robot, this controller decides which actions to take or which signals to send to the actuators [42]. The robot takes its actions u according to the controller (please note, actions in policy search context are represented with u instead of a). The robot controller can be stochastic, i.e., $\pi(u|s)$ or deterministic, i.e., $\pi(s)$. After the action execution the robot transitions to another state according to the probabilistic transition function $p(s_{t+1}|s_t, u_t)$. These states and actions of the robot form a trajectory $\tau = (s_0, u_0, s_1, u_1, ...)$. The corresponding return for the trajectory τ is represented as $R(\tau)$. The global utility of the robot is denoted as:

$$J(\theta) = \mathbb{E}_{\tau \sim \pi_\theta}[R(\tau)]. \tag{10}$$

Computing the expectation in $\mathbb{E}_{\tau \sim \pi_\theta}[R(\tau)]$ requires to run an infinite number of trajectories with the current controller. The way to go around this difficulty is to sample the expectation. After performing a finite set of trajectories, the return is computed over these trajectories. Thus, the goal is:

$$\theta^* = \underset{\theta}{\mathrm{argmax}}\, J(\theta) = \underset{\theta}{\mathrm{argmax}} \sum_\tau P(\tau, \theta) R(\tau) \tag{11}$$

where θ^* is the estimate of global performance and $P(\tau, \theta)$ is the probability of τ under policy π_θ.

Here RL addresses a black-box optimization problem in that the function which relates the performance to the policy parameters is unknown. There are two families of methods: direct policy search and gradient descent [42]. In direct policy search algorithms, approximate gradient descent is performed by "random trial then selection" methods, like genetic algorithms, evolution strategies, finite differences, cross entropy, etc. These algorithms need many samples and can escape from local minima if large enough variations are used. In gradient descent methods, a mathematical transformation is used so that policy gradient methods can be applied. In these methods, the policy gradient update is given by:

$$\theta_{k+1} = \theta_k + \alpha \nabla_\theta J(\theta) \tag{12}$$

where α is a learning rate, and the policy gradient is given by [17]:

$$\nabla_\theta J(\theta) = \sum_\tau \nabla_\theta P(\tau, \theta) R(\tau). \tag{13}$$

There are different methods to estimate the gradient $\nabla_\theta J(\theta)$, interested readers may refer to [17]. Policy-based methods have the advantage of being effective in high dimensional or continuous action spaces and having better convergence properties.

Some methods learn both policy and value functions. These methods are called actor-critic methods, where 'actor' is the learned policy that is trained using policy gradient with estimations from the critic, and 'critic' refers to the learned value function that evaluates the policy.

3.4. Deep Reinforcement Learning

Learning in RL progresses over discrete time steps by the agent interacting with the environment. Obtaining an optimal policy requires a considerable amount of interaction with the environment, which results in high memory and computational complexity. Therefore, the tabular approaches that represent state-value functions, $v_\pi(s)$, or state-action value functions, $q_\pi(s, a)$, as explicit tables are limited to low-dimensional problems, and they become unsuitable for large state spaces. A common way to overcome this limitation is to find a generalization for estimating state values by using a set of features in each state. In other words, the idea is to use a parameterized functional form with weight vector $w \in \mathbb{R}^d$ for representing $v_\pi(s)$ or $q_\pi(s, a)$ that are written as $\hat{v}(s; \theta)$ or $\hat{q}(s, a; \theta)$

instead of tables [14] (p. 161). Such approximate solution methods are called function approximators. The reduction of the state space by using the generalization capabilities of neural networks, especially deep neural networks, is becoming increasingly popular. Deep Learning (DL) has the ability to perform automatic feature extraction from raw data. DRL introduces DL to approximate the optimal policy and/or optimal value functions [14] (p. 192). Recently, there has been an increasing interest in using DL for scaling RL problems with high-dimensional state spaces.

The DQN method, first presented by Mnih et al. [43], combines Q-learning with Convolutional Neural Networks (CNN) for learning to play a wide variety of Atari games better than humans. In DQN, the agent's experiences $e_t = (s_t, a_t, r_t, s_{t+1})$ are stored at each time step t in a data set $D_t = \{e_1, ..., e_t\}$, so-called experience replay memory. Q-learning updates are applied on a mini-batch uniformly sampled from the experience replay memory. The Q-learning update is done using Equation (14):

$$L_i(\theta_i) = \mathbb{E}_{s,a,r,s' \sim U(D)}\left[\left(r + \gamma \max_{a'} Q(s', a'; \hat{\theta}_i) - Q(s, a; \theta_i)\right)^2\right] \tag{14}$$

where θ_i represents the parameters (weights) of the Q-network at iteration i and $\hat{\theta}_i$ represents the parameters used to compute the target network at iteration i. The target network parameters $\hat{\theta}_i$ are updated to the parameters θ_i after every C iterations.

4. Categorization of RL Approaches in Social Robotics Based on RL Type

In human-human communication, a communication medium is a means of conveying information to other people. It can be in different forms such as verbal, nonverbal, affective, and tactile. Human-robot interaction overlaps with human-human interaction to a certain extent. Furthermore, there can be an additional physical interface (i.e., a computer, a tablet, a smart game board, etc.) shared between the robot and the human. In the interaction between the robot and the human, information transmission is bidirectional, the robot and the human can be sender, receiver, or both. In the surveyed papers, we see all these communication channels being utilized, especially for the RL problem formulation. As it has already been stated in the introduction, one of the prominent characteristics of social robots is the ability to interact and communicate. Therefore, we provide two categorizations in this section: first we categorize the papers based on RL types, after which we provide a further discussion and categorization with respect to the utilized communication channels and interaction dynamics for the reward functions.

4.1. Bandit-Based Methods

Bandit-based methods can be considered as a simplified case of RL in which the next state does not depend on the action taken by the agent. Different bandit-based methods explored in social robotics [4,44–47], such as dueling bandit learning [44], k-armed bandit method (multi-armed bandit) [4,45,46], and Exponential-Weight Algorithm for Exploration and Exploitation (Exp3) algorithm [47].

4.1.1. Additional Physical Communication Medium between the Robot and the Human

Learning user preferences to personalize the user experience is used in customizing advertisements and search results. A similar approach was applied in HRI studies [4,44]. Whereas the customization is done in the background for personalized experiences in websites using users' clicks, it is adapted for social interactions by asking the user to select their preferences using the buttons. In other words, these studies use a physical communication medium between the robot and the human. Schneider and Kummert [44] investigated a dueling bandit learning approach for preference learning. The algorithm draws two or more actions, and the relative preference is used as reward. It is defined as follows: In each time step $t > 0$ a pair of arms $(k_t^{(1)}, k_t^{(2)})$ is selected and presented to the user, if the user prefers $k_t^{(1)}$ over $k_t^{(2)}$ then $w_t = 1$, and $w_t = 2$ otherwise where w_t is a noisy comparison result. The distribution of outcomes is represented by a preference matrix $P = [p_{ij}]_{KxK}$, here p_{ij} is the probability that the user preferred arm i over arm j. The participant provided pairwise comparisons via a button. In the work by Ritschel et al. [4], the robot adapted its linguistic style to the user's preferences. They defined the learning tasks as k-armed bandit problems. The adaptation was done based on explicit human feedback given via buttons in the form of numeric reward (-1, $+1$). The actions of the robot were a set of scripted utterances. Similarly, Ritschel et al. [46] used an additional medium between the robot and the user. They employed the social robot Reeti as a nutrition adviser, where a custom hardware was utilized to obtain the information about the selected drink [46]. Their custom hardware included an electronic vessel holder and a smart scale that could communicate with the robot. The problem was formalized as an k-armed bandit problem where the actions of the robot were scripted spoken advice. The reward was calculated from the amount of calories and quantity of the selected drink.

4.1.2. Verbal and Nonverbal Communication Plus an Interface

Social robots can use any natural communication channel, and benefit from different user interfaces. The studies [45–47] take advantage of a physical medium shared across the robot and the human to simplify the state space representations. Leite et al. [45] used a multi-armed bandit for empathetic supportive strategies in the context of a chess companion robot for children. The difference in the probabilities of the user being in a positive mood before and after employing supportive strategies was used as a reward. The child's affective state was calculated by using visual facial features (smile and gaze) and contextual features of the game (game evolution i.e winning/losing, chessboard configuration). Similarly, in the work by Gao et al. [47] the user's task-related parameters were monitored through the puzzle interface. The robot's behaviors were adapted by combining a decision tree model with the Exp3 [48]. The Exp3 algorithm maintains a list of weights for each of the actions, which are used for selecting the next action. The reward was the user's task performance in combination with the user's verbal feedback. The set of robot actions included four supportive behaviors to help the user to solve the puzzle game.

4.2. Model-Based and Model-Free Reinforcement Learning
Verbal Communication

Considering the challenge of modeling real-world human-robot interactions, the majority of papers included in this survey use model-free RL. Nevertheless, several recent works started to investigate model-based RL for HRI [49,50]. One of the challenges of real-world robot learning is the delayed reward. There is an assumption that the result of an agent's observations of its environment is available instantly. However, there can be a lag in human reaction to robot actions in HRI. When the reward of the robot depends on human responses, reward shaping can be useful for the robot to get more frequent feedback. Reward shaping is a technique that consists of augmenting the natural reward signal so that additional rewards are provided to make the learning process easier [51]. Studies in [49,50] presented methods including model-based RL and reward shaping for HRI. Tseng et al. [49] proposed a model-based RL strategy for a service robot learning the varying user needs and preferences, and adjusting its behaviors. The proposed reward model was used to shape the reward through human feedback by calculating temporal correlations of robot actions and human feedback. Concretely, they modeled human response time using a gamma distribution. This formulation was found to be effective (more cumulative reward collected) in dealing with delayed human feedback. The work by Martins et al. [50] presented a user-adaptive decision-making technique based on a simplified version of model-based RL and POMDP formulation. Three different reward functions were formulated, and compared in the experiments. Their entropy-based reward shaping mechanism devised using an information-based term. The purpose of using the information term was to increase the reward given for an action leading to unknown transitions, thereby encouraging the robot to investigate the impact of new actions on the user.

4.3. Value-Based Methods

In recent years, there has been an increasing interest in applying RL methods to social robotics with growing trend towards value-based methods. Q-learning, along with its different variations, is the most commonly used RL method in social robotics. The studies using Q-learning are [3,13,34,52–61]. These comprise studies using standard Q-learning [3,54,55,58,60,62], studies modify Q-learning for dealing with delayed reward [52], tuning the parameters for Q-learning such as α [13,34,52], dealing with decreasing human feedback over time [52], comparing their proposed algorithm with Q-learning [33,49,61,63,64], variation of Q-learning called Object Q-learning [64–66], combining Q-learning with fuzzy inference [67], SARSA [68,69], TD(λ) [70], MAXQ [33,71,72], R-learning [32], and Deep Q-learning [35,36,73,74].

4.3.1. Tactile Communication

When the user is involved in the learning process by providing feedback in the form of reward or guidance, the general approach is either using an additional interface or utilizing the sensory information such as internal (robot's onboard sensors) or external cameras and microphones. Nowadays, many social robots are equipped with tactile capabilities. However, the usage of the robots' touch sensors as a feedback mechanism has received relatively little attention in the context of RL in social robotics. Yet [52,53] benefited from the robot's tactile sensors instead of an additional interface between the user and the robot. Barraquand and Crowley [52] conducted five experiments with different modifications of the classical Q-learning algorithm. The human teacher provided feedback through tactile sensors of the Sony AIBO robot, caressing the robot for the positive feedback and tapping the robot for the negative feedback. The action space comprised two actions; bark and play. The first experiment was standard Q-learning with human reward. Since the human ceased giving feedback over time, they concluded that the learning rate α should be adapted. In the second experiment, they used the asynchronous Q-learning algorithm. In asynchronous Q-learning, the learning rate α may be different for different state-action pairs. The learning rate is decreased when the system encountered the same situations and actions. In relation to standard Q-learning this modification increased the effectiveness of the algorithm, i.e., it learned faster and forgot more slowly. Because the learning rate was much smaller when there was no feedback. To overcome the delayed reward, they considered to increase the effect of human-delivered positive reward in larger time frames and to decrease the effect of negative reward in a shorter time frame. The use of an eligibility trace with a heuristic for delayed reward was found to be more efficient than classical Q-learning (generalizing experience to cover similar situations). The authors noted that learning rate, reward propagation, and analogy (i.e., propagating information to similar states) can improve the effectiveness of learning from social interaction. Yang et al. [53] proposed a Q-learning based approach that combines homeostasis and IRL. The internal factors, i.e., the drives and motivations worked as a triggering mechanism to initiate the robot's services. However, the reward in the real-world experiments was given by the user touching the robot's head, left hand, and right hand to give positive, negative, and dispensable feedback, respectively [53]. The authors trained their model in a simulator and deployed it on the Pepper robot.

4.3.2. Additional Physical Communication Medium between the Robot and the Human

Since we identify social robots with interaction, the robot learning within a social scenario stands out in the surveyed papers. Alternatively, there are studies where social interaction is not the main concern however, the main purpose is training a social robot to do a task. As an example, a human teacher trains the agent through a GUI [26,30], speech and gestures [28,31]. In Suay and Chernova [26], human teacher trained a social robot. They performed experiments similar to those presented in [75] in a real-world scenario with the Nao robot [26]. The human trainer observed the robot in its environment via a webcam and provided reward based on the robot's past actions or anticipatory guidance for selecting future actions through a GUI. They conducted four sets of experiments (small state space and only reward, large state space and only reward, small state space and reward plus guidance, large state space and reward plus guidance) to investigate the effect of teacher guidance and state space size on learning performance in IRL. The task was object sorting and the size of state space depended on the object descriptor features. Their results showed that the guidance accelerated the learning by significantly decreasing the learning time and the number of states explored. They observed that human guidance helped the robot to reduce the action space and its positive effect was more visible in large state-space. In a similar vein, Suay et al. [30] conducted a user study in which 31 participants taught a Nao robot to catch the robotics toys by using one of three algorithms: Behavior Networks, IRL, and Confidence-Based Autonomy. The study compared the results of these algorithms in terms of algorithm usability and teaching performance by non-expert users. In IRL, the participants provided positive or negative feedback in the form of reward through an on-screen interface. In terms of teaching performance, users achieved better performance using Confidence-Based Autonomy, however, IRL was better of modelling user behavior. It has been noted in much of the literature that teaching with IRL requires more time than with other methods because users had the tendency to stop rewarding or to vary their reward strategy. This affected the training time, which is a drawback to this approach.

4.3.3. Verbal and Nonverbal Communication

We discuss different human feedback types in IRL in Section 5.1. When a human teacher trains an agent, the positive or negative feedback might convey several meanings, even lack of feedback can give information to the agent depending on the teacher's training strategy [76]. For example, Thomaz and Breazeal [31] realized that human trainers might have multiple intentions with the negative reward they are giving, such as the last taken action was bad and future actions should correct the current state. They performed experiments with two different platforms: the Leonardo robot learned pressing buttons and a virtual agent learned baking a cake (Sophie's kitchen). The virtual agent responded to the negative reward by taking an UNDO action, i.e., the opposite action. In the examples with the Leonardo robot, the human teacher provided verbal feedback. After negative feedback, the robot expected the human teacher to guide it through refining the example by using speech and gestures (collaborative dialog). Although the interactive Q-learning with the addition of UNDO behavior was tested only on the virtual agent, it is worth mentioning that the proposed algorithm was more efficient compared to standard IRL. It had several advantages such as robust exploration strategy, fewer states visited, fewer failures occurred and fewer action trials done for learning the task. Continuing along these lines, Thomaz and Breazeal [28] explored how self-exploration and human social guidance can be coupled for leveraging intrinsically motivated active learning. They called the presented approach socially guided exploration, in which the robot could learn by intrinsic motivations, however, it could also take advantage of a human teacher's guidance when available. The robot learner with human guidance generalized better to new starting states and reached the desired goal states faster than the self-exploration.

4.3.4. Higher Level Interaction Dynamics: Engagement

Social robots are expected to exhibit flexible and fluent face-to-face social conversation. The natural conversational abilities of social robots should not be only limited to short basic task related sentences. However, they should be able to engage users in the interactions with chat and entertainment, varying from storytelling to jokes together with human-like vocalizations and sounds. As an example, Papaioannou et al. [60] reported that users spent more time with the robot which can carry out small chat together with task-based dialogue compared to the robot that conversed only task-based dialogue. In their system, the agent was trained using the standard Q-learning algorithm with simulated users and tested with the Pepper robot where the robot assisted visitors of a shopping mall by providing information about and directions to the shops, current discounts in the shops, among other things. In the problem definition, states were represented with 12 features such as user engaged, task completed, distance, turn taking, etc. The action space consisted of 8 actions, A = [PerformTask, Greet, Goodbye, Chat, GiveDirections, Wait, RequestTask, RequestShop]. The reward was encoded as predefined numerical values based on task completion by the agent, including the engagement of the user. Another study considering user engagement is Keizer et al. [1], who applied a range of ML techniques in the presented system that included a modified iCat robot (with additional manipulator arms with grippers) and multimodal input sensors for tracking facial expressions, gaze behavior, body language and location of the users in the environment. The reward function was a weighted sum of task-related parameters. For each individual user i the reward function R_i was defined as $R_i = 350 \times TC_i - 2 \times W_i - TO_i - SP_i$. TC_i is short for Task Complete, and is a binary variable. W_i (Waiting) is a binary variable showing whether the user i is ready to order but not engaged with the system. TO_i stands for Task Ongoing and is a binary variable describing whether the user is interacting with the robot but has not been served. SP_i is short for Social Penalties and corresponds to several social penalties (e.g., while the user i is still talking to the system, it turns its attention to another user). An experimental evaluation compared a hand-coded and trained system. The authors reported that the trained system performed better and it was found to be faster at detecting user engagement than the hand-coded one, while the latter was more stable. In [55,57,59], the authors investigated the entertainment capabilities of social robots using RL. Ritschel et al. [57] presented a social-cues-driven Q-learning approach for adapting the Reeti robot to keep the user engaged during the interaction. The engagement of the user was estimated from the user's movement through the Kinect 2 sensor by using a Dynamic Bayesian Network. They used the change in the engagement as a reward in the storytelling scenario to adapt the robot's utterance based on the personality of the user. In similar fashion, the work by Weber et al. [59] incorporated social signals in the learning process, namely the participants' vocal laughs and visual smiles as reward. In the problem formulation, they used a two-dimensional vector containing probabilities of laughs and smiles for state representation, and the action space consisted of sounds, grimaces and three types of jokes. They used an average reward based on all samples from the punchline to the end with a predefined punchline for every joke. The human social signals were captured and processed by using the Social Signals Interpretation (SSI) framework [77]. Their purpose was to understand the user's humor preferences in an unobtrusive manner in order to improve the engagement skills of the robot. In a joke-telling scenario, the Reeti robot adapted its sense of humor (grimaces, sounds, three kinds of jokes and their combination) by using Q-learning with a linear function approximator. Likewise Addo and Ahamed [55] presented a joke telling scenario with a torso Nao robot for entertaining a human audience. They used Q-learning in which the actions of the robot were pre-classified jokes, and the numerical reward corresponded to affective states of the user. However, the affective states of the participants were captured by a self-reported feedback signal. After each joke, the human participant provided a verbal feedback (i.e., reward) such as "very funny", "funny", "indifferent" and "not funny".

4.3.5. Affective Communication: Facial Expressions

Human facial expressions are perhaps one of the richest and most powerful tools in social communication. Facial expressions analysis is commonly used in HRI for understanding users and enhancing their experience. Affective facial expressions can also facilitate robot learning in RL. Recently, it is becoming more popular to use off-the-shelf applications in social robotics for different perception and recognition modules. Affectiva software [78] analyzes facial expressions from videos or in real-time. The studies [58,68,69] used this software for affective child-robot interaction. In the work by Gordon et al. [68] a tutoring system for children was presented. The system included an Android tablet and the Tega robot setup integrated with the Affectiva software for facial emotion recognition. They used the SARSA algorithm where the reward was a weighted sum of valence and engagement. Both valence and engagement values were obtained from the Affectiva software. Similar to [68], Park et al. [58] used the Tega robot as a language learning companion for young children. A personalized policy was trained through 6–8 sessions of interaction by using a tabular Q-learning algorithm. The reward function was a weighted sum of engagement and learning gains of the child. The engagement was obtained from the Affectiva software. The learning gains in the reward function was represented as numerical values ([−100, 0, +50, +100]) depending on the lexical and syntactic complexity of the phrase relative to the child's level. Gamborino and Fu [69] presented an approach for socially assistive robots for children to support them in emotionally difficult situations using SARSA. In the proposed method, the human trainer selects the actions for the social robot RoBoHoN (small humanoid smartphone robot) through an interface with the purpose to improve the mood of the child depending on her/his current affective state. The affective state of the child was based on seven basic facial emotions and engagement obtained by the Affectiva software and stored in an input feature vector to classify the mood of the child as good or bad. The emotions were binarized as 1 or 0 depending on whether the value was greater or less than the average, respectively. The robot suggested a set of actions to the trainer. The aim was to suggest actions that would match with the trainer's action preferences. This way the agent would act independently, without feedback from the trainer. Another study using facial expressions is Zarinbal et al. [54], in which Q-learning was used for query-based scientific document summarization with a social robot. The problem formulation was as follows: In each state $S_t :< x_i, score^t(x_i) >$ a summary that consisted of M sentences was generated, where x_i is a sentence and $i = 1, 2, ..., M$. The scoring scheme was updated based on the human-delivered reward. The reward $r_t \in \{-1, 0, 1\}$ depended on the classified facial expressions: dislike, neutral and like. In state S_t, the robot presented the sentence x^* to the user and based on his reward r_t. The authors concluded that user feedback may improve the query-based text summarization.

4.3.6. Verbal Communication

The curse of dimensionality is a phenomenon that refers to problems with high dimensional data. Representing state and action spaces as explicit tables becomes impractical for large spaces. To overcome the problem of large state space, approximate solutions are used, one of them being fuzzy techniques. This approach is also explored for HRI, e.g., Chen et al. [67] and Patompak et al. [32] used fuzzification and fuzzy inference together with Q-learning. These works employed verbal communication in their user studies. Chen et al. [67] proposed a multi-robot system for providing services in a drinking-at-a-bar scenario. The authors used a modified Q-learning algorithm combined with fuzzy inference which was called information-driven fuzzy friend-Q (IDFFQ) learning for understanding and adapting the behaviors of the mentioned multi-robot system based on the emotion and intention of the user. The reward function was defined as $r = (r_t + r_h)/2$. Task completion r_t (i.e., robots selected the drink the user preferred) and the human's satisfaction with the robots' task performance r_h were predefined numerical values. Fuzzification of emotions was done using the triangular and trapezoidal membership function in the pleasure-arousal plane. They compared the proposed algorithm with their previous algorithm, Fuzzy Production Rule-based Friend-Q learning (FPRFQ) [79]. The authors noted that the current algorithm was superior in that it resulted in higher collected reward and faster response time of the robots. Patompak et al. [32] proposed a dynamic social force model for social HRI. The authors considered two interaction areas: a quality interaction area and a private area. The quality interaction area was defined as the distance from which the users can be engaged in high-quality interactions with robots. The proposed model was designed by a fuzzy inference system, the membership parameters were optimized by using the R-learning algorithm [80]. R-learning is an average reward RL approach; it does not discount future rewards [81]. They argued that R-learning was suitable for the scenario since they intended to take every interaction experience into account equally. In the real robot experiments, positive or negative verbal rewards were provided by the participants.

Another study that used verbal communication for the reward is [62]. In this study, a gesture recognition system categorized the body trunk patterns as towards (the person is facing the robot), neutral (the trunk is facing the robot between $3°$–$15°$ away), and away (orientation of the trunk is more than $15°$). The recognized gestures were interpreted as a person's accessibility level, which was used to determine the person's affective state. In the Q-learning-based decision-making system, the robot had drives and emotional states which were utilized for action selection. In particular, a state is represented as $s(y_H, y_R, d)$ where y_H is the accessibility level of the human, y_R is emotional state of the robot and d is the dominant drive. State transition probabilities, Q-values for each state, and reward for each transition were predetermined numerical values. The satisfaction of the robot's drives depended on the robot completing the task. In the experimental scenario, the Brian robot reminded the user about daily activities (eat, use the bathroom, go for a walk and take medication) and the user verbally stated 'yes' or 'no' after the robot's action, with 'no' meaning that the robot's drive is not satisfied and it will continue to try to satisfy the drive. The authors mentioned that the robot could use its drives in one or two iterations for the reminders except the drive related to using the bathroom. It was attributed to people potentially being uncomfortable with this reminder.

4.3.7. Higher Level Interaction Dynamics: Attention

Social robots have the potentials for information acquisition from both verbal and nonverbal communication. Not only can they gesture, maintain eye contact, and share attention with their users, but they can also estimate the users' non-verbal cues and behave accordingly. In this interaction, both actors can interpret verbal and nonverbal social cues to communicate effectively. For natural fluid HRI, robot non-verbal behaviors together with verbal communication are thoroughly discussed in [82]. These social cues do not only convey a basic message but also carry higher-level interaction dynamics such as attention, engagement, comfort, and so on. The following works highlight these in the context of RL in social robotics. Chiang et al. [56] proposed a Q-learning based approach for personalizing the human-like robot ARIO's interruption strategies based on the user's attention and the robot's belief in the person's awareness of itself. The authors called it the "robot's theory of awareness". They formulated the problem based on the user attention, which was referred to as a Human-Aware Markov Decision Process. The human attention was estimated with a trained Hidden Markov Model (HMM) from human social cues (face direction, body direction, and voice detection). The reward consisted in predefined numerical values based on the robot's theory of awareness of the user. The robot had six actions (gestures: head shake and arm wave; navigation: approach the user and move around; audio: make sound and call name) to draw the user's attention while the user was reading. The optimal policy converged after two hours of interaction. The robot developed personalized policies for each user depending on their interruption preferences. Another study considering human attention in their problem formulation is Hemminghaus and Kopp [3]. They used Q-learning to adapt the robot head Furhat's behavior in a memory game scenario. In the game, the robot assisted the participant by guiding their attention towards target objects in a shared spatial environment. In the proposed hierarchical approach, the high-level behavior was mapped to low-level behaviors, which could then be directly executed by the robot. The purpose of using Q-learning was to learn the execution of high-level behaviors through low-level behaviors. In the problem formulation, states were represented in terms of the user's gaze, user's speech, and game state. The game state represented the number of remaining card pairs in the game. The action space included actions such as speaking, gazing, etc. or a combination of those actions. The reward was designed as $r = r_{pos} - c$ if success $r = c.r_{neg}$ if no effect. The robot received a positive reward r_{pos} if the robot's action helped the user to find the correct pair. The robot received a negative reward

r_{neg} if the action had no effect on helping the user. c represents the cost of the chosen action in cases where the costs were determined manually. Moro et al. [61] is another study that considered the attention of the user. Their scenario was an assistive tea-making activity for older people with dementia. The authors proposed an algorithm involving Learning from Demonstration (LfD) and Q-learning for personalized robot behavior according to a user's cognitive abilities [61]. The Casper robot learned to imitate the combination of speech and gestures from a collected data set. The robot learns to select the suitable labeled behavior (i.e., speech and gestures initially learned from demonstrations) that is most likely to transition the user into the desired state, i.e., focused on the activity and completing the correct step. The reward function, $R(s, b_l^i)$, depended on b_l^i, the labeled behavior displayed by the robot, and the state s where $s = \{s_r, s_u\}$. Here, s_r represents a set of robot activity states, and s_u is the user state such that $s_u = \{s_{fnc}, s_{ac}\}$. In the user state, s_{fnc} represents the user functioning state which is one of five mental functioning states: focused, distracted, having a memory lapse, showing misjudgment, or being apathetic. The user activity state, s_{ac}, represents possible actions that can be performed by seniors with cognitive impairment: successfully completing a step, being idle, repeating a step, performing a step incorrectly, or declining to continue the activity. The robot was rewarded according to the state the user transitioned into—a positive reward if the user was focused and completed the activity, and a negative reward if the user transitioned to an undesirable state. The authors compared the proposed approach with Q-learning, and reported that the proposed approach required fewer interactions for convergence and fewer steps required to complete the tea-making activity. In all the papers explained above, the robot takes the users attention into account for deciding its actions. Shared attention refers to situations involving mutual gaze, gaze following, imperative pointing and declarative pointing. Da Silva and Francelin Romero [63] presented a robotic architecture for shared attention which included an artificial motivational system driving the robot's behaviors to satisfy its intrinsic needs, so-called necessities. The motivational system comprised necessity units that were implemented as a simple perceptron with recurrent connections. The input to the artificial motivational system was provided by a perception module used to detect the environmental state and to encode the state in first order logic with predicates. This module included face recognition with head pose estimation and a visual attention mechanism. The necessities of the robot were associated with a state-action pair in the training phase of the learning algorithm. The activation of a necessity unit was dependent on the input signal representing a stimulus detected from the environment (i.e., the perception module) and empirically defined parameters. They compared the performance of three different RL algorithms, namely contingency learning, Q-learning and Economic TG (ETG) methods for shared attention in social robotics. ETG is a relational RL algorithm that incorporates a tree-based method for storing examples [83]. Because ETG performed better in the simulation experiments, they decided to employ it in real-world experiments which entailed one of the authors interacting with the robotic head. The authors reported that the robot's corrected gaze index, which was defined as frequency of gaze shifts from the human to the location that the human is looking at, was increased over time during learning.

4.3.8. Affective Communication

Humans use affective communication consciously or unconsciously in their daily conversations by expressing feelings, opinions, or judgments. Social robots can facilitate their learning process through sensing and building representations of affective responses. This idea was used in [33,71,72]. In these studies, the socially assistive robot Brian 2.0 was employed as a social motivator by giving assistance, encouragement, and celebration in a memory game scenario. In the scenario, the participants interacted with the robot one-on-one with the objective to find the matching pictures in the memory card game (4×4 grid, 16 picture cards). The robot's behaviors were adapted using a MAXQ method to reduce the activity-induced stress in the user. The MAXQ approach is a hierarchical formulation, which accommodates a hierarchical decomposition of the target problem into smaller subproblems by decomposing the value function of an MDP into combinations of value functions of smaller integral MDPs [84]. The authors argued that the MAXQ algorithm was suitable for memory game scenarios due to its temporal abstraction, state abstraction, and sub-task abstraction. These abstractions also helped to reduce the number of Q-values that needed to be stored. The detailed system was presented in [33]. In their system, they used three different types of sensory information: a noise-canceling microphone for recognizing human verbal actions, an emWave earclip heart rate sensor for affective arousal level and a webcam for monitoring the activity state (depending upon whether matching card pairs were found or not). They used a two-stage training process involving offline training followed by online training. The purpose of the first stage was to determine the optimal behaviors for the robot with respect to the card game. The offline training was carried out on a human user simulation model created with the interaction data of ten participants. In the second stage, they aimed to personalize the robot according to the user's state (affective arousal and game state) for different participants in online interactions. The affective arousal and user activity state formed the user state (e.g., stressed: high arousal and not matching card, pleased: low arousal and matching card). The success of the robot's actions was subject to the improvement of a person's user state from a stressed state to a stress-free state.

4.4. Deep Reinforcement Learning

For natural interaction, it is important that social robots possess human-like social interaction skills, which requires features from high dimensional signals. In these cases, DRL can be useful. In fact, several researchers have begun to examine the applicability of DRL in social robotics [35,36,73,74,85–87].

4.4.1. Tactile Communication

One of the pioneering works using DRL in social robotics was presented by [36]. Here, a Pepper robot learned to choose among predefined actions for greeting people, based on visual input. In their work, they succeeded to map two different visual input sources, the Pepper robot's RGBD camera and the webcam, to discrete actions (waiting, looking towards the human, hand waving and handshaking) of the robot. The reward was provided by a touch sensor located on the robot's right hand to detect handshaking. The robot received a predefined numerical reward (1 or -0.1) based on a successful or unsuccessful handshake. A successful handshake was detected through the external touch sensor. The proposed multimodal DQN consists of two identical streams of CNN for action-value function estimation—one for grayscale frames and another for depth frames. The grayscale and depth images were processed independently, and the Q-values from both streams were fused for selecting the best possible action. This method comprised two phases: the data generation phase and the training phase. In the data generation phase, the Pepper robot interacted with the environment and collected data. After this phase, the training phase began. This two-stage algorithm was useful in that it did not pause the interaction for training. Qureshi et al. [36] used 14 days of interaction data where each day of the experiment corresponded to one episode. The same authors applied a variation of DQN, the Multimodal Deep Attention Recurrent Q-Network (MDARQN) [73], to the same handshaking scenario in [36]. In their previous study, the robot was unable to indicate its attention. For adding perceptibility to the robot's actions, a recurrent attention model was used, which enabled the Q-network to focus on certain parts of the input image. Similar to their previous work [36], two identical Q-networks were used (one for grayscale frames and one for depth frames). Each Q-network consisted of convnets, a Long Short-term Memory (LSTM) network, and an attention network [88]. The convnets were used to transform visual frames into feature vectors. The network transforms an input image into D-dimensional L feature vectors, each of them representing a part of the image $a_t = \{a_t^1, ..., a_t^L\}, a_t^l \in \mathbb{R}^D$. This feature vector was provided as an input to the attention network for generating the annotation vector $z \in \mathbb{R}^D$. The annotation vector z_t

is the dynamic representation of a part of an input image at time t. z_t is computed with $z_t = \sum_{l=1}^{L} \beta_t^l a_t^l$. The LSTM network used the annotation vector z_t for computing the next hidden state. Each of the streams of the MDARQN model were trained by using the back-propagation method. The outputs from the two streams were normalized separately and averaged to create output Q-values of MDARQN. As in their previous work, handshake detection was used for the reward function (-0.1 for unsuccessful handshakes and 1 for successful handshakes). The horizontal and vertical axes of the input image were divided into five subregions, and the Q-network enabled to focus on certain parts of the input image. The attention mechanism of the robot used the annotation vector z_t to determine the pixel location to direct maximal attention to the input image. This region selection provided computational benefits by reducing the number of training parameters. Another work from the same authors Qureshi et al. [74] proposed an intrinsically motivated DRL approach for the same handshaking scenario. The proposed method utilized three basic events to represent the current state of the interaction, i.e., eye contact, smile, and handshake. These event occurrences were predicted at the next time step according to the state-action pair by a neural network called Pnet. Another neural network called Qnet was employed for action selection policy guided by the intrinsic reward. The reward was determined based on the prediction error of Pnet, i.e., the error between actual occurrences of events $e(t + 1)$ and Pnet's prediction $\hat{e}(t + 1)$. An OpenCV-based event detector module provided the labels for three events (i.e., actual event occurrence). The Qnet was a dual stream deep convolutional neural network mapping pixels to q-values of the actions (wait, look towards human, wave hand, and shake hand). Pnet was a multi-label classifier which was trained to minimize the prediction error between $\hat{e}$ and e by using the Binary Cross Entropy (BCE) loss function. The reward consisted in predetermined numerical values depending on the prediction error between e and $\hat{e}$. They investigated the impact of three different reward functions named strict, neutral and kind. In all reward functions, if all three events are predicted successfully by Pnet, Qnet receives a reward of 1, if all events are predicted wrong then Qnet gets a reward of -0.1. If one or two events are predicted correctly then different reward functions penalize differently, with the strict reward having the highest penalties. The authors reported that the reward functions with more positive reward on incorrect predictions yielded more socially acceptable behavior. They compared the collected total reward from 3 days of experiments in a public place, each day following a different policy (random policy, Qnet policy, and the previously employed method [36]). The current proposed model led to more human-like behaviors, according to the human evaluators.

4.4.2. No Communication Medium

Another study using the Pepper robot and DQN was presented by Cuayáhuitl [35]. In their scenario, human participants played a 'Noughts and Crosses' game with two different grids (small and big) against the Pepper robot. They used a CNN for recognizing game moves, i.e., hand-writing on the grid. These visual perceptions and the verbal conversations of the participant were given as an input to their modified DQN. The author modified the Deep Q-Learning with Experience Replay [43] by adding the identification of the worst action set $\hat{A}$. $\hat{A}$ included actions with $min(r(s,a) < 0 \ \forall a \in A)$ and A is the set of actions leading to win the game. The action selection was done with $\max_{a \in A \setminus \hat{A}} Q(s, a; \theta)$.

In other words, the proposed DQN algorithm refines the action set at each step to make the agent learn to infer the effects of its actions (such as selecting the actions that lead to winning or to avoid losing). The reward consisted in predefined numerical values based on the performance of the robot in the game. Therefore, this study does not use any communication medium for reward formulation. The robot received the highest reward in the cases 'about to win' or 'winning', whereas the robot received the lowest reward in the cases 'about to lose' or 'losing'.

4.4.3. Nonverbal Communication

Expressive robot behaviors including facial expressions, gestures, and posture are found to be useful to express the robots' internal states, goals, and desires [89]. To date, several studies have investigated the production of expressive robot behaviors using DRL, including gaze [85,86] and facial expressions [87]. Lathuilière et al. [85] modeled Q-learning with a Long Short Term Memory (LSTM) to fuse audio and visual data for controlling the gaze of the robotic head to direct it towards the targets of interest. The reward function was defined as $R_t = F_{t+1} + \alpha \sum_{t+1}$ where $\alpha \geq 0$ serves as an adjustment parameter. If the speech sources lie within the camera's field of view, large α values return large rewards, i.e, α permits to give importance to speaking persons. The reward function includes face reward F_t ($\alpha = 0$) and speaker reward ($\alpha > 0$). The number of visible people (face reward) and the presence of speech sources in the camera field of view (speaker reward) were observed from the temporal sequence of camera and microphone observations. The proposed DRL model was trained on a simulated environment with simulated people moving and speaking, and on the publicly available AVDIAR dataset. In this offline training, they compared the reward obtained with four different networks: early fusion and late fusion of audio and video data, as well as only audio data and only video data. The authors emphasized the importance of audio-visual fusion in the context of gaze control for HRI. They reported that the proposed method outperformed the handcrafted strategies. Lathuilière et al. [86] extended the study presented in [85] by investigating the impact of the discount factor, the window size (number of past observations affects the decision), and LSTM network size. They reported that in the experiments with AVDIAR dataset, high discount factors were prone to overfit, whereas in the simulated environment low discount factors resulted in worse performance. Using smaller window sizes accelerated the training, however, larger window sizes performed better in simulated environment. Changing the LSTM size did not make a substantial difference in the results. In a similar vein, Churamani et al. [87] utilized visual and audial data for enabling the Nico robot to express empathy towards the users. They focused on both recognizing the emotions of the user and generating emotions for the robot to display. The presented model consisted of three modules: an emotion perception module, an intrinsic emotion module, and an emotion expression module. For the perception module, both the visual and audio channels were used to train a Growing-When-Required (GWR) Network. For the emotion expression module, they used a Deep Deterministic Policy Gradient (DDPG) based actor-critic architecture. The reward was the symmetry of the eyebrows and mouth in offline pre-training, whereas in online training the reward was provided by the participant deciding whether the expressed facial expression was appropriate. The Nico robot expressed its emotions through programmable LED displays in the eyebrow and mouth area.

4.5. Policy-Based Methods

Higher Level Interaction Dynamics: Comfort

In the domain of socially assistive robotics, the robots are expected to be adaptive to their users to some extent, by using social interaction parameters (for example, the interaction distance, the speed of motion and utterances) regarding to the task, to the users' comfort and personality. Several studies [90–92] examined the Policy Gradient Reinforcement Learning (PGRL) for adapting the robot behaviors using social interaction parameters. Mitsunaga et al. [90,91] presented a study where the Robovie II robot adjusted its behaviors (i.e., proxemics zones, eye contact ratio, waiting time between utterance and gesture, motion speed) according to comfort and discomfort signals of humans (i.e., body re-positioning amount and the time spent gazing at the robot).These signals were used as reward. The goal of the robot was to minimize these signals, thereby reducing experienced discomfort in the human interactant. In [92], an ActiveMedia Pioneer 2-DX mobile robot adapted its personality by changing the interaction distance, speed and frequency of motions, and vocal content (what and how the robot says things). The purpose of this adaptation was to improve the user's task performance. Their reward function was based on user performance, defined as the number of performed exercises. Specifically, the number of performed exercises over the previous 15 s was computed every second and results were averaged over a 60 s period to produce the final evaluation for each policy. They used a threshold for the reward function (7 exercises in the first 10 min) and a time range to adjust the fatigue incurred by the participant. The participant's performance was tracked by the robot through a light-weight motion capture system worn by the participant.

5. Categorization of RL Approaches in Social Robotics Based on Reward

We now present a review of the literature but with focus on the reward function. Designing the reward function is perhaps the most crucial step in the implementation of an RL framework. One of the main contributions of this paper is a categorization of different types of reward functions that are used in RL and social robotics. The categorization is given in Figure 4.

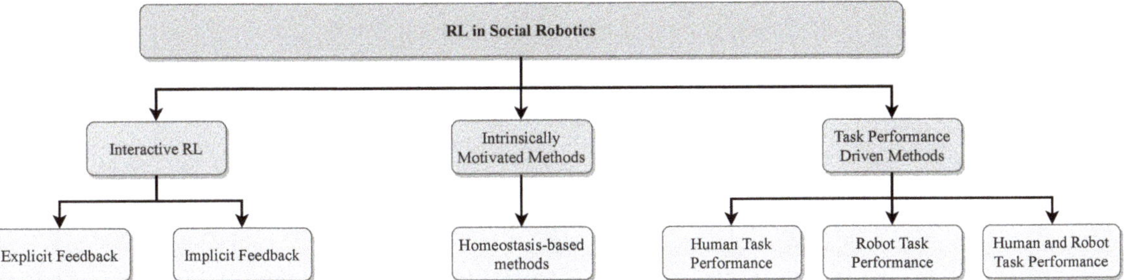

Figure 4. Reinforcement Learning approaches in social robotics.

As we have already discussed the used RL methods in Section 4, they are not included here. Moreover, the evaluation methodologies are also discussed in a separate section (see Section 6).

5.1. Interactive Reinforcement Learning

Different approaches have been proposed for incorporating the human assistance in the learning process of artificial agents, including learning from human feedback [24,76] and learning from demonstration. Learning from demonstration is beyond the scope of this paper, we focus on learning from human feedback. In traditional RL, the agent receives environmental reward from a predefined reward function. Interactive RL makes use of human feedback in the learning process in combination with or without environmental reward. Interactive RL framework is given in Figure 5. Integrating human feedback with RL can be accomplished in different ways, such as via evaluative feedback [93], corrective feedback [94] or guidance [95].

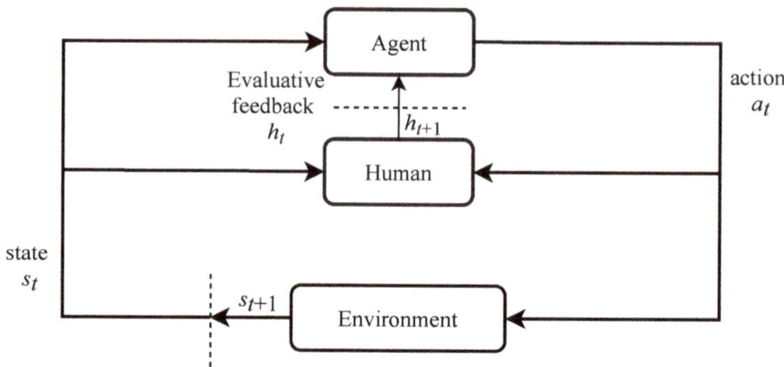

Figure 5. Interaction in Interactive Reinforcement Learning (reproduced from [96]).

Li et al. [96] discuss different interpretations of human evaluative feedback in interactive reinforcement learning (referred to as human-centered RL throughout the paper). They distinguish between three types of human evaluative feedback: interactive shaping, learning from categorical feedback and learning from policy feedback. In interactive shaping, human feedback is interpreted as numeric reward, and this reward can be myopic i.e., $\gamma = 0$ [93] or non-myopic i.e., γ is different from 0 [97]. Human feedback might be erroneous when the task is repetitive. Moreover, human teachers tend to give less frequent feedback (e.g., due to boredom and fatigue) as the learning progresses. Modeling human feedback has been found to be an efficient strategy when the meaning of human-delivered feedback is ambiguous [76]. Loftin et al. [76] developed a probabilistic model of human teacher's feedback. They interpret human feedback as categorical feedback, considering that human teachers may have different feedback strategies. In their work, depending on the human teacher's training strategy, a lack of feedback can convey information about the agent's behavior. Human training strategies are categorized into four groups: reward-focused strategy (positive reward for correct actions and no feedback for incorrect actions), punishment-focused strategy (no feedback for correct actions and punishment for incorrect actions), balanced strategy (positive reward for correct actions and punishment for incorrect actions) and inactive strategy (the human teacher rarely provides feedback). Corrective feedback can be categorized under policy feedback. As an example, Celemin and Ruiz-del Solar [94] presented a framework named COACH (COrrective Advice Communicated by Humans) which uses human corrective feedback in the action domain as binary signals (i.e., increase or decrease the magnitude of the current action). In their comparison with classical reinforcement learning approaches, they showed that RL agents can benefit from human feedback, i.e., learning progresses faster [94]. When the agent learns both human feedback and environment reward, the human feedback can be used to guide the agent's exploration [95]. The guidance includes both providing feedback on past actions and guiding the agent in the learning process through future-directed rewards. Human guidance can reduce the action space by narrowing down the action choices [98], which speeds up the training process by accelerating the convergence towards an optimal policy.

In the context of HRI, the human can be in the learning loop by way of varying types of inputs, such as providing feedback via a GUI (e.g., by button or mouse clicks). Alternatively, the feedback can be delivered more naturally, via emotions, gestures and speech. Therefore, this category comprises two subcategories: (1) explicit feedback, when the feedback is direct, provided through an interface such as ratings, and labels; (2) implicit feedback, if the human feedback is spontaneous behavior or reactions such as non-verbal cues and social signals. The terms "explicit feedback" and "implicit feedback" are adopted from Schmidt [99]'s "implicit interaction" study in human-computer interaction. For a quick summary of the studies, see Table 1.

Table 1. Summary of Interactive Reinforcement Learning approaches in social robotics.

Reference	Subcategory	Type of RL	Reward	Social Robot
Barraquand et al. [52]	Explicit feedback	Q-learning	User provides reward by using robot's tactile sensors	Aibo
Suay et al. [26,30]	Explicit feedback	Q-learning	Human teacher delivers reward or guidance through a GUI	Nao
Knox et al. [29]	Explicit feedback	TAMER	Human teacher provides reward by using a remote	Nexi
Yang et al. [53]	Explicit feedback	Q-learning	User gives reward by touching robot's tactile sensors	Pepper
Schneider et al. [44]	Explicit feedback	Dueling bandit	User provides feedback through a button	Nao
Churamani et al. [87]	Explicit feedback	DDPG	User gives reward whether robot's expression is appropriate to affective context of dialogue	Nico
Tseng et al. [49]	Explicit feedback	Modified R-Max	User provides reward through a software	ARIO
Gamborino et al. [69]	Explicit feedback	SARSA	User's transition between bad and good mood state	RoBoHoN
Ritschel et al. [4]	Explicit feedback	k-armed bandit	User gives reward via buttons	Reeti
Patompak et al. [32]	Implicit feedback	R-learning	Verbal reward by the user based on robot's social distance	Pepper
Thomaz et al. [31]	Implicit feedback	Q-learning	Human teacher provides reward or guidance	Leonardo
Thomaz et al. [28]	Implicit feedback	Q-learning	Human teacher provides guidance through speech or gestures	Leonardo
Gruneberg et al. [100,101]	Implicit feedback	Not specified	Human teacher's smile and frown	Nao
Addo et al. [55]	Implicit feedback	Q-learning	Verbal reward of the user	Nao
Zarinbal et al. [54]	Implicit feedback	Q-learning	Human teacher's facial expressions	Nao
Mitsunaga et al. [90,91]	Implicit feedback	PGRL	Discomfort signals of the user	Robovie II
Leite et al. [45]	Implicit feedback	Multi-armed bandit	User's affective cues and task-related features	iCat
Chiang et al. [56]	Implicit feedback	Q-learning	Numerical values based on attention and engagement levels of the user	ARIO
Gordon et al. [68]	Implicit feedback	SARSA	Weighted sum of facial valence and engagement	Tega

Table 1. Cont.

Reference	Subcategory	Type of RL	Reward	Social Robot
Ritschel et al. [57]	Implicit feedback	Q-learning	Change in user engagement	Reeti
Weber et al. [59]	Implicit feedback	Q-learning	Vocal laughter and visual smiles	Reeti
Park et al. [58]	Implicit feedback	Q-learning	Weighted sum of engagement and learning	Tega
Ramachandran et al. [102]	Implicit feedback	POMDP	Engagement level of the user	Nao
Martins et al. [50]	Implicit feedback	Model-based RL and POMDP	The robot's actions' impact on the user	GrowMu

5.1.1. Explicit Feedback

In the explicit feedback approach, the feedback of the human teacher is given by direct manipulations and generally through an artificial interface. The human teacher observes the agent's actions and environment states and subsequently provides feedback to the agent through a graphical user interface (GUI) or through the robot's (touch) sensors. In this approach, the feedback from the human teacher is noiseless and direct in the form of numerical values provided via a button, a Graphical User Interface (GUI), or through the robot's touch sensors. In general, the main purpose of the interaction is to teach the robot to do something in this category. Unlike the explicit feedback category, in the implicit feedback category, the majority of studies include a social scenario such as robot tutoring, robots supporting the human in a game, etc. The studies under this category are [4,26,29,30,44,52,53].

5.1.2. Implicit Feedback

Human social signals are widely used as reward in social human-robot interaction. The most commonly used signals are human emotions, as these have a great influence on decision-making [103]. Computational models of emotions have been studied by many researchers as part of the agent's decision making architecture, by modelling the RL agents with emotions or incorporating human emotions as an input to the learning process. As an example, Moerland et al. [104] surveyed RL studies focusing on agent/robot emotions. Since emotions also play an important role in communication and social robots [7], there exist various studies considering these aspects for RL and social robotics. In the implicit feedback approach, the agent learns from spontaneous natural behavior and reactions of the interactant, i.e., emotions, speech, gestures, etc. This type of feedback is noisy and indirect. In other words, in this approach, human feedback requires pre-processing and the quality of the feedback depends on the perception and recognition algorithms being used. Unlike explicit feedback, the implicit feedback is not provided directly through an interface. Instead, the human's emotions or verbal instructions serve as reward or guidance signals. The studies in this category are [28,31,32,45,50,55–57,59,68,90,91,100–102].

5.2. Methods Using Intrinsic Motivation

It is a common approach to examine the biological and psychological decision-making mechanisms and to use a similar method for autonomous systems. One such approach consists in combining intrinsic motivation with reinforcement learning. Intrinsic motivation is a concept in psychology, which denotes the internal natural drive to explore the environment, as well as gain new knowledge and skills. The activities are performed for inherent satisfaction rather than external rewards [105]. Researchers have proposed computational approaches that use intrinsic motivation [106]. In intrinsically motivated RL, the main idea is using intrinsic motivations as a form of reward [107]. There are different intrinsic motivation models within the RL framework [20]. However, in social robotics, the idea of maintaining the internal needs of the robot (detailed in Section 5.2) has received much attention [13,34,63–66,108]. One exception is [74], in which prediction error of social event occurrences was used as intrinsic motivation. For a quick summary, see Table 2.

Homeostasis-Based Methods

Homeostasis, as defined by Cannon [109], refers to a continuous process of maintaining an optimal internal state in the physiological condition of the body for survival. Berridge [110] explains homeostasis motivation with a thermostat example that behaves as a regulatory system by continuously measuring the actual room temperature and comparing it with a predefined set point, and activating the air conditioning system if the measured temperature deviates from the predefined set point. In the same manner, the body maintains its internal equilibrium through a variety of voluntary and involuntary processes and behaviors. The homeostasis-based RL in social robotics is presented in [13,34,64–66,108]. These studies introduced a biologically inspired approach that depends on homeostasis. The robot's goal was to keep its well-being as high as possible while considering both internal and external circumstances. The common theme in these studies is that the robot has motivations and drives (needs), where each drive has a connection with a motivation as in Equation (15).

$$if \ D_i < L_d \ then \ M_i = 0$$
$$if \ D_i \geqslant L_d \ then \ M_i = D_i + w_i \tag{15}$$

Motivations whose drives are below the activation levels do not initiate a robot behavior. This was formulated as $if \ D_i < L_d \ then \ M_i = 0$ where D_i is a drive, L_d the activation level, and M_i is the related motivation. The motivation depends on two factors: the associated drive and the presence of an external stimulus, this was formulated as $if \ D_i \geqslant L_d \ then \ M_i = D_i + w_i$ where w_i is the related external stimulus. These motivations serve as action stimulation to satiate the drives. A drive can be seen as a deficit that leads the agent to take action in order to alleviate this deficit and maintain an internal equilibrium. The ideal value for a drive is zero, corresponding to the absence of need. The robot learns how to act in order to maintain its drives within an acceptable range, i.e., to maintain its well-being. The well-being of the robot was defined as:

$$Wb = Wb_{ideal} - \sum_i \alpha_i D_i \tag{16}$$

where Wb_{ideal} is the value of the well-being when all drives are satiated, and α_i is the set of the personality factors that weight the importance of each drive. The variation of the robot's well-being is used as reward signal and calculated with the Equation (17)

$$\Delta Wb = Wb_t - Wb_{t-1} \tag{17}$$

i.e., the difference between the current well-being Wb_t and the well-being in the previous step Wb_{t-1}.

In several works [64–66], a variation of the traditional Q-learning algorithm was used in addition to the homeostasis-based approach. In all of these, the authors referred to the proposed algorithm as Object Q-learning. In this approach, there are actions associated with each object in the environment, and the robot considers its state in relation to every object independently. Thus, there is an assumption that an action execution in relation to a certain object does not influence the state of the robot in relation to other objects. However, in reality, an action execution may create collateral effects. In other words, an action associated with a particular object, e.g., approaching it, may affect the robot's state in relation to other objects, e.g., moving away from them. The update of Q-values accounted for these collateral effects. The purpose of this simplification was to reduce the number of states during the learning process. In their experiments, to reduce the state space, the robot learned what to do with each object without considering its relation to other objects. In other words, they assumed that an action execution associated with a certain object will not affect the state of the robot in relation to the rest of the objects. The proposed algorithm was implemented on the social robot Maggie that lived in a laboratory and interacted with several objects in the environment (e.g., a music player, a docking station, or humans). Castro-González et al. [65] appears to be closely linked to the other papers discussed here with one difference being that a discrete emotion, fear, was used as one of the motivations. Unlike other motivation-drive pairs, no drive was associated with the 'fear motivation' (i.e., fear is not a deficiency of any need). 'Fear motivation was' linked to dangerous situations (that can cause damage the robot) and directed the robot to a secure state. As an example, the motivation 'social' was not updated if the user who occasionally hit the robot was around. For a quick summary, refer to Table 2.

Table 2. Summary of Intrinsically Motivated Methods in social robotics.

Reference	Subcategory	Type of RL	Reward	Social Robot
Malfaz et al. [108]	Homeostasis based	Q-learning	Wellbeing of the robot	Maggie
Castro-Gonzalez et al. [64–66]	Homeostasis based	Object Q-learning	Variation of robot's wellbeing	Maggie
Maroto et al. [13]	Homeostasis based	Q-learning	Maximization of robot's well-being	Mini
Perula et al. [34]	Homeostasis based	Q-learning	Well-being of the robot	Mini
Da Silva et al. [63]	-	Economic TG	Generated on the basis of internal state estimate	Robotic head
Qureshi et al. [74]	-	DQN	Prediction error of an action conditional prediction network	Pepper

5.3. Methods Driven by Task Performance

Task performance denotes the effectiveness with which an agent performs a given task, and the performance metrics can vary for different tasks. In these methods, the design of the reward function is based on task-driven measures, which often include some problem-specific information, especially the task performance of the robot, task performance of the human, or both. For a quick summary, see Table 3.

5.3.1. Human Task Performance Driven Methods

In these human task performance driven methods, the reward function is based on the user's success in performing a task related to the interaction with the robot. The studies in this category are [47,92].

5.3.2. Robot Task Performance Driven Methods

In these methods, the reward design depends on the robot's task performance. Robot behaviors that satisfy the user's preferences, accurate completion of the task, finishing the task within a desired amount of time, visiting certain states, and robot actions that benefit or satisfy the user are examples for task performance measures. The studies in this category are [1,3,35,36,46,60,62,67,70,85,86].

Table 3. Summary of Task performance driven methods in social robotics.

Reference	Subcategory	Type of RL	Reward	Social Robot
Tapus et al. [92]	Human task performance	PGRL	User performance	Pioneer 2-DX
Gao et al. [47]	Human task performance	Multi-arm bandit	User task performance and user's verbal feedback	Pepper
Chan et al. [71,72]	Human and robot task performance	MAXQ	Success of the robot's actions in helping or improving user's affect and task performance	Brian 2.0
Chan et al. [33]	Human and robot task performance	MAXQ	Task performance of human and robot	Brian 2.0
Moro et al. [61]	Human and robot task performance	Q-learning	Numerical numbers based on robot's performance on user's activity state	Casper
Nejat et al. [62]	Robot task performance	Q-learning	User provides verbal feedback	Brian
Ranatunga et al. [70]	Robot task performance	TD(λ)	Head and eye kinematic scheme of the robot	Zeno
Keizer et al. [1]	Robot task performance	Monte-Carlo control	The robot's performance as a bartender	iCat
Qureshi et al. [36]	Robot task performance	Multimodal DQN	Numerical values based on robot's handshake success	Pepper
Papaioannou et al. [60]	Robot task performance	Q-learning	Task completion of the robot	Pepper
Qureshi et al. [73]	Robot task performance	MDARQN	Numerical values based on robot's handshaking success	Pepper
Hemminghaus et al. [3]	Robot task performance	Q-learning	Robot's task performance and execution cost of the robot's action	Furhat
Chen et al. [67]	Robot task performance	Q-learning	Numerical values based on correctly completed tasks	Mobile robots
Ritschel et al. [46]	Robot task performance	n-armed bandit	Robot's performance at convincing user to select healthy drink	Reeti
Lathuiliere et al. [85,86]	Robot task performance	DQN	Number of observed faces and presence of speech sources in the visual field	Nao
Cuayahuitl [35]	Robot task performance	DQN	Numerical values based on robot's performance in the game	Pepper

5.3.3. Human and Robot Task Performance Driven Methods

In the previous two sections, we listed the studies using task performance of the robot and human as reward signal. There are also studies that use a combination of the human's and the robot's task performance as reward signal. As an example, in [33,72] the robot received the highest reward if the user completed the task successfully. The robot also received reward for its actions that were suitable for the current situation. Likewise, in [61], the robot was rewarded based on actions that transitioned the user into a desirable state (e.g., completing the activity). Other papers in this category are [33,71,72].

6. Evaluation Methodologies

The past decade has seen a rapid growth of social robotics in diverse uncontrolled environments such as homes, schools, hospitals, shopping centers, or museums. In this review, we have seen various application domains in a range of fields including therapy [3], eldercare [62], entertainment [59], navigation [32], healthcare [44], education [58], personal robots [13], and rehabilitation [92]. Research in the field of social robotics and human-robot interaction becomes crucial as more and more robots are entering our lives. This brings many challenges as social robots are required to deal with dynamic and stochastic elements in social interaction in addition to the challenges in robotics. Besides these challenges, validation of social robotics systems with users necessitates efficient evaluation methodologies. Recent studies underline the importance of evaluation and assessment methodologies in HRI [111]. However, developing a standardized evaluation procedure still remains a difficult task. Furthermore, in RL-based robotic systems, there is a need to explore various human-level factors (personal preferences, attitudes, emotions, etc.) to assure that the learned policy leads to better HRI. Additionally, how can we evaluate whether the learned policy conveys the intended social skill(s)? As an example, in Qureshi et al. [36,73,74]'s study, the model performance on a test dataset was evaluated by three volunteers who judged if the robot's action was an appropriate one for the current scenario. In [87], there both annotators and participants rated whether the robot was able to associate the facial expressions with the conversation context. The independent annotators' ratings were higher than the participants', which, as the authors argued, might be explained by discrepancies between the participants' actual expressed emotion and the intended emotion. In such cases, additional sensory information could be useful for validating that the adaptive robot behaviors lead to better HRI. For example, Park et al. [58] analyzed the body poses and electrodermal activity (EDA) of the participants to check their correlation with participant's engagement. This kind of approach could be used to support subjective evaluations. A comparative evaluation methodology considering both the learned policy and the user's experience about the interaction is another way of evaluation. As an example [32,33,56,90,91] presented the policy for each participant as well as a discussion on the effectiveness of the robot behavior on the user based on their comments and subjective evaluations.

The papers in the scope of this manuscript used different evaluation and assessment methodologies for their algorithms and for their systems with users. We identify three types of evaluation methodologies: (1) an evaluation from the algorithm point of view, (2) evaluation and assessment of user experience-related subjective measures, and (3) evaluation of both learning algorithm-related factors and user experience-related factors. Several studies only reported the self-rated questionnaire results [45] or user opinions [55]. There are also studies which do not include any evaluation, and only a short discussion regarding the learned policy [53,57,100,101].

The cumulative collected reward over time is the most commonly used evaluation method. As learning progresses, the frequency of negative rewards is expected to decrease and positive rewards are expected to increase. Thus, the cumulative reward and comparing the reward across different settings and variations of algorithms are one of the measures for evaluating the efficiency of learning [49,50,52,85,86]. The evolution of the learning algorithm over time (e.g., the evolution of Q values) is another evaluation method. Several studies presented only the learning evolution of their system without mentioning how a participant would perceive the learned robot behaviors [13,34,61,63–66,108]. Comparison of user experiences (e.g., learning gains of children) for adaptive and non-adaptive robot is another way of evaluation [68,102]. We also see evaluation by using only interaction related objective measures such as the frequency of turn-taking and dialogue duration with the robot [60]. Task-related evaluation measures (i.e., the number of moves needed to solve a game with an adaptive versus a random robot) together with Q-matrix [3], or average task success and average reward [35] are used. In some IRL studies, the purpose is only teaching a robot. In these studies, evaluation metrics are training time [26,29], or training related parameters (e.g., the amount of positive and negative feedback) [28].

Studies reporting both subjective user opinions and algorithm related measures are [30,44,46,59,92]. Interaction related objective measures such as interaction duration, distance to the robot, preferred motion speed of the robot in combination with questionnaires are other measures for evaluating the efficiency of the learned policy. Studies also use a comparison of different algorithms in terms of average steps, average reward, average execution time together with questionnaires [67], and the number of times the preferences of the trainer match with the agent's action [69], reward and discussion of observations from the experiments [46], questionnaires and task-related parameters (e.g., time to complete the task) [47].

7. Discussion

In this paper, we present the RL approaches in social robotics. In virtual game environments (e.g., Atari, Go, etc.) which are commonly used testbeds for RL implementations, the goal is well defined (e.g., getting higher scores, accomplishing a game level, or winning the game). In social robotics, the goal is not that clear. Still, we argue that social robots could provide a unique potential testbed for RL implementations in real-world scenarios, in a sense that they can deal with transparency issues by showing their internal states through social cues (e.g., facial expressions, gaze, speech, LEDs on their body, tablet). In Section 5, we presented RL approaches based on reward types. IRL with implicit reward is the most widely used approach in social robotics since human social cues occur naturally during the interaction. However, the change in social cues can be slow, which leads to sparse reward. A combination of the reward approaches presented Section 5, namely intrinsically motivated methods, IRL with implicit feedback, and task performance-driven methods could be an approach to deal with the sparse reward problem. This way the robot could receive a reward even when there is no dramatic change in social cues or the task is not completed in one step. Similar to the homeostasis-based approaches, combining emotional models for robots' decision-making mechanisms could be helpful. The interested reader may refer to [104] which presents a thorough analysis and discussion of computational emotional models incorporated within RL. Th sparse reward problem is not the only problem in real-world social HRI. We continue to the discussion with the proposed solutions for real-world RL problems in Section 7.1. Later on, we present possible interesting future directions in Section 7.2.

7.1. Proposed Solutions to Real-World RL Problems

RL is a powerful and versatile algorithmic tool and has been shown to perform better than humans in simulated environments [43] However, the progress on applying RL methods to real-world systems is not so advanced yet. This is due to the complexity of the real-world. Dulac-Arnold et al. [112] discuss nine challenges of realizing RL on real-world systems. Here, we discuss these challenges and how some papers tackled them in real-world HRI with social robots.

The first mentioned challenge is "training off-line from the fixed logs of an external behavior policy". This challenge applies to HRI since users would not tolerate the long pauses and action delays of the social robot. As an example, Qureshi et al. [36] suggested an approach where they divided training into two stages. In the first stage, the robot interacts with the environment and gathers data, whereas in the second stage the robot rests and trains.

The second challenge is "learning on the real system from limited samples". This challenge is especially valid for HRI since the interaction time with the users is limited in controlled lab experiments. Moreover, users get bored and tired with longer interaction duration. As mentioned [112] exploration must be limited. As an example, in [13,34], exploration and exploitation phases are separated. A predefined duration is set for the exploration phase, in which the robot runs through all possible states and actions. Moreover, they also decreased the learning rate α throughout the exploration phase to increase the importance of previously learned information as the learning progresses. In the exploitation phase, they set α to 0. As mentioned in [112], for improving the sample efficiency expert demonstrations can be beneficial to avoid learning from scratch. For example, Moro et al. [61] combined LfD with Q-learning for a Casper robot helping older people in a tea making scenario. Another mentioned solution was model-based RL, of which we see two examples in social robotics [49,50]. In addition, long-term interactions (several sessions [58,68,102]) are important for HRI and could be beneficial for RL to collect samples.

The third challenge is "high-dimensional continuous state and action spaces". In the context of social robotics, the problem also needs to be simplified due to the low onboard computational power of most platforms. That might be another reason for a small set of actions in the reviewed papers. To overcome this challenge we see several approaches. As an example, human guidance was found to effective [26], as well as Object Q-learning [64–66] and action elimination [35].

The fourth challenge is "safety constraints that should never or at least rarely be violated". The mentioned approaches for this challenge in [112] include imposing safety constraints during the training. In the current literature, social robot interactions are generally conducted in a controlled laboratory environment and the researchers are available to intervene if any problems occur. Therefore, this challenge seems to get little attention.

The fifth challenge is "tasks that may be partially observable, alternatively viewed as non-stationary or stochastic". We see several attempts in social robotics to deal with this challenge such as in POMDP based approaches [50,102], and in DRL where several frames are stacked together for incorporating the history of the agent observations. Another mentioned approach to deal with this challenge was using recurrent networks which were applied in [63].

The sixth challenge is "reward functions that are unspecified, multi-objective, or risk-sensitive". Some papers that use simulated environments for training and testing on real-world interactions. In these papers, there are different reward functions for the simulated world and the real-world. Generally, the real-world reward functions are simplified to one parameter such as feedback of the user or predefined numeric numbers, whereas the simulated world reward functions are more complex including several parameters.

The seventh challenge is "system operators who desire explainable policies and actions". This is particularly valid for social robotics, since ambiguous robot behaviors might affect the user's willingness to interact again. Moreover, if the human trains the robot, the intention and internal state of the robot becomes crucial for the success of the training. As an example, Knox et al. [29] discussed the transparency challenges and their effect on the training time. Thomaz and Breazeal [28] observed that participants had a tendency to wait for eye contact with the robot before saying the next utterance while training the robot. These kinds of social cues on the robot could be used for explaining its actions and internal states.

The eighth challenge is "inference that must happen in real-time at the control frequency of the system". The real-world is slower than the simulated world both in reaction and data generation. To deal with this challenge, several researchers used an additional interface between the robot and the human, so that the inference is received from the interface rather than robot control.

The ninth challenge is "large and/or unknown delays in the system actuators, sensors, or rewards". We see several approaches to deal with this challenge, as an example [52] considered to increase the effect of human-delivered positive reward in larger time frames and to decrease the effect of negative reward in a shorter time frame. Another approach was estimating reward from natural human feedback using the gamma distribution [49].

7.2. Future Outlook

There are still many interesting potential problems and open questions to be solved in RL for social robotics. Applications on physically embodied robots are limited due to the enormous challenge of complexity and uncertainty in real-world social interactions. The increased prevalence of RL in physical social robots will shed further light on this topic. Another unanswered question is how RL-based social robotics may include the generation of reward signals from ambiguous or conflicting sources of implicit feedback, and how learned skills can be transferred to different robots. Further work could also investigate larger state-action spaces, as current studies are mostly limited to a small sets.

Despite the fact that there are goal-oriented approaches for social robot learning [113,114], in the current literature, the social robot that learns through RL has only one goal, such as performing a single task and optimizing a single reward function. However, in many real-world scenarios, a robot may need to perform a diverse set of tasks. As an example, socially assistive robots designed with the purpose of assisting older people in their houses may need to accomplish several tasks such as medication reminders, detecting issues, informing caregivers, and managing plans. Multi-goal RL enables an agent to learn multiple goals, hence the agent can generalize the desired behavior and transfer skills to unseen goals and tasks [115]. This has been applied on robotic manipulation tasks in a simulated environment [115]. However, applying the multi-goal RL framework to social robots would be a fruitful area for future work.

Another interesting future direction might be the application of multi-objective RL in social robotics. The task efficiency and user satisfaction can be two objectives where the robot would try to maximize both objectives by formalizing the problem as a multi-objective MDP. As an example, Hao et al. [116] presents a multi-objective weighted RL in which the agent had two objectives: minimizing the cost of service execution and eliminating the user's negative emotions. We refer the interested reader to the survey on multi-objective decision making for a more detailed explanation of the topic [117].

Recent developments in the field of deep neural networks have led to an increasing interest in DRL. Applying DRL in social robotics has also received recent attention, however, studies focused on small sets of actions and single task scenarios. In this regard, social robots with larger sets of actions would be a promising area for further work. Another future direction can be a further investigation of hyper-parameters of RL in social robotics. This was briefly discussed in [1], as an example, in turn-based interactions relatively small discount factors (i.e., $0.7 \leqslant \gamma \leqslant 0.95$) are more common, whereas for the frame-based interactions with rather long trajectories, higher discount factors seem to be more suitable (i.e., $\gamma \geqslant 0.99$). In deep networks, the selection of different hyper-parameters affects the accuracy of the algorithm [118]. This also applies to DRL, Lathuilière et al. [86] presented several experiments to evaluate the impact of some of the principal parameters of their deep network structure.

Thus far, model-free RL learning a value function or a policy through trial and error is the most commonly used approach in social robotics. However, model-based RL that focuses on learning a transition model of the environment serving as a simulation remains to be further explored. In particular, having a user model can ease the learning process. Although it is difficult to model human reactions, having a model can play a crucial role in reducing the number of required interactions in the real-world. The model-based approach can also help with the problem of hardware depreciation which may arise in model-free RL in robotics because of the considerable amount of interaction time. Simulating the interaction environment can ease the training without manual interventions and a need for maintenance. Nonetheless, transferring the learned policies in simulation directly to the physical robot may not be trivial due to undermodeling and uncertainty about system dynamics [15]. A common limitation is that most of the works are not generalizable, i.e., utilizing the knowledge learned by one robot on the other or utilizing the task knowledge for other tasks. The Google AI team trained a model-based Deep Planning Network (PlaNet) agent which achieved six different tasks (i.e., cartpole swing-up, cartpole balance, finger spin, cheetah run, etc.) [119]. A similar approach for a physical social robot would be an interesting future direction.

RL problems are formalized as MDPs in fully observable environments. However, in the case of HRI not all of the required observations are available, due to the underlying effect of psychological states on human behavior. It has been demonstrated that POMDPs are able to model the uncertainties and inherent interaction ambiguities in real-world HRI scenarios [120]. Hausknecht and Stone [121] proposed a method that couples a Long Short Term Memory with a Deep Q-Network to handle the noisy observations characteristic of POMDPs. A similar approach would be useful in social robotics problems to better capture the dynamics of the environment. We included two examples of POMDP approaches in social robotics, [50,102]. Further investigation would constitute an interesting line of research.

8. Conclusions

In this work, we give an overview of the work on RL in social robotics. We surveyed the literature and presented a thorough analysis of RL approaches in social robotics. Social robots have two important characteristics: physical embodiment and interaction/communication capabilities. Therefore, we included studies with physically embodied robots. Moreover, we categorize the papers based on the used RL type. In this categorization, we discuss and group the papers based on the communication medium used for reward formulation. Considering the importance of designing the reward function, we also categorize the papers based on the nature of the reward. The evaluation methods of the papers are also grouped by whether or not they use subjective and algorithmic metrics. We then provide a discussion in the view of real-world RL challenges and proposed solutions. The points that remain to be explored, including the approaches that have thus far received less attention are also given in the discussion section. To conclude, despite tremendous leaps in computing power and advances in learning methods, we are still a long way from general-purpose, robust, and versatile social robots that can learn several skills from naive users with real-world interactions. In spite of the immediate challenges, we see steady progress of RL applications in social robotics with an increasing interest in recent years.

Author Contributions: N.A. was the main author responsible for conducting literature research, methodology definition, and paper writing. A.L. supervised the study, and has been involved in structuring and writing the paper. All authors have read and agreed to the published version of the manuscript.

Funding: This research was funded by European Union's Horizon 2020 research and innovation program under the Marie Skłodowska-Curie grant agreement No 721619 for the SOCRATES project.

Institutional Review Board Statement: Not applicable.

Informed Consent Statement: Not applicable.

Data Availability Statement: Not applicable.

Conflicts of Interest: The authors declare no conflict of interest.

References

1. Keizer, S.; Ellen Foster, M.; Wang, Z.; Lemon, O. Machine Learning for Social Multiparty Human–Robot Interaction. *ACM Trans. Interact. Intell. Syst.* **2014**, *4*, 14:1–14:32. [CrossRef]
2. de Greeff, J.; Belpaeme, T. Why robots should be social: Enhancing machine learning through social human-robot interaction. *PLoS ONE* **2015**, *10*, e0138061. [CrossRef]
3. Hemminghaus, J.; Kopp, S. Towards Adaptive Social Behavior Generation for Assistive Robots Using Reinforcement Learning. In Proceedings of the 2017 ACM/IEEE International Conference on Human-Robot Interaction (HRI 2017), Vienna, Austria, 6–9 March 2017; pp. 332–340. [CrossRef]
4. Ritschel, H.; Seiderer, A.; Janowski, K.; Wagner, S.; André, E. Adaptive linguistic style for an assistive robotic health companion based on explicit human feedback. In Proceedings of the 12th ACM International Conference on PErvasive Technologies Related to Assistive Environments, Rhodes, Greece, 5–7 June 2019; pp. 247–255.
5. Sutton, R.S.; Barto, A.G. *Introduction to Reinforcement Learning*; MIT Press: Cambridge, UK, 1998; Volume 2.

6. Barto, A.G.; Sutton, R.S.; Watkins, C. *Learning and Sequential Decision Making*; University of Massachusetts Amherst: Amherst, MA, USA, 1989.

7. Fong, T.; Nourbakhsh, I.; Dautenhahn, K. A survey of socially interactive robots. *Robot. Auton. Syst.* **2003**, *42*, 143–166. [CrossRef]

8. Breazeal, C. Toward sociable robots. *Robot. Auton. Syst.* **2003**, *42*, 167–175. [CrossRef]

9. Duffy, B.R. Anthropomorphism and the social robot. *Robot. Auton. Syst.* **2003**, *42*, 177–190. [CrossRef]

10. Bartneck, C.; Forlizzi, J. A design-centred framework for social human-robot interaction. In Proceedings of the 13th IEEE International Workshop on Robot and Human Interactive Communication (RO-MAN 2004), Kurashiki, Japan, 20–22 September 2004; pp. 591–594.

11. Hegel, F.; Muhl, C.; Wrede, B.; Hielscher-Fastabend, M.; Sagerer, G. Understanding social robots. In Proceedings of the 2009 Second International Conferences on Advances in Computer-Human Interactions, Cancun, Mexico, 1–7 February 2009; pp. 169–174.

12. Yan, H.; Ang, M.H.; Poo, A.N. A survey on perception methods for human–robot interaction in social robots. *Int. J. Soc. Robot.* **2014**, *6*, 85–119. [CrossRef]

13. Maroto-Gómez, M.; Castro-González, Á.; Castillo, J.; Malfaz, M.; Salichs, M. A bio-inspired motivational decision making system for social robots based on the perception of the user. *Sensors* **2018**, *18*, 2691. [CrossRef]

14. Sutton, R.S.; Barto, A.G. *Reinforcement Learning: An Introduction*; A Bradford Book: Cambridge, MA, USA, 2018.

15. Kober, J.; Bagnell, J.A.; Peters, J. Reinforcement learning in robotics: A survey. *Int. J. Robot. Res.* **2013**, *32*, 1238–1274. [CrossRef]

16. Kormushev, P.; Calinon, S.; Caldwell, D. Reinforcement learning in robotics: Applications and real-world challenges. *Robotics* **2013**, *2*, 122–148. [CrossRef]

17. Deisenroth, M.P.; Neumann, G.; Peters, J. A survey on policy search for robotics. *Found. Trends® Robot.* **2013**, *2*, 388–403.

18. Garcıa, J.; Fernández, F. A comprehensive survey on safe reinforcement learning. *J. Mach. Learn. Res.* **2015**, *16*, 1437–1480.

19. Bhagat, S.; Banerjee, H.; Ho Tse, Z.T.; Ren, H. Deep reinforcement learning for soft, flexible robots: Brief review with impending challenges. *Robotics* **2019**, *8*, 4. [CrossRef]

20. Oudeyer, P.Y.; Kaplan, F. How can we define intrinsic motivation. In Proceedings of the 8th International Conference on Epigenetic Robotics: Modeling Cognitive Development in Robotic Systems, Brighton, UK, 30–31 July 2008; pp. 93–101.

21. Thomaz, A.; Hoffman, G.; Cakmak, M. Computational human-robot interaction. *Found. Trends Robot.* **2016**, *4*, 105–223.

22. Thomaz, A.L.; Breazeal, C.; Barto, A.G.; Picard, R. Socially Guided Machine Learning. 2006. Available online: https://scholarworks.umass.edu/cs_faculty_pubs/183 (accessed on 2 February 2020).

23. Holzinger, A.; Plass, M.; Holzinger, K.; Crişan, G.C.; Pintea, C.M.; Palade, V. Towards interactive Machine Learning (iML): Applying ant colony algorithms to solve the traveling salesman problem with the human-in-the-loop approach. In Proceedings of the International Conference on Availability, Reliability, and Security (ARES 2016), Salzburg, Austria, 31 August–2 September 2016; pp. 81–95.

24. Knox, W.B.; Stone, P. Reinforcement learning from simultaneous human and MDP reward. In Proceedings of the 11th International Conference on Autonomous Agents and Multiagent Systems (AAMAS '12), Valencia, Spain, 4–8 June 2012; pp. 475–482.

25. Isbell, C.; Shelton, C.R.; Kearns, M.; Singh, S.; Stone, P. A social reinforcement learning agent. In Proceedings of the Fifth International Conference on Autonomous Agents, (AGENTS '01), Montreal, QC, Canada, 28 May–1 June 2001; pp. 377–384.

26. Suay, H.B.; Chernova, S. Effect of human guidance and state space size on interactive reinforcement learning. In Proceedings of the 20th IEEE International Workshop on Robot and Human Communication (RO-MAN 2011), Atlanta, GA, USA, 31 July–3 August 2011; pp. 1–6.

27. Thomaz, A.L.; Hoffman, G.; Breazeal, C. Reinforcement learning with human teachers: Understanding how people want to teach robots. In Proceedings of the 15th IEEE International Symposium on Robot and Human Interactive Communication (RO-MAN 2006), Hatfield, UK, 6–8 September 2006; pp. 352–357.

28. Thomaz, A.L.; Breazeal, C. Experiments in socially guided exploration: Lessons learned in building robots that learn with and without human teachers. *Connect. Sci.* **2008**, *20*, 91–110. [CrossRef]

29. Knox, W.B.; Stone, P.; Breazeal, C. Training a Robot via Human Feedback: A Case Study. In *Social Robotics*; Herrmann, G., Pearson, M.J., Lenz, A., Bremner, P., Spiers, A., Leonards, U., Eds.; Springer International Publishing: Cham, Switzerland, 2013; pp. 460–470.

30. Suay, H.B.; Toris, R.; Chernova, S. A Practical Comparison of Three Robot Learning from Demonstration Algorithm. *Int. J. Soc. Robot.* **2012**, *4*, 319–330. [CrossRef]

31. Thomaz, A.L.; Breazeal, C. Asymmetric Interpretations of Positive and Negative Human Feedback for a Social Learning Agent. In Proceedings of the 16th IEEE International Symposium Robot and Human Interactive Communication (RO-MAN 2007), Jeju, Korea, 26–29 August 2007; pp. 720–725. [CrossRef]

32. Patompak, P.; Jeong, S.; Nilkhamhang, I.; Chong, N.Y. Learning Proxemics for Personalized Human–Robot Social Interaction. *Int. J. Soc. Robot.* **2019**, *12*, 267–280. [CrossRef]

33. Chan, J.; Nejat, G. Social Intelligence for a Robot Engaging People in Cognitive Training Activities. *Int. J. Adv. Robot. Syst.* **2012**, *9*. [CrossRef]

34. Pérula-Martínez, R.; Castro-Gonzalez, A.; Malfaz, M.; Alonso-Martín, F.; Salichs, M.A. Bioinspired decision-making for a socially interactive robot. *Cogn. Syst. Res.* **2019**, *54*, 287–301. [CrossRef]

35. Cuayáhuitl, H. A data-efficient deep learning approach for deployable multimodal social robots. *Neurocomputing* **2019**, *396*, 587–598. [CrossRef]

36. Qureshi, A.H.; Nakamura, Y.; Yoshikawa, Y.; Ishiguro, H. Robot gains social intelligence through multimodal deep reinforcement learning. In Proceedings of the 16th IEEE-RAS International Conference on Humanoid Robots, Humanoids, Cancun, Mexico, 15–17 November 2016; pp. 745–751. [CrossRef]

37. Zhang, H.; Yu, T. Taxonomy of Reinforcement Learning Algorithms. In *Deep Reinforcement Learning: Fundamentals, Research and Applications*; Dong, H., Ding, Z., Zhang, S., Eds.; Springer: Singapore, 2020; pp. 125–133. [CrossRef]

38. Bellman, R. On the theory of dynamic programming. *Proc. Natl. Acad. Sci. USA* **1952**, *38*, 716. [CrossRef]

39. Rummery, G.A.; Niranjan, M. *On-line Q-Learning Using Connectionist Systems*; University of Cambridge, Department of Engineering: Cambridge, UK, 1994; Volume 37.

40. Watkins, C.J.C.H. Learning from Delayed Rewards. 1989. Available online: http://www.cs.rhul.ac.uk/~chrisw/new_thesis.pdf (accessed on 7 December 2019).

41. Gosavi, A. Boundedness of iterates in Q-learning. *Syst. Control Lett.* **2006**, *55*, 347–349. [CrossRef]

42. Sigaud, O.; Stulp, F. Policy search in continuous action domains: An overview. *Neural Netw.* **2019**, *113*, 28–40. [CrossRef] [PubMed]

43. Mnih, V.; Kavukcuoglu, K.; Silver, D.; Rusu, A.A.; Veness, J.; Bellemare, M.G.; Graves, A.; Riedmiller, M.; Fidjeland, A.K.; Ostrovski, G.; et al. Human-level control through deep reinforcement learning. *Nature* **2015**, *518*, 529–533. [CrossRef] [PubMed]

44. Schneider, S.; Kummert, F. Exploring embodiment and dueling bandit learning for preference adaptation in human-robot interaction. In Proceedings of the 26th IEEE International Symposium on Robot and Human Interactive Communication (RO-MAN 2017), Lisbon, Portugal, 28 August–1 September 2017; pp. 1325–1331.

45. Leite, I.; Pereira, A.; Castellano, G.; Mascarenhas, S.; Martinho, C.; Paiva, A. Modelling empathy in social robotic companions. In Proceedings of the International Conference on User Modeling, Adaptation, and Personalization (UMAP 2011), Girona, Spain, 11–15 July 2011; pp. 135–147.

46. Ritschel, H.; Seiderer, A.; Janowski, K.; Aslan, I.; André, E. Drink-o-mender: An adaptive robotic drink adviser. In Proceedings of the 3rd International Workshop on Multisensory Approaches to Human-Food Interaction, Boulder, CO, USA, 16–20 October 2018; pp. 1–8.

47. Gao, Y.; Barendregt, W.; Obaid, M.; Castellano, G. When robot personalisation does not help: Insights from a robot-supported learning study. In Proceedings of the 27th IEEE International Symposium on Robot and Human Interactive Communication (RO-MAN 2018), Nanjing, China, 27–31 August 2018; pp. 705–712.

48. Bubeck, S.; Cesa-Bianchi, N. Regret Analysis of Stochastic and Nonstochastic Multi-armed Bandit Problems. *Found. Trends®* *Mach. Learn.* **2012**, *5*, 1–122. [CrossRef]

49. Tseng, S.H.; Liu, F.C.; Fu, L.C. Active Learning on Service Providing Model: Adjustment of Robot Behaviors Through Human Feedback. *IEEE Trans. Cogn. Dev. Syst.* **2018**, *10*, 701–711. [CrossRef]

50. Martins, G.S.; Al Tair, H.; Santos, L.; Dias, J. αPOMDP: POMDP-based user-adaptive decision-making for social robots. *Pattern Recognit. Lett.* **2019**, *118*, 94–103. [CrossRef]

51. Wiewiora, E. Reward Shaping. In *Encyclopedia of Machine Learning*; Sammut, C., Webb, G.I., Eds.; Springer: Boston, MA, USA, 2010; pp. 863–865.

52. Barraquand, R.; Crowley, J.L. Learning Polite Behavior with Situation Models. In Proceedings of the 3rd ACM/IEEE International Conference on Human Robot Interaction (HRI 2008), Amsterdam, The Netherlands, 12–15 March 2008; pp. 209–216. [CrossRef]

53. Yang, C.; Lu, M.; Tseng, S.; Fu, L. A companion robot for daily care of elders based on homeostasis. In Proceedings of the 56th Annual Conference of the Society of Instrument and Control Engineers of Japan (SICE 2017), Kanazawa, Japan, 19–22 September 2017; pp. 1401–1406. [CrossRef]

54. Zarinbal, M.; Mohebi, A.; Mosalli, H.; Haratinik, R.; Jabalameli, Z.; Bayatmakou, F. A New Social Robot for Interactive Query-Based Summarization: Scientific Document Summarization. In Proceedings of the International Conference on Interactive Collaborative Robotics (ICR 2019), Istanbul, Turkey, 20–25 August 2019; pp. 330–340.

55. Addo, I.D.; Ahamed, S.I. Applying affective feedback to reinforcement learning in ZOEI, a comic humanoid robot. In Proceedings of the 23rd IEEE International Symposium on Robot and Human Interactive Communication (RO-MAN 2014), Edinburgh, UK, 25–29 August 2014; pp. 423–428.

56. Chiang, Y.S.; Chu, T.S.; Lim, C.; Wu, T.Y.; Tseng, S.H.; Fu, L.C. Personalizing robot behavior for interruption in social human-robot interaction. In Proceedings of the 2014 IEEE International Workshop on Advanced Robotics and its Social Impacts (ARSO 2014), Evanston, IL, USA, 11–13 September 2014; pp. 44–49. [CrossRef]

57. Ritschel, H.; Baur, T.; André, E. Adapting a Robot's linguistic style based on socially-aware reinforcement learning. In Proceedings of the 26th IEEE International Symposium on Robot and Human Interactive Communication (RO-MAN 2017), Lisbon, Portugal, 28 August–1 September 2017; pp. 378–384. [CrossRef]

58. Park, H.W.; Grover, I.; Spaulding, S.; Gomez, L.; Breazeal, C. A model-free affective reinforcement learning approach to personalization of an autonomous social robot companion for early literacy education. In Proceedings of the AAAI Conference on Artificial Intelligence, Honolulu, HI, USA, 27 January–1 February 2019; Volume 33, pp. 687–694.

59. Weber, K.; Ritschel, H.; Aslan, I.; Lingenfelser, F.; André, E. How to Shape the Humor of a Robot—Social Behavior Adaptation Based on Reinforcement Learning. In Proceedings of the International Conference on Multimodal Interaction, ICMI 2018, Boulder, CO, USA, 16–20 October 2018; pp. 154–162. [CrossRef]

60. Papaioannou, I.; Dondrup, C.; Novikova, J.; Lemon, O. Hybrid chat and task dialogue for more engaging hri using reinforcement learning. In Proceedings of the 26th IEEE International Symposium on Robot and Human Interactive Communication (RO-MAN 2017), Lisbon, Portugal, 28 August–1 September 2017; pp. 593–598.

61. Moro, C.; Nejat, G.; Mihailidis, A. Learning and Personalizing Socially Assistive Robot Behaviors to Aid with Activities of Daily Living. *ACM Trans. Hum. Robot. Interact.* **2018**, *7*. [CrossRef]

62. Nejat, G.; Ficocelli, M. Can I be of assistance? The intelligence behind an assistive robot. In Proceedings of the 2008 IEEE International Conference on Robotics and Automation (ICRA 2008), Pasadena, CA, USA, 19–23 May 2008; pp. 3564–3569.

63. Da Silva, R.R.; Francelin Romero, R.A. Modelling Shared Attention Through Relational Reinforcement Learning. *J. Intell. Robot. Syst.* **2012**, *66*, 167–182. [CrossRef]

64. Castro-González, Á.; Malfaz, M.; Gorostiza, J.F.; Salichs, M.A. Learning behaviors by an autonomous social robot with motivations. *Cybern. Syst.* **2014**, *45*, 568–598. [CrossRef]

65. Castro-González, Á.; Malfaz, M.; Salichs, M.A. An autonomous social robot in fear. *IEEE Trans. Auton. Ment. Dev.* **2013**, *5*, 135–151. [CrossRef]

66. Castro-González, Á.; Malfaz, M.; Salichs, M.A. Learning the Selection of Actions for an Autonomous Social Robot by Reinforcement Learning Based on Motivations. *Int. J. Soc. Robot.* **2011**, *3*, 427–441. [CrossRef]

67. Chen, L.; Wu, M.; Zhou, M.; She, J.; Dong, F.; Hirota, K. Information-Driven Multirobot Behavior Adaptation to Emotional Intention in Human–Robot Interaction. *IEEE Trans. Cogn. Dev. Syst.* **2018**, *10*, 647–658. [CrossRef]

68. Gordon, G.; Spaulding, S.; Westlund, J.K.; Lee, J.J.; Plummer, L.; Martinez, M.; Das, M.; Breazeal, C. Affective personalization of a social robot tutor for children's second language skills. In Proceedings of the 30th AAAI Conference on Artificial Intelligence, Phoenix, AZ, USA, 12–17 February 2016.

69. Gamborino, E.; Fu, L.C. Interactive Reinforcement Learning based Assistive Robot for the Emotional Support of Children. In Proceedings of the 18th International Conference on Control Automation and Systems (ICCAS 2018), Pyeong Chang, Korea, 17–20 October 2018; pp. 708–713.

70. Ranatunga, I.; Rajruangrabin, J.; Popa, D.O.; Makedon, F. Enhanced Therapeutic Interactivity Using Social Robot Zeno. In Proceedings of the 4th International Conference on PErvasive Technologies Related to Assistive Environments (PETRA 2011), Crete, Greece, 25–27 May 2011; pp. 1–6 . [CrossRef]

71. Chan, J.; Nejat, G. Minimizing task-induced stress in cognitively stimulating activities using an intelligent socially assistive robot. In Proceedings of the 20th International Symposium on Robot and Human Interactive Communication (RO-MAN 2011), Atlanta, GA, USA, 31 July–3 August 2011; pp. 296–301.

72. Chan, J.; Nejat, G. A learning-based control architecture for an assistive robot providing social engagement during cognitively stimulating activities. In Proceedings of the 2011 IEEE International Conference on Robotics and Automation, Shanghai, China, 9–13 May 2011; pp. 3928–3933.

73. Qureshi, A.H.; Nakamura, Y.; Yoshikawa, Y.; Ishiguro, H. Show, attend and interact: Perceivable human-robot social interaction through neural attention Q-network. In Proceedings of the IEEE International Conference on Robotics and Automation (ICRA 2017), Singapore, 29 May–3 June 2017; pp. 1639–1645.

74. Qureshi, A.; Nakamura, Y.; Yoshikawa, Y.; Ishiguro, H. Intrinsically motivated reinforcement learning for human–robot interaction in the real-world. *Neural Netw.* **2018**, *107*, 23–33. [CrossRef] [PubMed]

75. Thomaz, A.; Breazeal, C. Adding guidance to interactive reinforcement learning. In Proceedings of the 20th Conference on Artificial Intelligence (AAAI 2006), Boston, MA, USA, 16–20 July 2006.

76. Loftin, R.; Peng, B.; MacGlashan, J.; Littman, M.L.; Taylor, M.E.; Huang, J.; Roberts, D.L. Learning behaviors via human-delivered discrete feedback: Modeling implicit feedback strategies to speed up learning. *Auton. Agents Multi Agent Syst.* **2016**, *30*, 30–59. [CrossRef]

77. Wagner, J.; Lingenfelser, F.; Baur, T.; Damian, I.; Kistler, F.; André, E. The social signal interpretation (SSI) framework: Multimodal signal processing and recognition in real-time. In Proceedings of the 21st ACM International Conference on Multimedia, Barcelona, Spain, 21–25 October 2013; pp. 831–834.

78. McDuff, D.; Mahmoud, A.; Mavadati, M.; Amr, M.; Turcot, J.; Kaliouby, R.E. AFFDEX SDK: A cross-platform real-time multi-face expression recognition toolkit. In Proceedings of the 2016 CHI Conference Extended Abstracts on Human Factors in Computing Systems, San Jose, CA, USA, 7–12 May 2016; pp. 3723–3726.

79. Chen, L.F.; Liu, Z.T.; Dong, F.Y.; Yamazaki, Y.; Wu, M.; Hirota, K. Adapting multi-robot behavior to communication atmosphere in humans-robots interaction using fuzzy production rule based friend-Q learning. *J. Adv. Comput. Intell. Intell. Inform.* **2013**, *17*, 291–301. [CrossRef]

80. Schwartz, A. A Reinforcement Learning Method for Maximizing Undiscounted Rewards. In Proceedings of the 10th International Conference on Machine Learning (ICML 1993), Amherst, MA, USA, 27–29 June 1993; Volume 298, pp. 298–305.

81. Mahadevan, S. Average reward reinforcement learning: Foundations, algorithms, and empirical results. *Mach. Learn.* **1996**, *22*, 159–195. [CrossRef]

82. Mavridis, N. A review of verbal and non-verbal human–robot interactive communication. *Robot. Auton. Syst.* **2015**, *63*, 22–35. [CrossRef]

83. Da Silva, R.R.; Policastro, C.A.; Romero, R.A. Relational reinforcement learning applied to shared attention. In Proceedings of the 2009 International Joint Conference on Neural Networks (IJCNN 2009), Atlanta, GA, USA, 14–19 June 2009; pp. 2943–2949.

84. Dietterich, T.G. Hierarchical reinforcement learning with the MAXQ value function decomposition. *J. Artif. Intell. Res.* **2000**, *13*, 227–303. [CrossRef]

85. Lathuilière, S.; Massé, B.; Mesejo, P.; Horaud, R. Deep Reinforcement Learning for Audio-Visual Gaze Control. In Proceedings of the IEEE/RSJ International Conference on Intelligent Robots and Systems (IROS 2018), Madrid, Spain, 1–5 October 2018; pp. 1555–1562.

86. Lathuilière, S.; Massé, B.; Mesejo, P.; Horaud, R. Neural network based reinforcement learning for audio–visual gaze control in human–robot interaction. *Pattern Recognit. Lett.* **2019**, *118*, 61–71. [CrossRef]

87. Churamani, N.; Barros, P.; Strahl, E.; Wermter, S. Learning Empathy-Driven Emotion Expressions using Affective Modulations. In Proceedings of the International Joint Conference on Neural Networks (IJCNN 2018), Rio de Janeiro, Brazil, 8–13 July 2018; pp. 1–8. [CrossRef]

88. Xu, K.; Ba, J.; Kiros, R.; Cho, K.; Courville, A.; Salakhudinov, R.; Zemel, R.; Bengio, Y. Show, attend and tell: Neural image caption generation with visual attention. In Proceedings of the International Conference on Machine Learning (ICML 2015), Lille, France, 6–11 July 2015; pp. 2048–2057.

89. Breazeal, C. Role of expressive behaviour for robots that learn from people. *Philos. Trans. R. Soc. B Biol. Sci.* **2009**, *364*, 3527–3538. [CrossRef]

90. Mitsunaga, N.; Smith, C.; Kanda, T.; Ishiguro, H.; Hagita, N. Robot behavior adaptation for human-robot interaction based on policy gradient reinforcement learning. *J. Robot. Soc. Jpn.* **2006**, *24*, 820–829. [CrossRef]

91. Mitsunaga, N.; Smith, C.; Kanda, T.; Ishiguro, H.; Hagita, N. Adapting robot behavior for human–robot interaction. *IEEE Trans. Robot.* **2008**, *24*, 911–916. [CrossRef]

92. Tapus, A.; Ţăpuş, C.; Matarić, M.J. User—Robot personality matching and assistive robot behavior adaptation for post-stroke rehabilitation therapy. *Intell. Serv. Robot.* **2008**, *1*, 169. [CrossRef]

93. Knox, W.B.; Stone, P. Interactively shaping agents via human reinforcement: The TAMER framework. In Proceedings of the 5th International Conference on Knowledge Capture, Redondo Beach, CA, USA, 1–4 September 2009; pp. 9–16.

94. Celemin, C.; Ruiz-del Solar, J. An interactive framework for learning continuous actions policies based on corrective feedback. *J. Intell. Robot. Syst.* **2019**, *95*, 77–97. [CrossRef]

95. Thomaz, A.L.; Breazeal, C. Teachable robots: Understanding human teaching behavior to build more effective robot learners. *Artif. Intell.* **2008**, *172*, 716–737. [CrossRef]

96. Li, G.; Gomez, R.; Nakamura, K.; He, B. Human-centered reinforcement learning: A survey. *IEEE Trans. Hum. Mach. Syst.* **2019**, *49*, 337–349. [CrossRef]

97. Knox, W.B.; Stone, P. Framing reinforcement learning from human reward: Reward positivity, temporal discounting, episodicity, and performance. *Artif. Intell.* **2015**, *225*, 24–50. [CrossRef]

98. Thomaz, A.L.; Breazeal, C. Reinforcement Learning with Human Teachers: Evidence of Feedback and Guidance with Implications for Learning Performance. In Proceedings of the 21st National Conference on Artificial intelligence (AAAI 2006), Boston, MA, USA, 16–20 July 2006; Volume 6, pp. 1000–1005.

99. Schmidt, A. Implicit human computer interaction through context. *Pers. Technol.* **2000**, *4*, 191–199. [CrossRef]

100. Grüneberg, P.; Suzuki, K. A lesson from subjective computing: Autonomous self-referentiality and social interaction as conditions for subjectivity. In Proceedings of the AISB/IACAP World Congress 2012: Computational Philosophy, Part of Alan Turing Year, Birmingham, UK, 2–6 July 2012; pp. 18–28.

101. Grüneberg, P.; Suzuki, K. An approach to subjective computing: A robot that learns from interaction with humans. *IEEE Trans. Auton. Ment. Dev.* **2013**, *6*, 5–18. [CrossRef]

102. Ramachandran, A.; Sebo, S.S.; Scassellati, B. Personalized Robot Tutoring using the Assistive Tutor POMDP (AT-POMDP). In Proceedings of the 33rd AAAI Conference on Artificial Intelligence (AAAI 2019), Honolulu, HI, USA, 27 January–1 February 2019; pp. 8050–8057.

103. Lerner, J.S.; Li, Y.; Valdesolo, P.; Kassam, K.S. Emotion and decision making. *Annu. Rev. Psychol.* **2015**, *66*, 799–823. [CrossRef]

104. Moerland, T.M.; Broekens, J.; Jonker, C.M. Emotion in reinforcement learning agents and robots: A survey. *Mach. Learn.* **2018**, *107*, 443–480. [CrossRef]

105. Ryan, R.M.; Deci, E.L. Intrinsic and extrinsic motivations: Classic definitions and new directions. *Contemp. Educ. Psychol.* **2000**, *25*, 54–67. [CrossRef] [PubMed]

106. Oudeyer, P.Y.; Kaplan, F. What is intrinsic motivation? A typology of computational approaches. *Front. Neurorobot.* **2009**, *1*, 6. [CrossRef] [PubMed]

107. Chentanez, N.; Barto, A.G.; Singh, S.P. Intrinsically motivated reinforcement learning. In Proceedings of the Advances in Neural Information Processing Systems (NIPS 2004), Vancouver, BC, Canada, 13–18 December 2004; pp. 1281–1288.

108. Malfaz, M.; Castro-González, Á.; Barber, R.; Salichs, M.A. A biologically inspired architecture for an autonomous and social robot. *IEEE Trans. Auton. Ment. Dev.* **2011**, *3*, 232–246. [CrossRef]

109. Cannon, W.B. *The Wisdom of the Body*; W.W. Norton & Company, Inc.: New York, NY, USA, 1939.

110. Berridge, K.C. Motivation concepts in behavioral neuroscience. *Physiol. Behav.* **2004**, *81*, 179–209. [CrossRef]
111. Sim, D.Y.Y.; Loo, C.K. Extensive assessment and evaluation methodologies on assistive social robots for modelling human–robot interaction–A review. *Inf. Sci.* **2015**, *301*, 305–344. [CrossRef]
112. Dulac-Arnold, G.; Mankowitz, D.; Hester, T. Challenges of real-world reinforcement learning. *arXiv* **2019**, arXiv:1904.12901.
113. Lockerd, A.; Breazeal, C. Tutelage and socially guided robot learning. In Proceedings of the IEEE/RSJ International Conference on Intelligent Robots and Systems (IROS 2004), Sendai, Japan, 28 September–2 October 2004; Volume 4, pp. 3475–3480.
114. Liu, L.; Li, B.; Chen, I.M.; Goh, T.J.; Sung, M. Interactive robots as social partner for communication care. In Proceedings of the IEEE International Conference on Robotics and Automation (ICRA 2014), Hong Kong, China, 31 May–7 June 2014; pp. 2231–2236.
115. Bai, C.; Liu, P.; Zhao, W.; Tang, X. Guided goal generation for hindsight multi-goal reinforcement learning. *Neurocomputing* **2019**, *359*, 353–367. [CrossRef]
116. Hao, M.; Cao, W.; Liu, Z.; Wu, M.; Yuan, Y. Emotion Regulation Based on Multi-objective Weighted Reinforcement Learning for Human-robot Interaction. In Proceedings of the 12th Asian Control Conference (ASCC 2019), Kitakyushu-shi, Japan, 9–12 June 2019; pp. 1402–1406.
117. Roijers, D.M.; Vamplew, P.; Whiteson, S.; Dazeley, R. A Survey of Multi-objective Sequential Decision-making. *J. Artif. Intell. Res.* **2013**, *48*, 67–113. [CrossRef]
118. Zhang, X.; Yao, L.; Huang, C.; Sheng, Q.Z.; Wang, X. Intent Recognition in Smart Living through Deep Recurrent Neural Networks. In Proceedings of the International Conference on Neural Information Processing, Guangzhou, China, 14–18 November 2017; pp. 748–758.
119. Hafner, D.; Lillicrap, T.; Fischer, I.; Villegas, R.; Ha, D.; Lee, H.; Davidson, J. Learning Latent Dynamics for Planning from Pixels. In Proceedings of the 36th International Conference on Machine Learning (ICML 2019), Long Beach, CA, USA, 9–15 June 2019; pp. 2555–2565.
120. Kostavelis, I.; Giakoumis, D.; Malassiotis, S.; Tzovaras, D. A POMDP Design Framework for Decision Making in Assistive Robots. In Proceedings of the International Conference on Human-Computer Interaction (HCI 2017), Vancouver, BC, Canada, 9–14 July 2017; pp. 467–479.
121. Hausknecht, M.; Stone, P. Deep Recurrent Q-Learning for Partially Observable MDPs. In Proceedings of the 2015 AAAI Fall Symposium on Sequential Decision Making for Intelligent Agents (AAAI-SDMIA15), Arlington, VA, USA, 12–14 November 2015.

Letter

A Framework for Human-Robot-Human Physical Interaction Based on N-Player Game Theory

Rui Zou [†], **Yubin Liu** *, **Jie Zhao and Hegao Cai**

State Key Laboratory of Robotics and Systems, Harbin Institute of Technology, Harbin 150001, China;
18b908014@stu.hit.edu.cn (R.Z.); jzhao@hit.edu.cn (J.Z.); hgcai@hit.edu.cn (H.C.)
* Correspondence: liuyubin@hit.edu.cn
† Current address: 92 Xidazhi Street, Nangang District, State Key Laboratory of Robotics and Systems, Harbin Institute of Technology, Harbin 150001, China.

Received: 31 July 2020; Accepted: 1 September 2020; Published: 3 September 2020

Abstract: In order to analyze the complex interactive behaviors between the robot and two humans, this paper presents an adaptive optimal control framework for human-robot-human physical interaction. N-player linear quadratic differential game theory is used to describe the system under study. N-player differential game theory can not be used directly in actual scenerie, since the robot cannot know humans' control objectives in advance. In order to let the robot know humans' control objectives, the paper presents an online estimation method to identify unknown humans' control objectives based on the recursive least squares algorithm. The Nash equilibrium solution of human-robot-human interaction is obtained by solving the coupled Riccati equation. Adaptive optimal control can be achieved during the human-robot-human physical interaction. The effectiveness of the proposed method is demonstrated by rigorous theoretical analysis and simulations. The simulation results show that the proposed controller can achieve adaptive optimal control during the interaction between the robot and two humans. Compared with the LQR controller, the proposed controller has more superior performance.

Keywords: physical human-robot interaction; game theory; adaptive optimal control; robot control

1. Introduction

In the past decade, physical human-robot interaction has attracted the attention of the research community due to the urgent requirement for robot technology in unstructured environment [1–4]. Physical human-robot interaction combines the advantages of humans and robots, which means that humans are good at reasoning and problem solving with high flexibility, while robots perform well in terms of execution as well as guaranteeing the accuracy of task execution [5,6]. The combination of these advantages has led to the wide application of physical human-robot interaction, such as teleoperation [7,8], collaborative assembly [9,10], and collaborative transportation [11–13].

Two types of specific human robot interaction strategies have been widely studied: co-activity type of interaction strategy and master-slave control strategy [14,15]. Co-activity type of interaction strategy is used in typical rehabilitation robots that help limb movement training or intelligent industrial systems that support heavy objects to resist gravity, where robots completely ignore human users' behaviors [16,17]. In contrast, the master–slave control strategy is used in the teleoperated robots or force extender exoskeletons use where robots completely follow the control of human users [18]. However, these strategies can only be used for specific interactive behaviors, the general framework for analyzing various interactive behaviors between robot and humans is still missing [19,20].

It has been pointed out that game theory can be used as a general framework to analyze complex interactive behaviors between multiple agents because different combinations of individual cost functions and different optimization objectives can be used to describe various interactive behaviors in game theory [21]. In [22], the human and the robot were been regarded as two agents and game theory was used in order to analyze the performance of the two agents. In [23], the optimal control was obtained for a given game with a linear system cost function by solving the coupled Riccati equation. In [24], an optimal control algorithm was developed for human-robot collaboration by solving the Riccati equation in each loop. In [25–28], policy iteration was used to solve the Nash equilibrium solution in order to improve the calculation speed. In [29], cyber-physical human systems was modeled via an interplay between reinforcement learning and game theory. In [30], haptic shared control for human-robot collaboration was modeled by a game-theoretical approach. In [31], human-like motion planning was studied based on game theoretic decision making. In [32], cooperative game was used for human-robot collaborative manufacturing. In [33], a bayesian framework was proposed for nash equilibrium inference in human-robot parallel play. In [19], non-cooperative differential game theory was used to model human-robot interaction system that results in a variety of interaction strategies. However, the above studies only consider two agents, that is, the interaction between one human and one robot. Therefore, the aforementioned methods are not suitable for human-robot-human physical interaction where more than one human interact with one robot physically. It is worth noting that the physical interaction between one robot and two humans will bring greater advantages such as operating larger loads, improving the flexibility and robustness of the system [28,34–37]. These greater advantages are brought by the team collaboration between the robot and two humans. To the authors' acknowledgment, no literature have researched the problem of the physical interaction between one robot and two humans based on game theory.

In the paper, a general adaptive optimal control framework for human-robot-human physical interaction is proposed based on N-player game theory. Accordingly, the robot and two humans can interact with each other optimally by learning each other's control. N-player differential game theory was used to model the human-robot-human interaction system in order to analyze the complex interactive behaviors between the robot and two humans. In N-player differential game theory, humans' control objectives are assumed to be knowledge [38,39]. However, N-player differential game theory can not be used directly in actual scenerie since the robot cannot know humans' control objectives in advance. In order to let the robot know humans' control objectives, the paper presents an online estimation method to identify unknown humans' control objectives based on the recursive least squares algorithm. Subsequently, the Nash equilibrium solution of the multi-human robot physical interaction is obtained by solving the coupled Riccati equation to achieve coupled optimization. Finally, the effectiveness of the proposed method is demonstrated by rigorous theoretical analysis and simulation experiments. This paper makes the following four contributions.

(1) N-player differential game theory is firstly used to model the human-robot-human interaction system.
(2) An online estimation method to identify unknown humans' control objectives based on the recursive least squares algorithm is presented.
(3) A general adaptive optimal control framework for human-robot-human physical interaction is propose based on (1) and (2).
(4) The effectiveness of the proposed method is demonstrated by rigorous theoretical analysis and simulation experiments.

The remainder of this paper is organized, as follows: Section 2 models the human-robot-human physical interaction system based on N-player differential game theory. Section 3 establishes an adaptive optimal control law, and the control performance of the system is analyzed theoretically. Section 4 verifies the effectiveness of the proposed method through simulation experiments. Finally, Section 5 concludes this work.

2. Problem Formulation

2.1. System Description

The system considered contains two humans and one robot. An example scenario is shown in Figure 1, where the robot and the humans collaborate to perform an object transporting task. In this shared control task, when the control objectives of humans' change, the robot should recognize the humans' control objectives and response adaptively and optimally. The forces exerted by the humans on the object are measured by force sensors at the interaction point. It is worth noting that the humans' control objectives are unknown to the robot.

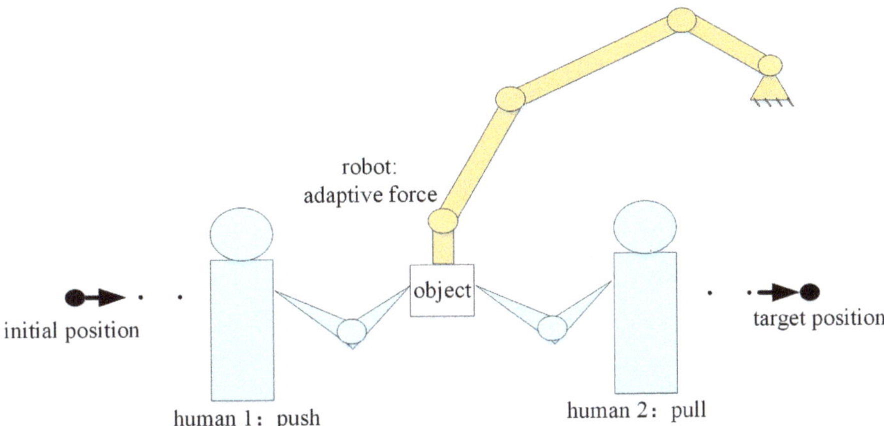

Figure 1. A scenario where the humans and the robot collaborate to perform an object transporting task.

The forward kinematics of the robot are described as

$$x(t) = \phi(q(t)) \tag{1}$$

where $x(t) \in \mathbb{R}^m$ and $q(t) \in \mathbb{R}^n$ are the positions in Cartesian space and joint space respectively, m and n are degrees of freedom. Derivation of Equation (1) with time can be obtained

$$\dot{x}(t) = J(q(t))\dot{q}(t) \tag{2}$$

where $J(q(t)) \in \mathbb{R}^{m \times n}$ is the Jacobian matrix.

The following impedance model is given in Cartesian space

$$M_d \ddot{x}(t) + C_d \dot{x}(t) = u(t) + f_1(t) + f_2(t) \tag{3}$$

where $M_d \in \mathbb{R}^{m \times m}$ is the desired inertial matrix, $C_d \in \mathbb{R}^{m \times m}$ is the damping matrix, $u(t) \in \mathbb{R}^m$ is the control input in the Cartesian space [40–42], $f_1(t) \in \mathbb{R}^n$ is the contact force between object and human 1, $f_2(t) \in \mathbb{R}^n$ is the contact force between object and human 2.

To track a common and fixed target $x_d \in \mathbb{R}^m$ ($\dot{x}_d \in \mathbb{R}^m$) in cooperative object transporting task, Equation (3) can be transformed, as following

$$M_d(\ddot{x}(t) - \ddot{x}_d) + C_d(\dot{x}(t) - \dot{x}_d) = u(t) + f_1(t) + f_2(t). \tag{4}$$

In order to ease the design of the control, Equation (4) can be rewritten as the following state-space form

$$\dot{z} = Az + B_1 u + B_2 f_1 + B_3 f_2$$

$$z = \begin{bmatrix} x(t) - x_d \\ \dot{x}(t) \end{bmatrix}, A = \begin{bmatrix} 0_m & 1_m \\ 0_m & -M_d^{-1} C_d \end{bmatrix}$$

$$B_1 = B_2 = B_3 = B = \begin{bmatrix} 0_m \\ M_d^{-1} \end{bmatrix} \tag{5}$$

where 0_m and 1_m denote $m \times m$ zero and unit matrices, respectively.

2.2. Problem Formulation

According to non-cooperative differential game theory, in the paper, the interaction between the robot and the humans is described as a game between N players (in this paper, $N = 3$) [43]. In the game, each player will minimize their respective cost function

$$\Gamma \equiv \int_{t_0}^{\infty} z^T Q z + u^T u d_t$$

$$\Gamma_1 \equiv \int_{t_0}^{\infty} z^T Q_1 z + f_1^T f_1 d_t$$

$$\Gamma_2 \equiv \int_{t_0}^{\infty} z^T Q_2 z + f_2^T f_2 d_t$$

$$Q = \begin{bmatrix} Q_{01} & 0_{n \times n} \\ 0_{n \times n} & Q_{02} \end{bmatrix} \tag{6}$$

$$Q_1 = \begin{bmatrix} Q_{11} & 0_{n \times n} \\ 0_{n \times n} & Q_{12} \end{bmatrix}$$

$$Q_2 = \begin{bmatrix} Q_{21} & 0_{n \times n} \\ 0_{n \times n} & Q_{22} \end{bmatrix}$$

where $\Gamma, \Gamma_1, \Gamma_2$ are cost functions of the robot, human 1, and human 2, respectively, Q, Q_1, Q_2 are state weights matrices of the robot, human 1 and human 2, respectively. Each player achieves the cooperative object transporting task by minimizing the error to the target while minimizing their own costs. Q, Q_1, Q_2 contain two components corresponding to position regulation and velocity, respectively. Q_{01}, Q_{11}, Q_{21} correspond to position regulation and Q_{02}, Q_{12}, Q_{22} correspond to velocity.

In [27], the N-player game has been studied if the cost functions are known. However, Γ_1, Γ_2 are unknown to the robot because they are determined by the humans. Therefore, a method is proposed in the paper to estimate Γ_1, Γ_2 in order to achieve adaptive optimal control and, thus, the human-robot-human cooperative object transporting task.

2.3. N-Player Differential Game Theory

Based on the differential game theory of linear systems, for N-player game the following linear differential equation [43] is considered:

$$\dot{z} = Az + B_1 u_1 + \cdots + B_N u_N, z(0) = z_0. \tag{7}$$

Each player has a quadratic cost function that they want to minimize:

$$\Gamma_i = \int_0^{\infty} z^T Q_i z + u_i^T u_i d_t, i = 1, \cdots, N \tag{8}$$

Different types of multi-agent behaviors are defined in game theory, which can be achieved through different concepts of game equilibrium [44,45]. In this paper, Nash equilibrium is considered. In the sense of Nash equilibrium, each player minimizes their cost function:

$$u_i = -\eta_i z, \eta_i = B_i^T P_i$$

$$(A - \sum_{j \neq i}^{N} B_i \eta_i)^T P_i + P_i (A - \sum_{j \neq i}^{N} B_i \eta_i)_i + Q_i - P_i B_i B_i^T P = 0, i = 1, \cdots, N \tag{9}$$

where N is equal to 3 in this paper. In the sense of Nash equilibrium, the humans and the robot minimizes their own cost function:

$$u = -\alpha z$$
$$\alpha = B^T P_r \tag{10a}$$

$$f_1 = -\beta z$$
$$\beta = B^T P_1 \tag{10b}$$

$$f_2 = -\gamma z$$
$$\gamma = B^T P_1 \tag{10c}$$

$$A_r^T P_r + P_r A_r + Q - P_r B B^T P_r = 0_{2n}, A_r = A - B\beta - B\gamma \tag{10d}$$

$$A_1^T P_1 + P_1 A_1 + Q - P_1 B B^T P_1 = 0_{2n}, A_1 = A - B\alpha - B\gamma \tag{10e}$$

$$A_2^T P_2 + P_2 A_2 + Q - P_2 B B^T P_2 = 0_{2n}, A_2 = A - B\alpha - B\beta \tag{10f}$$

where $\alpha \equiv \begin{bmatrix} \alpha_e, \alpha_v \end{bmatrix}$ is the feedback gain of the robot, $\beta \equiv \begin{bmatrix} \beta_e, \beta_v \end{bmatrix}$ is the feedback gain of the human 1, $\gamma \equiv \begin{bmatrix} \gamma_e, \gamma_v \end{bmatrix}$ is the feedback gain of of the human 2. $\alpha_e, \beta_e, \gamma_e$ are the position error gains, $\alpha_v, \beta_v, \gamma_v$ are the velocity gains, P_r, P_1, P_2 are the solutions of the above well-known Riccati equation consisting of Equation (10d–f). The robot and the humans influence each other through A_r, A_1, and A_2 in order to achieve the interactive control and the coupling optimization.

β, γ are unknown to the robot. Therefore, we aim to propose a method to estimate them in the following section.

3. Adaptive Optimal Control

A recursive least squares algorithm with forgetting factors is used in this paper to get the estimate $\hat{\beta}, \hat{\gamma}$ of β, γ in order to estimate the feedback gains of the humans in real time and avoid the data saturation phenomenon caused by the standard least squares algorithm [46]. Subsequently, the estimate $\hat{Q}_1, \hat{Q}_2$ of Q_1, Q_2 can be obtained using Equation (10e,f).

Equation (10e) is used as the model for identification. For convenience, we let $\theta_1 = -\beta^T, y_1 = f_1^T$, $W = z^T$. Subsequently, Equation (10b) can be rewritten as

$$y_1 = W\theta_1. \tag{11}$$

The feedback gain of the human 1 are estimated by minimizing the total prediction error

$$J_1 = \int_0^t exp(-\lambda_1 t) \|y_1(s) - W(s)\hat{\theta}_1\|^2 ds \tag{12}$$

where λ_1 is the constant forgetting factor. The update rule of the parameter θ_1 can be obtained as

$$\dot{\hat{\theta}}_1 = -PW^T e_1$$
$$\dot{P} = \lambda_1 P - PW^T W P \tag{13}$$
$$e_1 = \hat{y}_1 - y_1.$$

The estimated error of $\hat{\theta}_1$ is

$$e_{\theta_1}(t) = exp(-\lambda_1 t)P(t)P^{-1}(0)e_{\theta_1}(0). \tag{14}$$

Thus, the estimate $\hat{\beta}$ can be obtained as

$$\hat{\beta} = -\hat{\theta}_1^T. \tag{15}$$

Similarly, we let $\theta_2 = -\gamma^T$, $y_2 = f_2^T$, $W = z^T$. Afterwards, Equation (10c) can be rewritten as

$$y_2 = W\theta_2. \tag{16}$$

The feedback gain of the human 2 are estimated by minimizing the total prediction error

$$J_2 = \int_0^t exp(-\lambda_2 t)\|y_2(s) - W(s)\hat{\theta}_2\|^2 ds \tag{17}$$

where λ_2 is the constant forgetting factor. The update rule of the parameter θ_2 can be obtained as

$$\begin{aligned} \dot{\hat{\theta}}_2 &= -PW^T e_2 \\ \dot{P} &= \lambda_2 P - PW^T WP \\ e_1 &= \hat{y}_2 - y_2. \end{aligned} \tag{18}$$

The estimated error of $\hat{\theta}_2$ is

$$e_{\theta_2}(t) = exp(-\lambda_2 t)P(t)P^{-1}(0)e_{\theta_2}(0). \tag{19}$$

Thus, the estimate $\hat{\gamma}$ can be obtained as

$$\hat{\gamma} = -\hat{\theta}_2^T. \tag{20}$$

Equations (13), (15), (18) and (20) are critical, because they enable each agent to recognize their partners' control objectives and use Equation (10a–f) to adjust their own control.

In order to ensure the performance of cooperative object transporting task, we let

$$Q + Q_1 + Q_2 \equiv C \tag{21}$$

where C is the total weight. The cooperative object transporting task fixes the task performance through the total weight C and uses Equation (21) to share the the effort between 2 humans and the robot. Equation (21) makes the proposed controller be able to adjust the contributions between the humans and the robot and makes the humans and the robot take complementary roles as well.

The control architecture is shown in Figure 2.

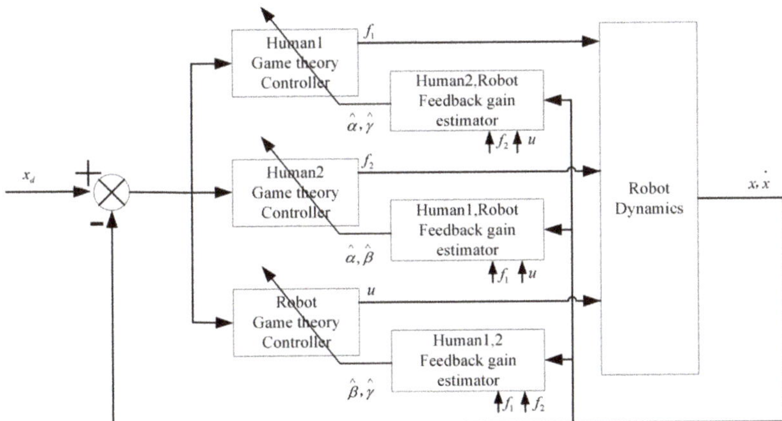

Figure 2. Control Architecture.

A pseudo-code summarizes the implementation procedures of the proposed method as Algorithm 1.

Algorithm 1 Adaptive optimal control algorithm based on N-player game

---

Input: Current state z, target x_d.
Output: Robot's control input u, estimated the humans' cost function state weight $\hat{Q}_1, \hat{Q}_2$ in Equation (10e,f).
Begin
    Define x_d, initialize $Q, \hat{Q}_1, \hat{Q}_2, u, f_1, f_2, \hat{z}, \alpha, \hat{\beta}, \hat{\gamma}, P_r, \hat{P}_1, \hat{P}_2$, set λ_1 in Equation (13), λ_2 in Equation (18), C in Equation (21), the terminal time t_f of one trial.
    While $t < t_f$ do
        Measure the position $x(t)$, velocity $\dot{x}(t)$, and form z.
        Update $\hat{\beta}$ using Equations (13) and (15), Update $\hat{\gamma}$ using Equations (18) and (20).
        Solve the Riccati equation in Equation (10d) to obtain P, and calculate the robot's control input u.
        Calculate estimated the humans' cost function state weights $\hat{Q}_1, \hat{Q}_2$ in Equation (10e,f) using the Riccati equation.
        Compute robot's cost function state weight Q according to Equation (21).

Theorem 1. *Consider the robot dynamics shown in Equation (5). If the robot and the humans estimate the parameters of their partners' controller and adjust their own control according to Equations (10a–f), (13), (15), (18), (20) and (21), then the following conclusions will be drawn:*

- *The closed-loop system is stable, and $z, \alpha, \hat{\beta}, \hat{\gamma}, u$ are bounded.*
- *$\lim_{x \to \infty} \hat{Q}_1 = Q_1$, $\lim_{x \to \infty} \hat{Q}_2 = Q_2$, which indicate that $\hat{Q}_1, \hat{Q}_2$ converge to the correct values Q_1, Q_2, if z is persistently exciting.*
- *The Nash equilibrium is achieved for th human-robot-human interaction system.*

Proof of Theorem 1. $\hat{\beta}, \hat{\gamma}$ influence u, f_1, f_2, z as following:

$$\dot{\hat{z}} = A\hat{z} + B\hat{u} + Bf_1 + B_2. \tag{22}$$

By subtracting Equation (5) from Equation (22), we have

$$\dot{e}_z = Ae_z + B(\hat{u} - u) + Be_{f_1} + Be_{f_2} \tag{23}$$

where $e_z = \hat{z} - z$. By considering Equation (10a–c), we have

$$\dot{e}_z = (A - B\alpha)e_z + Be_{\theta_1}^T z + Be_{\theta_2}^T z. \tag{24}$$

Consider the Lyapunov function candidate as following

$$W = \frac{1}{2}z^T z + \frac{1}{2}e_{\theta_1}^T e_{\theta_1} + \frac{1}{2}e_{\theta_2}^T e_{\theta_2} + \frac{\chi}{2}e_z^T e_z \tag{25}$$

where $\chi = min(\frac{2(\lambda_1 - \rho)\pi}{\varphi^2\|B\|^2}, \frac{2(\lambda_2 - \rho)\pi}{\varphi^2\|B\|^2})$, with ρ being the upper bound of the maximum eigenvalue of $\dot{P}P^{-1}$, π being the lower bound of the minimum eigenvalue of $B\alpha - A$, φ being the upper bound of $\|z\|$.

When considering function $V = \frac{1}{2}z^T z$ and differentiating V with respect to time, we obtain

$$\dot{V} = z^T \dot{z} = -z^T(B\alpha + B\beta + B\gamma - A)z. \tag{26}$$

According to Equation (10d), $B\alpha + B\beta + B\gamma - A$ is positive definite if Q is positive definite, it follows $\lim_{t\to\infty}\|z\| = 0$. Therefore, z is bounded and we define φ as the upper bound of $\|z\|$. By differentiating Equation (25), with respect to time, and considering Equations (14), (19) and (24), we obtain

$$
\begin{aligned}
\dot{W} =& z^T\dot{z} + e_{\theta_1}^T\dot{e}_{\theta_1} + e_{\theta_2}^T\dot{e}_{\theta_2} + \chi e_z^T\dot{e}_z \\
=& -z^T(B\alpha + B\beta + B\gamma - A)z \\
& -\lambda_1 e_{\theta_1}^T e_{\theta_1} + e_{\theta_1}^T\dot{P}P^{-1}e_{\theta_1} - \lambda_2 e_{\theta_2}^T e_{\theta_2} + e_{\theta_2}^T\dot{P}P^{-1}e_{\theta_2} \\
& -\chi e_z^T(B\alpha - A)e_z + \chi e_z^T Be_{\theta_1}^T z + \chi e_z^T Be_{\theta_2}^T z \\
\leq& -z^T(B\alpha + B\beta + B\gamma - A)z \\
& -\lambda_1\|e_{\theta_1}\|^2 + \rho\|e_{\theta_1}\|^2 - \lambda_2\|e_{\theta_2}\|^2 + \rho\|e_{\theta_2}\|^2 \\
& -\chi\pi\|e_z\|^2 + \chi\varphi\|B\|\|e_z\|\|e_{\theta_1}\| + \chi\varphi\|B\|\|e_z\|\|e_{\theta_2}\| \\
=& -z^T(B\alpha + B\beta + B\gamma - A)z \\
& -(\sqrt{\lambda_1 - \rho}\|e_{\theta_1}\| - \sqrt{\frac{\chi\pi}{2}}\|e_z\|)^2 \\
& -2\sqrt{\lambda_1 - \rho}\sqrt{\frac{\chi\pi}{2}}\|e_{\theta_1}\|\|e_z\| + \chi\varphi B\|\|e_z\|\|e_{\theta_1}\| \\
& -(\sqrt{\lambda_2 - \rho}\|e_{\theta_2}\| - \sqrt{\frac{\chi\pi}{2}}\|e_z\|)^2 \\
& -2\sqrt{\lambda_2 - \rho}\sqrt{\frac{\chi\pi}{2}}\|e_{\theta_2}\|\|e_z\| + \chi\varphi\|B\|\|e_z\|\|e_{\theta_2}\| \\
\leq& -z^T(B\alpha + B\beta + B\gamma - A)z \\
& +(-2\sqrt{\lambda_1 - \rho}\sqrt{\frac{\chi\pi}{2}} + \chi\varphi\|B\|)\|e_z\|\|e_{\theta_1}\| \\
& +(-2\sqrt{\lambda_2 - \rho}\sqrt{\frac{\chi\pi}{2}} + \chi\varphi\|B\|)\|e_z\|\|e_{\theta_2}\| \\
\leq& 0
\end{aligned}
\tag{27}
$$

According to Equations (26) and (27), we have $\lim_{t\to\infty}\|z\| = 0$, $\lim_{t\to\infty}\|e_z\| = 0$. Therefore, $z(t)$ is bounded and $\lim_{t\to\infty}\|\dot{e}_z\| = 0$. According to Equation (27), we have $\lim_{t\to\infty}\|e_{\theta_1}\| = 0$, $\lim_{t\to\infty}\|e_{\theta_2}\| = 0$. Because of $e_{\theta_1} = \hat{\theta}_1 - \theta_1 = (-\hat{\beta})^T - (-\beta)^T = \beta^T - \hat{\beta}^T$, $e_{\theta_2} = \hat{\theta}_2 - \theta_2 = (-\hat{\gamma})^T - (-\gamma)^T = \gamma^T - \hat{\gamma}^T$, we can obtain $\lim_{t\to\infty}\|\beta^T - \hat{\beta}^T\| = 0$, $\lim_{t\to\infty}\|\gamma^T - \hat{\gamma}^T\| = 0$. β, γ are assumed to be bounded, since they are the feedback gains of the humans. Therefore, $\hat{\beta}, \hat{\gamma}$ are also bounded. According to Equation (10a–c), P_1, P_2 are also bounded. According to Equation (10d), A_r is bounded. Therefore, P, α and u are bounded.

According to Equation (10e), we can calculate the estimated errors $e_{Q_1} = \hat{Q}_1 - Q_1, e_{Q_2} = \hat{Q}_2 - Q_2$. e_{Q_1}, e_{Q_2} are due to the errors e_P, e_{P_1}, e_{P_2}. Because e_P, e_{P_1}, e_{P_2} converge to zero, we have $\lim_{t \to \infty} \|e_{Q_1}\| = 0$, $\lim_{t \to \infty} \|e_{Q_2}\| = 0$, that is $\lim_{t \to \infty} \hat{Q}_1 = Q_1, \lim_{t \to \infty} \hat{Q}_2 = Q_2$.

Multiplying Equation (10d) by $\hat{z}^T$ on the left side and by $\hat{z}$ on the right side, and considering Equation (13), we have

$$
\begin{aligned}
0 =& \hat{z}^T Q \hat{z} + \hat{z}^T P_r B B^T P_r \hat{z} + \hat{z}^T P_r \dot{\hat{z}} \\
& + \hat{z} P_r \dot{\hat{z}}^T + \hat{z}^T P_r H e_z + \hat{z} P_r H e_z^T \\
\equiv& \hat{\sigma}.
\end{aligned}
\tag{28}
$$

Considering $\lim_{t \to \infty} e_z = 0, \lim_{t \to \infty} \dot{e}_z = 0$, we can obtain

$$
\begin{aligned}
\lim_{t \to \infty} \sigma \equiv& \lim_{t \to \infty} (z^T Q Z + z^T P_r B B^T P_r z \\
& + z^T P_r \dot{z} + z P_r \dot{z}^T) \\
=& 0.
\end{aligned}
\tag{29}
$$

Similarly, we can obtain

$$
\begin{aligned}
\lim_{t \to \infty} \sigma_1 \equiv& \lim_{t \to \infty} (z^T Q_1 Z + z^T P_1 B B^T P_1 z \\
& + z^T P_1 \dot{z} + z P_1 \dot{z}^T) \\
=& 0 \\
\lim_{t \to \infty} \sigma_2 \equiv& \lim_{t \to \infty} (z^T Q_2 Z + z^T P_2 B B^T P_2 z \\
& + z^T P_2 \dot{z} + z P_2 \dot{z}^T) \\
=& 0.
\end{aligned}
\tag{30}
$$

$\lim_{t \to \infty} \sigma = 0, \lim_{t \to \infty} \sigma_1 = 0$ and $\lim_{t \to \infty} \sigma_2 = 0$ indicate that the Nash equilibrium is achieved for the human-robot-human interaction system. $\square$

4. Simulations and Results

4.1. Experimental Design and Ssimulation Settings

With the development of the robot technology, in the future, robots will enter our homes and become a member of family in our daily lives. In our daily lives, we often need to carry various objects. Some objects (e.g., objects with smaller size and lower weight) can be successfully carried by one human; some objects (e.g., objects with medium size and medium weight) need to be carried successfully by two humans; some objects (e.g., objects with larger size and higher weight) can be carried successfully by three or more humans. Consider one scenario: In our home, we have an object (such as a table with a relatively larger size and higher weight) that need to be carried by three humans. However, there are only two humans in the home. In this case, we can let the robot help us carry the object together with the two humans. The robot can play the same role as one human. A simulation is conducted with CoppeliaSim in order to verify the control performance of the controller proposed in this paper. The version of CoppeliaSim that we used is CoppeliaSim 4.0.0 (CoppeliaSim Edu, Windows). Figure 3 demonstrates the CoppeliaSim simulation scenario of cooperative object transporting task. The humans cooperate with the robot to transport the object between -10 cm and $+10$ cm back and forth along the horizontal direction.

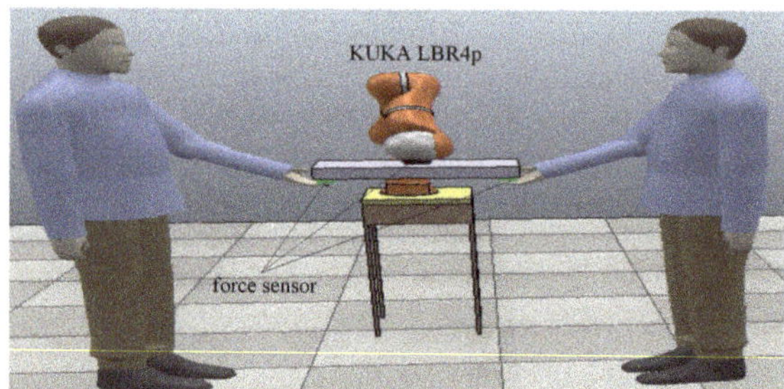

Figure 3. Simulation of cooperative object transporting task. The humans cooperate with the robot to transport the object back and forth between −10 cm and +10 cm along the horizontal direction. The forces that are exerted by the humans on the object are measured by force sensors at the interaction point.

The controller that is proposed in this paper implements interactive control because every agent considers the control of other partners. In order to present the advantages of the proposed controller, we compare the proposed controller with the linear quadratic regulators (LQR) optimal controller. The LQR controller can be obtained by setting $A_r = A$, $A_1 = A$, $A_2 = A$ in Equation (10d–f). The LQR controller allows each agent to form its own control input optimally, but it ignores the controls of other partners. Let $Q = Q_1 = Q_2 = diag(100, 0)$.

The cost functions of the humans usually change during the physical human-robot-human interaction. The robot needs to identify the change and adaptively adjust its own cost function in order to complete the cooperative object transporting task. In order to verify the ability of the robot to adaptively interact with two humans when humans' cost functions change, we simulated a scenario where the robot cooperated with the humans to perform an object transporting task. The task performance is achieved by setting the value of C in Equation (21). Let $C = diag(300, 0)$. The cost functions of the human 1 and the human 2 change randomly according to $Q_1 = diag(50, 0) + \rho \cdot diag(50, 0)$, $Q_2 = diag(50, 0) + \rho \cdot diag(50, 0)$ (ρ is a uniformly distributed random number between $[0, 1]$).

The human-robot-human cooperative object transporting task can be fulfilled with less effort with the proposed controller. In order to make this affirmation, we made a comparison with a human-robot cooperative object transporting task. In simulation of the human-robot-human cooperative object transporting task, we let $Q = Q_1 = Q_2 = diag(100, 0)$. In simulation of the human-robot cooperative object transporting task, we let $Q = diag(100, 0)$, $Q_1 = diag(100, 0)$, $Q_2 = diag(0, 0)$.

We assume that the humans and the robot do not have prior knowledge of each other (thus, initially $\hat{\alpha} \equiv 0$, $\hat{\beta} \equiv 0$, $\hat{\gamma} \equiv 0$). The control input of the robot are generated by Equations (5), (10a–f), (13), (15), (18) and (20). The simulated interaction forces f_1, f_2 of the human 1 and the human 2 are generated by a similar set of equations. The simulation time is 40 s. Let the inertia of the robot $M_d = 6$ kg, the damping of the robot $C_d = -0.2 \ N \cdot m^{-1}$ [19], the real-time least squares algorithm forgetting factor $\lambda_1 = \lambda_2 = 0.95$. Simulation time step is 0.005 s.

4.2. Results

Figure 4 depicts the change in position of the end effector with respect to time. The results plotted in Figure 4 is a smooth curve that looks like a sinusoidal signal. This smooth curve is determined by Equation (3). In Equation (3), $u(t), f_1(t), f_2(t)$ are iteratively calculated by our proposed controller based on game theory. Due to the fact that the humans and the robot do not transport the object at a constant speed using our method, the end effector follows a curve signal rather than a straight line signal. As can be seen from Figure 4, the end effector can reach the target position with the proposed controller which means that the cooperative object transporting task is successfully fulfilled. In contrast, the end effector can not reach the target position with the LQR controller, which means that the cooperative object transporting task is not successfully fulfilled. The reason why the cooperative object transporting task can be successfully fulfilled with the proposed controller rather than with the LQR controller is that the proposed controller considers the interaction with other partners. When one partner decreases effort, the other partners will gradually increase their efforts to ensure the successful fulfillment of the cooperative object transporting task. In contrast, the LQR controller does not consider the interaction with other partners, so the cooperative object transporting task cannot be guaranteed to be successfully fulfilled.

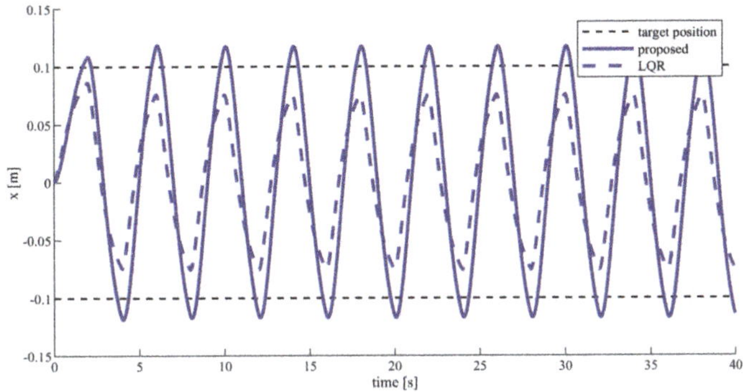

Figure 4. The end effector position value.

In Figure 5, we can see that the estimated humans' feedback gains converge to the real values in a few seconds. This means that the humans' feedback gains can be successfully estimated by the proposed method.

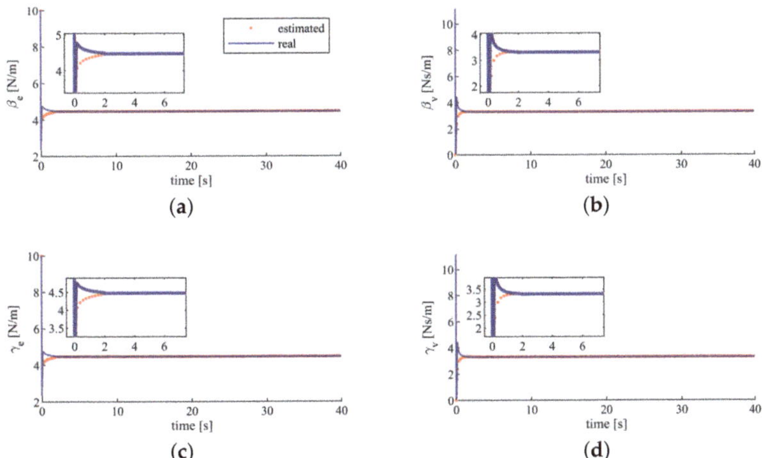

Figure 5. Control gains of humans. (**a**) the position error feedback gain of the human 1. (**b**) the velocity feedback gain of the human 1. (**c**) the position error feedback gain of the human 2. (**d**) the velocity feedback gain of the human 2.

Figure 6 demonstrates that fulfilling the cooperative object transporting task requires larger control gains β, γ with the LQR controller compared with the controller proposed in this paper. It means that accomplishing the same task requires less effort using the proposed controller. This is because that the proposed controller considers the interaction with other partners and calculates the minimal effort for the humans and the robot to complete the task. In contrast, the LQR controller doesn't consider the interaction with other partners, so the humans and the robot only minimize their own cost function and may, therefore, require larger effort.

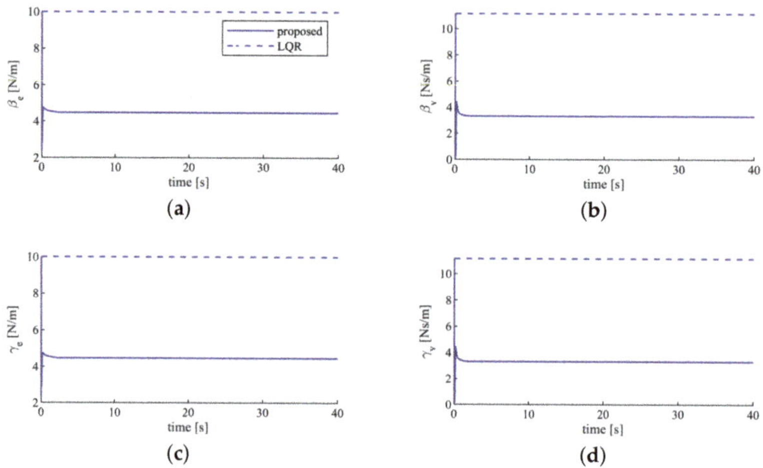

Figure 6. Humans' control gains (**a**) the position error feedback gain of the human 1. (**b**) the velocity feedback gain of the human 1. (**c**) the position error feedback gain of the human 2. (**d**) the velocity feedback gain of the human 2.

The feedback gains are affected by the state weights of the cost functions. In order to verify the advantages of the proposed controller when the state weights vary, we let Q_1 vary from 0 to $10Q$ with $Q_2 = diag(100,0)$ and let Q_2 vary from 0 to $10Q$ with $Q_1 = diag(100,0)$ respectively. It can be seen from Figure 7 that accomplishing the same task always requires less effort using the proposed controller. We can also see that the difference between the control gains with our proposed controller and the control gains with LQR controller becoming smaller when Q_1/Q or Q_2/Q increases, this is because that the robot's relative influence decreases.

From Figures 4–7, we can conclude that the human-robot-human cooperative object transporting task can be fulfilled with less effort and the system can be kept stable using the proposed controller.

It can be seen from Figure 8 that, when the cost functions of the human 1 and the human 2 change, the cost function of the robot will also change adaptively. When the sum of the state weights of the human 1 and the human 2 $Q_1 + Q_2$ increases, the state weight of the Robot Q decreases accordingly. Conversely, when the sum of the state weights of the human 1 and the human 2 $Q_1 + Q_2$ decreases, the state weight of the robot Q increases accordingly. The reason why the robot can change adaptively is that we set the constant C value in Equation (21). Equation (21) makes the proposed controller able to adjust the contributions between the humans and the robot and makes the humans and the robot take complementary roles as well.

Figure 9 shows that, using the proposed controller, the adaptive cooperative object transporting task can be fulfilled and the system can be kept stable.

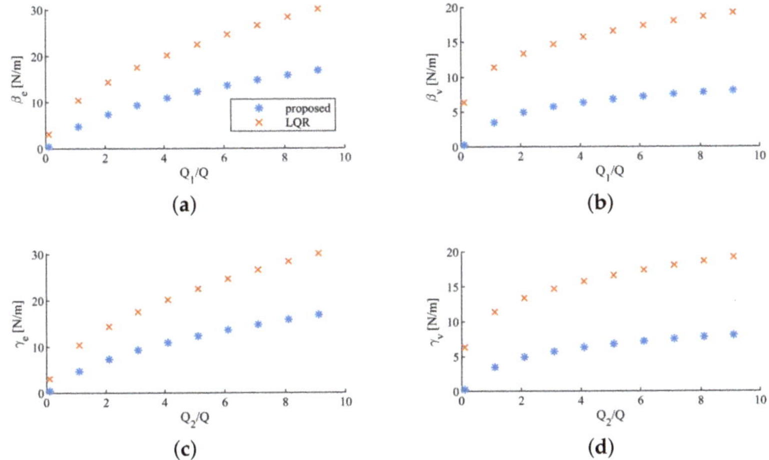

Figure 7. Control gains for different values of humans' state weights. (**a**) and (**b**) the state weight of the human 1 vary. (**c**) and (**d**) the state weight of the human 2 vary.

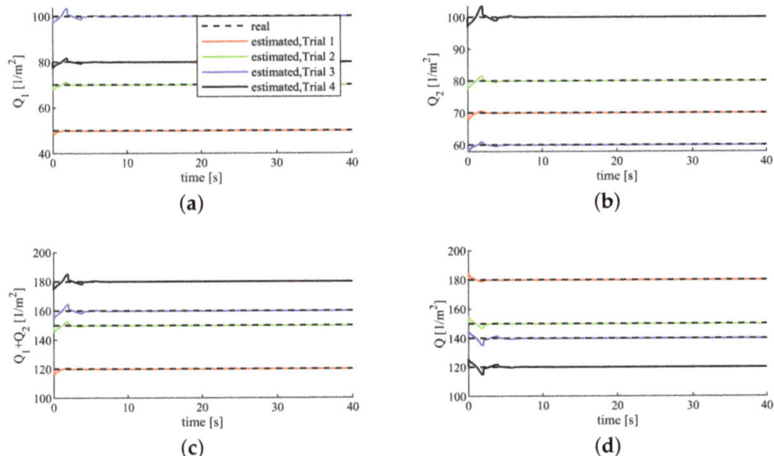

Figure 8. Humans' state weights. (**a**) the state weight of the human 1. (**b**) the state weight of the human 2. (**c**) the sum of the state weights of the human 1 and human 2. (**d**) the state weight of the robot.

From Figures 8 and 9, we can conclude that the adaptive cooperative object transporting task can be fulfilled with the proposed controller. During the physical interaction, the robot can successfully identify the change of each human's cost function, and then adaptively adjust its own cost function to achieve interactive optimal control.

Figure 10 demonstrates that fulfilling the human-robot-human cooperative object transporting task requires smaller control gains β_e, β_v as compared with the human-robot cooperative object transporting task. It means that accomplishing the same task requires less effort by means of the human-robot-human physical interaction. This is because the human-robot-human cooperative object transporting task considers the interaction with more partners (two partners) and calculates minimal effort for the humans and the robot to complete the task. In contrast, the human-robot cooperative object transporting task consider the interaction with less partners (only one partner), so the human and the robot may therefore require larger effort.

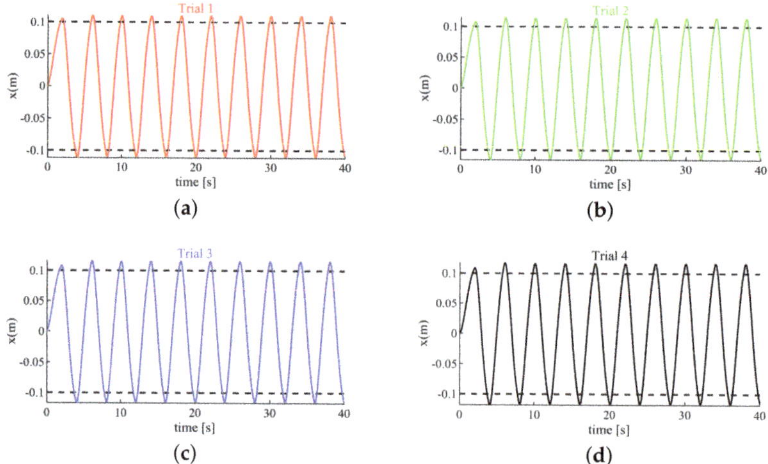

Figure 9. The end effector position value. (**a**) The end effector position value in Trial 1. (**b**) The end effector position value in Trial 2. (**c**) The end effector position value in Trial 3. (**d**) The end effector position value in Trial 4.

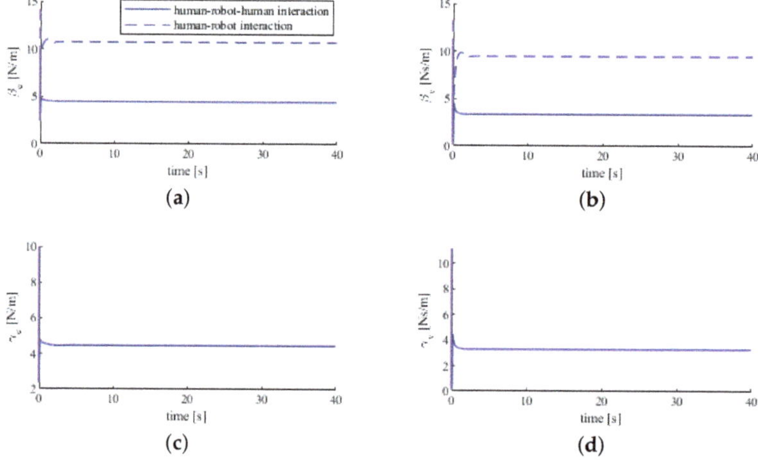

Figure 10. Humans' control gains. The dashed lines correspond to the human-robot cooperative object transporting task. The solid lines correspond to the human-robot-human cooperative object transporting task. (**a**) the position error feedback gain of the human 1. (**b**) the velocity feedback gain of the human 1. (**c**) the position error feedback gain of the human 2. (**d**) the velocity feedback gain of the human 2.

5. Conclusions

In this paper, the human-robot-human physical interaction problem has been studied. An adaptive optimal control framework for the human-robot-human physical interaction has been proposed based on N-player game theory. The recursive least squares algorithm based on forgetting factor has been used to identify unknown control parameters of the humans online. The performance of the controller proposed in this paper has been verified by simulations of cooperative object transporting task. The simulation results show that the proposed controller can achieve adaptive optimal control during the interaction between the robot and two humans and keep the system stable. Compared with the LQR controller, the proposed controller has more superior performance. Compared with the human-robot physical interaction, accomplishing the same cooperative object transporting task requires less effort by means of the human-robot-human physical interaction based on the approach proposed in the paper. Although this paper only conducts simulations on the physical interaction between one robot and two humans, it is worth mentioning that the framework that is proposed in this paper has the potential to be generalized to the situation where multiple robots physically interact with multiple humans. As future work, we will extend the framework to the interaction between multiple robots and multiple humans.

Author Contributions: R.Z. conceived the original ideas, designed all the experiments, and subsequently drafted the manuscript. Y.L. provided supervision and funding support for the project. J.Z. provided supervision and funding support for the project. H.C. provided supervision. All authors have read and agreed to the published version of the manuscript.

Funding: This work was supported by the Major Research Plan of the National Natural Science Foundation of China under Grant 91948201.

Conflicts of Interest: The authors declare no conflict of interest.

References

1. De Santis, A.; Siciliano, B.; De Luca, A.; Bicchi, A. An atlas of physical human-robot interaction. *Mech. Mach. Theory* **2008**, *43*, 253–270. [CrossRef]
2. Carolina, P.; Angelika, P.; Martin, B. A survey of environment-, operator-, and task-adapted controllers for teleoperation systems. *Mechatronics* **2010**, *20*, 787–801.
3. Losey, D.P.; McDonald, C.G.; Battaglia, E.; O'Malley, M.K. A review of intent detection, arbitration, and communication aspects of shared control for physical human-robot interaction. *Appl. Mech. Rev.* **2018**, *70*, 010804. [CrossRef]
4. Aslam, P.; Jeha, R. Safe physical human robot interaction-past, present and future. *J. Mech. Sci. Technol.* **2008**, *22*, 469.
5. Li, Y.; Ge, S.S. Human–robot collaboration based on motion intention estimation. *IEEE-ASME Trans. Mechatron.* **2013**, *19*, 1007–1014. [CrossRef]
6. Li, Y.; Ge, S.S. Force tracking control for motion synchronization in human-robot collaboration. *Robotica* **2016**, *34*, 1260–128. [CrossRef]
7. Sandra, H.; Martin, B. Human-oriented control for haptic teleoperation. *Proc. IEEE* **2012**, *100*, 623–647.
8. Chen, Z.; Huang, F.; Yang, C.; Yao, B. Adaptive fuzzy backstepping control for stable nonlinear bilateral teleoperation manipulators with enhanced transparency performance. *IEEE Trans. Ind. Electron.* **2019**, *67*, 746–756. [CrossRef]
9. Liu, C.; Masayoshi, T. Modeling and controller design of cooperative robots in workspace sharing human-robot assembly teams. In Proceedings of the IROS 2014, Chicago, IL, USA, 14–18 September 2014; pp. 1386–1391.
10. Zanchettin, A.M.; Casalino, A.; Piroddi, L.; Rocco, P. Prediction of human activity patterns for human-robot collaborative assembly tasks. *IEEE Trans. Ind. Inform.* **2018**, *15*, 3934–3942. [CrossRef]
11. Alexander, M.; Martin, L.; Ayse, K.; Metin, S.; Cagatay, B.; Sandra, H. The role of roles: Physical cooperation between humans and robots. *Int. J. Robot. Res.* **2012**, *31*, 1656–1674.
12. Costa, M.J.; Dieter, C.; Veronique, L.; Johannes, C.; El-Houssaine, A. A structured methodology for the design of a human-robot collaborative assembly workplace. *Int. J. Adv. Manuf. Technol.* **2019**, *102*, 2663–2681.
13. Daniel, N.; Jan, K. A problem design and constraint modelling approach for collaborative assembly line planning. *Robot. Comput. Integr. Manuf.* **2019**, *55*, 199–207.
14. Selma, M.; Sandra, H. Control sharing in human-robot team interaction. *Annu. Rev. Control* **2017**, *44*, 342–354.

15. Mahdi, K.; Aude, B. A dynamical system approach to task-adaptation in physical human-robot interaction. *Auton. Robot.* **2019**, *43*, 927–946.
16. Roberto, C.; Vittorio, S. Rehabilitation Robotics: Technology and Applications. In *Rehabilitation Robotics*; Colombo, R., Sanguineti, V., Eds.; Academic Press: London, UK, 2018; pp. xix–xxvi.
17. Colgate, J.E.; Decker, P.F.; Klostermeyer, S.H.; Makhlin, A.; Meer, D.; Santos-Munne, J.; Peshkin, M.A.; Robie, M. Methods and Apparatus for Manipulation of Heavy Payloads with Intelligent Assist Devices. U.S. Patent 7,185,774, 6 March 2007.
18. Zoss, A.B.; Kazerooni, H.; Chu, A. Biomechanical design of the Berkeley lower extremity exoskeleton (BLEEX). *IEEE-ASME Trans. Mechatron.* **2006**, *11*, 128–138. [CrossRef]
19. Li, Y.; Carboni, G.; Gonzalez, F.; Campolo, D.; Burdet, E. Differential game theory for versatile physical human-robot interaction. *Nat. Mach. Intell.* **2019**, *1*, 36–43. [CrossRef]
20. Li, Y.; Tee, K.P.; Yan, R.; Chan, W.L.; Wu, Y. A framework of human-robot coordination based on game theory and policy iteration. *IEEE Trans. Robot.* **2016**, *32*, 1408–1418. [CrossRef]
21. Nathanaël, J.; Themistoklis, C.; Etienne, B. A framework to describe, analyze and generate interactive motor behaviors. *PLoS ONE* **2012**, *7*, e49945.
22. Li, Y.; Tee, K.P.; Yan, R.; Chan, W.L.; Wu, Y.; Limbu, D.K. Adaptive optimal control for coordination in physical human-robot interaction. In Proceedings of the 2015 IEEE/RSJ International Conference on Intelligent Robots and Systems (IROS), Hamburg, Germany, 28 September–3 October 2015; pp. 20–25.
23. Kirk, D.E. Optimal control theory: An introduction. In *Optimal Control Theory*; Dover Publications: Mineola, NY, USA, 2004.
24. Li, Y.; Tee, K.P.; Chan, W.L.; Yan, R.; Chua, Y.; Limbu, D.K. Continuous role adaptation for human-robot shared control. *IEEE Trans. Robot.* **2015**, *31*, 672–681. [CrossRef]
25. Lewis, F.L.; Vrabie, D. Reinforcement learning and adaptive dynamic programming for feedback control. *IEEE Circuits Syst. Mag.* **2009**, *9*, 32–50. [CrossRef]
26. Vamvoudakis, K.G.; Lewis, F.L. Multi-player non-zero-sum games: Online adaptive learning solution of coupled Hamilton–Jacobi equations. *Automatica* **2011**, *47*, 1556–1569. [CrossRef]
27. Zhang, H.; Wei, Q.; Liu, D. An iterative adaptive dynamic programming method for solving a class of nonlinear zero-sum differential games. *Automatica* **2011**, *47*, 207–214. [CrossRef]
28. Liu, D.; Li, H.; Wang, D. Online synchronous approximate optimal learning algorithm for multi-player non-zero-sum games with unknown dynamics. *IEEE Trans. Syst. Man Cybern. Syst.* **2014**, *44*, 1015–1027. [CrossRef]
29. Albaba, B.M.; Yildiz, Y. Modeling cyber-physical human systems via an interplay between reinforcement learning and game theory. *Annu. Rev. Control* **2019**, *48*, 1–21. [CrossRef]
30. Music, S.; Hirche, S. Haptic Shared Control for Human-Robot Collaboration: A Game-Theoretical Approach. In Proceedings of the 21st IFAC World Congress, Berlin, Germany, 12–17 July 2020.
31. Turnwald, A.; Wollherr, D. Human-like motion planning based on game theoretic decision making. *Int. J. Soc. Robot.* **2019**, *11*, 151–170. [CrossRef]
32. Liu, Z.; Liu, Q.; Xu, W.; Zhou, Z.; Pham, D.T. Human-robot collaborative manufacturing using cooperative game: Framework and implementation. *Procedia CIRP* **2018**, *72*, 87–92. [CrossRef]
33. Bansal, S.; Xu, J.; Howard, A.; Isbell, C. A Bayesian Framework for Nash Equilibrium Inference in Human-Robot Parallel Play. *arXiv* **2020**, arXiv:2006.05729.
34. Antonelli, G.; Chiaverini, S.; Marino, A. A coordination strategy for multi-robot sampling of dynamic fields. In Proceedings of the 2012 IEEE International Conference on Robotics and Automation, Saint Paul, MN, USA, 14–18 May 2012; pp. 1113–1118.
35. Yan, Z.; Jouandeau, N.; Cherif, A.A. A survey and analysis of multi-robot coordination. *Int. J. Adv. Robot. Syst.* **2013**, *10*, 399. [CrossRef]
36. Martina, L.; Alessandro, M.; Stefano, C. A distributed approach to human multi-robot physical interaction. In Proceedings of the 2019 IEEE International Conference on Systems, Man and Cybernetics (SMC), Bari, Italy, 6–9 October 2019.
37. Kim, W.; Marta, L.; Balatti, P.; Wu, Y.; Arash, A. Towards ergonomic control of collaborative effort in multi-human mobile-robot teams. In Proceedings of the IEEE/RSJ International Conference on Intelligent Robots and Systems, Macau, China, 3–8 November 2019.
38. Starr, A.W.; Ho, Y.-C. Nonzero-sum differential games. *J. Optim. Theory Appl.* **1969**, *3*, 184–206. [CrossRef]

39. Fudenberg, D.; Tirole, J. Noncooperative game theory for industrial organization: An introduction and overview. *Handb. Ind. Organ.* **1989**, *1*, 259–327.

40. Hegan, N. Impedance Control: An Approach To Manipulation: Part I-Theory Part II-Implementation Part III-Applications. *J. Dyn. Syst. Meas. Control* **1985**, *107*, 1–24. [CrossRef]

41. Blank, A.A.; Okamura, A.M.; Whitcomb, L.L. Task-dependent impedance and implications for upper-limb prosthesis control. *Int. J. Robot. Res.* **2014**, *33*, 827–846. [CrossRef]

42. Vogel, J.; Haddadin, S.; Jarosiewicz, B.; Simeral, J.D.; Bacher, D.; Hochberg, L.R.; Donoghue, J.P.; van der Smagt, P. An assistive decision-and-control architecture for force-sensitive hand–arm systems driven by human–machine interfaces. *Int. J. Robot. Res.* **2015**, *34*, 763–780. [CrossRef]

43. Basar, T.; Olsder, G.J. *Dynamic Noncooperative Game Theory*, 2nd ed.; Society for Industrial and Applied Mathematics; The Math Works Inc.: Natick, MA, USA, 1999.

44. Shima, T.; Rasmussen, S. UAV cooperative decision and control: Challenges and practical approaches. In *UAV Cooperative Decision and Control*; SIAM: Philadelphia, PA, USA, 2009.

45. Hudas, G.; Vamvoudakis, K.G.; Mikulski, D.; Lewis, F.L. Online adaptive learning for team strategies in multi-agent systems. *J. Def. Model. Simul.* **2012**, *9*, 59–69. [CrossRef]

46. Tan, H.J.; Chan, S.C.; Lin, J.Q.; Sun, X. A New Variable Forgetting Factor-Based Bias-Compensated RLS Algorithm for Identification of FIR Systems With Input Noise and Its Hardware Implementation. *IEEE Trans. Circuits Syst. I Regul. Pap.* **2019**, *67*, 198–211. [CrossRef]

MDPI
St. Alban-Anlage 66
4052 Basel
Switzerland
Tel. +41 61 683 77 34
Fax +41 61 302 89 18
www.mdpi.com

Sensors Editorial Office
E-mail: sensors@mdpi.com
www.mdpi.com/journal/sensors